Lecture Notes in Computer Science 7872

Commenced Publication in 1973
Founding and Former Series Editors:
Gerhard Goos, Juris Hartmanis, and Jan van Leeuwen

T0216666

Zhi-Hua Zhou   Fabio Roli
Josef Kittler (Eds.)

Multiple Classifier Systems

11th International Workshop, MCS 2013
Nanjing, China, May 15-17, 2013
Proceedings

 Springer

Volume Editors

Zhi-Hua Zhou
Nanjing University
National Key Laboratory for Novel Software Technology
Nanjing, 210023, China
E-mail: zhouzh@nju.edu.cn

Fabio Roli
University of Cagliari
Department of Electrical and Electronic Engineering
09123 Cagliari, Italy
E-mail: roli@diee.unica.it

Josef Kittler
University of Surrey
Centre for Vision, Speech and Signal Processing
Guildford, Surrey, GU2 7XH, UK
E-mail: j.kittler@surrey.ac.uk

ISSN 0302-9743   e-ISSN 1611-3349
ISBN 978-3-642-38066-2   e-ISBN 978-3-642-38067-9
DOI 10.1007/978-3-642-38067-9
Springer Heidelberg Dordrecht London New York

Library of Congress Control Number: 2013936264

CR Subject Classification (1998): I.5, I.2.6, H.2.8, I.4, H.3

LNCS Sublibrary: SL 6 – Image Processing, Computer Vision, Pattern Recognition, and Graphics

Typesetting: Camera-ready by author, data conversion by Scientific Publishing Services, Chennai, India

Printed on acid-free paper

Springer is part of Springer Science+Business Media (www.springer.com)

Preface

This volume contains the papers presented at the 11th International Workshop on Multiple Classifier Systems (MCS 2013) held in Nanjing, China, during May 15–17, 2013. This was the 11th edition of the well-established series of meetings providing a leading international forum for the discussion of issues in multiple classifier systems and ensemble methods. The aim of the workshop is to bring together researchers from diverse communities concerned with this topic, including pattern recognition, machine learning, neural network, data mining, and statistics.

MCS 2013 received 59 full submissions. The Program Committee consisting of 45 experts carefully reviewed the submissions, with the help of external reviewers. Based on the reviews, 34 papers were selected for presentation at the workshop and inclusion in the proceedings. The workshop program and this volume were significantly enhanced by two invited talks given by world renowned experts: Bin Yu (UC Berkeley, USA) and Marcello Pelillo (Università Ca' Foscari Venezia, Italy).

This workshop would not have been possible without the help of many individuals and organizations. First of all, we would like to thank the Program Committee members and reviewers, for their great efforts in providing insightful comments on the submissions. We also wish to thank all the authors who have submitted their recent work to the workshop. The management of the papers, including the preparation of this proceedings volume, was done with the EasyChair conference management system. Special thanks go to the Local Arrangements and Publicity Chairs, Ming Li, Yang Yu, and Giorgio Fumera, for their outstanding contribution to the organization of MCS 2013.

This workshop was organized by the LAMDA Group of the National Key Laboratory for Novel Software Technology, Nanjing University, China, the Center for Vision, Speech and Signal Processing of the University of Surrey, UK, and the Department of Electrical and Electronic Engineering of the University of Cagliari, Italy. We thank the International Association for Pattern Recognition (IAPR), IEEE Computer Society Nanjing Chapter, and the National Science Foundation of China for their support.

May 2013

Zhi-Hua Zhou
Fabio Roli
Josef Kittler

Organization

Program Chairs

Zhi-Hua Zhou   Nanjing University, China
Fabio Roli   University of Cagliari, Italy
Josef Kittler   University of Surrey, UK

Local Arrangements Chair

Ming Li   Nanjing University, China

Publicity Chairs

Yang Yu   Nanjing University, China
Giorgio Fumera   University of Cagliari, Italy

Program Committee

Jon Atli Benediktsson, Iceland   G. Martinez-Munoz, Spain
Giorgio Fumera, Italy   Robi Polikar, USA
Cesare Furlanello, Italy   Juan J. Rodriguez, Spain
Joydeep Ghosh, USA   Lior Rokach, Israel
Lawrence Hall, USA   Arun Ross, USA
Tin Kam Ho, USA   Carlo Sansone, Italy
Nathan Intrator, Israel   Friedhelm Schwenker, Gemany
Philip Kegelmeyer, USA   Giovanni Seni, USA
Krzysztof Kryszczuk, Switzerland   Giorgio Valentini, Italy
Ludmila Kuncheva, UK   Terry Windeatt, UK
Guo-Zheng Li, China   Xin Yao, UK
Richard Maclin, USA   Yang Yu, China

Additional Reviewers

Battista Biggio   John Korecki
Luca Didaci   Nan Li
Benjamin Geiger   Xu-Ying Liu
Pedram Ghamisi   Gian Luca Marcialis
Xudong Kang   Leandro Minku

Supported by

National Science Foundation of China
IAPR (International Association for Pattern Recognition)
IEEE Computer Society Nanjing Chapter

Table of Contents

Improving Simple Collaborative Filtering Models Using Ensemble Methods

Ariel Bar[1], Lior Rokach[1], Guy Shani[1], Bracha Shapira[1], and Alon Schclar[2]

[1] Department of Information Systems Engineering
Ben-Gurion University of the Negev, Beer-Sheva, Israel
{arielba,liorrk,shanigu,bshapira}@bgu.ac.il
[2] School of Computer Science, Academic College of Tel Aviv-Yaffo
P.O.B. 8401, Tel Aviv 61083, Israel
alonschc@mta.ac.il

Abstract. In this paper we examine the effect of applying ensemble learning to the performance of collaborative filtering methods. We present several systematic approaches for generating an ensemble of collaborative filtering models based on a single collaborative filtering algorithm (single-model or homogeneous ensemble). We present an adaptation of several popular ensemble techniques in machine learning for the collaborative filtering domain, including bagging, boosting, fusion and randomness injection. We evaluate the proposed approach on several types of collaborative filtering base models: k-NN, matrix factorization and a neighborhood matrix factorization model. Empirical evaluation shows a prediction improvement compared to all base CF algorithms. In particular, we show that the performance of an ensemble of simple (weak) CF models such as k-NN is competitive compared with a single strong CF model (such as matrix factorization) while requiring an order of magnitude less computational cost.

Keywords: Recommendation Systems, Collaborative Filtering, Ensemble Methods.

1   Introduction

Collaborative Filtering is perhaps the most successful and popular method for providing predictions over user preferences, or recommending items. For example, in recent Netflix competitions, CF models were shown to provide the most accurate models. However, many of these methods require a very long training time in order to achieve high performance. Indeed, researchers suggest more and more complex models, with better accuracy, at the cost of higher computational effort.

Ensemble methods suggest that a combination of many simple identical models can achieve a performance of a complex model, at a lower training computation time. Various ensemble methods create a set of varying models using the same basic algorithm automatically, without forcing the user to explicitly learn a single set of model parameters that perform the best. The predictions of the resulting models are combined by, e.g., voting among all models. Indeed, ensemble methods have shown in

Z.-H. Zhou, F. Roli, and J. Kittler (Eds.): MCS 2013, LNCS 7872, pp. 1–12, 2013.

many cases the ability to achieve accuracy competitive with complex models. In this paper we investigate the applicability of a set of ensemble methods to a wide set of CF algorithms. We explain how to adapt CF algorithms to the ensemble framework in some cases, and how to use CF algorithms without any modifications in other cases. We run an extensive set of experiments, varying the parameters of the ensemble. We show that, as in other Machine Learning problems, ensemble methods over simple CF models achieve competitive performance with a single, more complex CF model at a lower cost.

2   Background and Related Work

Collaborative Filtering (CF) [1] is perhaps the most popular and the most effective technique for building recommendation systems. This approach predicts the opinion that the active user will have on items or recommends the "best" items to the active user, by using a scheme based on the active user's previous likings and the opinions of other, like-minded, users. The CF prediction problem is typically formulated as a triplet (U, I, R), where:

- U is a set of M users taking values form $\{u_1, u_2 \ldots u_m\}$.
- I is a set of N items taking values from $\{i_1, i_2 \ldots i_n\}$
- R - The ratings matrix, is a collection of historical rating records (each record contains a user id ($u \in U$), an item id ($i \in I$), and the rating that u gave to $i - r_{u,i}$.

A rating measures the preference by user u to item i, where high values mean stronger preferences. One main challenge of CF algorithms is to give an accurate prediction, denoted by $\hat{r}_{u,i}$ to the unknown entries in the ratings matrix, which is typically very sparse. Popular examples of CF methods include k-NN models [1-2] and Matrix Factorization models [3].

Ensemble is a machine learning approach that uses a combination of identical models in order to improve the results obtained by a single model. Unlike hybridization methods [4] in recommender systems that combine different types of recommendation models (e.g. a CF model and a content based model), the base models which construct the ensemble are based on a single learning algorithm.

Most improvements of collaborative filtering models either create more sophisticated models or add new enhancements to known ones. These methods include approaches such matrix factorization [3],[5], enriching models with implicit data[6], enhanced k-NN models [7], applying new similarity measures [8], or applying momentum techniques for gradient decent solvers [5].

In [9] the data sparsity problem of the ratings' matrix was alleviated by imputing the matrix with artificial ratings, prior to building the CF model. Ten different machine learning models were evaluated for the data imputing task, including an ensemble classifier (a fusion of several models). In two different experiments the ensemble approach provided lower MAE (mean absolute error). Note that this ensemble approach is a sort of hybridization method.

The framework presented in [10] describes three matrix factorization techniques, different in their parameters and constraints solving the matrix formation optimization problem. The best results (minimum RMSE – root mean square error) were achieved by an ensemble model which was constructed as a simple average of the three matrix factorization models.

Recommendations of several k-NN models are combined in [11] to improve MAE. The suggested model was a fusion between the User-Based CF approach and Item-Based CF approach. In addition the paper suggests lazy Bagging learning approach for computing the user-user, or item-item similarities.

In [12] a modified version of the AdaBoost.RT ensemble regressor (AdaBoost [13] variant designed for regression tasks) was shown to improve the RMSE measure of a neighborhood matrix factorization model. The authors demonstrate that adding more regressors to the ensemble reduces the RMSE (the best results were achieved with 10 models in the ensemble).

A heterogeneous ensemble model which blends five state-of-the-art CF methods was proposed in [14]. The hybrid model was superior to each of the base models. The parameters of the base methods were chosen manually.

The main contribution of this paper is a systematic framework for applying ensemble methods to CF methods. We employ automatic methods for generating an ensemble of collaborative filtering models based on a single collaborative filtering algorithm (homogeneous ensemble). We demonstrate the effectiveness of this framework by applying several ensemble methods to various base CF methods. In particular, we show that the performance of an ensemble of simple (weak) CF models such as k-NN is competitive compared with a single strong CF model (such as matrix factorization) while requiring an order of magnitude less computational cost.

3   Ensemble Framework

The proposed framework consists of two main components: (a) the ensemble method; and (b) the base CF algorithm. We investigate four common ensemble methods: Bagging, Boosting (a variant of AdaBoost), Fusion and Randomness Injection. These methods were chosen due to their improved accuracy when applied to classification problems, and the diversity in their mechanisms.

The Bagging and AdaBoost ensembles require the base algorithm to handle datasets in which samples may appear several times, or datasets where weights are assigned to the samples (equivalent conditions). Most of the base CF algorithms assume that each rating appears only once, and that all ratings have the same weight. In order to enable application of Bagging and Boosting, we modify the base CF algorithms to handle recurring and weighted samples. We evaluate four different base (modified) CF algorithms: k-NN User-User Similarity; k-NN Item-Item Similarity; Matrix Factorization (three variants of this algorithm) and Factorized Neighborhood. The first three algorithms are simpler, having a relatively low accuracy and rapid training time, while the last two are more complex, having better performance and higher training cost.

4   Ensemble Methods For CF

4.1   Bagging

The Bagging approach (Fig.1) [15] generates k different bootstrap samples (with re-placement) of the original dataset where each sample is used to construct a different CF prediction model. Each bootstrap sample (line 2) is in the size of the original rat-ing data set. The base prediction algorithm is applied to each bootstrap sample (line 3) producing k different prediction models. The ensemble model is a simple average over all the base ones (line 5).

Input:
- T – Training dataset of ratings $<U,I,R>$
- K – ensemble size.
- $BaseCF$– the Base CF prediction algorithm.

Output: *Bagging ensemble*

Method:
1. **for** i= 1 to K do:
2.   Create a random bootstrap sample T_i, by sampling T with replacemen
3.   Apply the $BaseCF$ to T_i and construct M_i.
4. **end for**

The prediction rule of the model is:

$$r_{ui}^{\wedge} = \sum_{}^{K} r_{u_i}^{M_i} / K$$

Fig. 1. Bagging algorithm for CF

4.2   Boosting

AdaBoost [15] is perhaps one of the most popular boosting algorithms in machine learning. In this approach, weights are assigned to each rating tuple, while an iterative process constructs a series of K models. After model M^i is learned, the weights are updated to allow the subsequent model, M_{i+1}, focus on the tuples that were poorly predicted by M_i. The ensemble model combines the predictions of each individual model via a weighted average according the accuracy of each model.

In this work we evaluated several variants of the AdaBoost.RT [16] algorithm. Ini-tial experiments with the original algorithm resulted with poor accuracy models; for all evaluated configurations, the ensemble model either had a negligible accuracy improvement or even an overall accuracy decrease compared to the original base model; Thus, we replaced the original relative error function with a pure absolute one as presented in Fig.2 (line 6 in the pseudo code). This modification resulted with im-provement of the original algorithm. As suggested in the original work, we initialize δ the demarcating threshold criteria to be the AE (the model error) of the original data-set. During the calibration process of the algorithm we evaluated different values for "n" (line 8 in the pseudo code), controlling the distribution function. The best results were achieved when we set n=1. We noticed that both the original and modified algorithms were highly unstable with n=2 or 3, as the accuracy of the final ensemble model decreased substantially.

Input:
- T – Training dataset of ratings $<U,I,R>$
- K – the ensemble size.
- *BaseCF* – the Base CF prediction algorithm
- δ – Threshold (0 $<\delta<$ *the rating score range*) for demarcating correct and incorrect predictions

Output: *AdaBoost.RT model*

Method:
1. Assign iteration number $t=1$
2. Assign initial distribution for each tuple in R:
   $D_t(r_{ui}) = 1/|R|$
3. **while** $t \leq K$ Do
4.   Apply *BaseCF* to T with distribution D_t, and construct the model M_t.
5.   **for each** rating $r_{ui} \in R$
6.   calculate $AE_t(r_{ui}) = | r_{ui}^{\wedge M_t} - r_{ui} |$
7.   calculate error rate of iteration t:

$$\varepsilon_t = \sum_{ui:AE_t(r_{ui})>\delta} D_t(r_{ui})$$

8.   Set $\beta_t = \varepsilon_t^n$
9.   Update distribution D_{t+1} as: //Z_t = normalization factor

$$D_{t+1}(r_{ui}) = \frac{D_t(r_{ui})}{Z_t} \times \begin{cases} \beta_t & if\ AE_t(r_{ui}) \leq \delta \\ 1 & otherwise \end{cases}$$

10.   Set t= t+1
11. **end while**

The prediction rule of the model is:

$$\hat{r_{ui}} = \sum_{i=1}^{K} \log(\frac{1}{\beta_i}) \cdot r_{ui}^{\wedge Mi} / \sum_{i=1}^{K} \log(\frac{1}{\beta_i})$$

Fig. 2. AdaBoost.RT algorithm for CF

4.3 Fusion

A straightforward way to construct an ensemble is to take a specific prediction algorithm, and use it several times on the same dataset, but each time with different initial parameters [17]. This process constructs different models, which can later be combined together by e.g. averaging. For the k-NN algorithms, we applied the following three fusion schemas:

1. Fusion by similarity metric - we combined the predictions of a two k-NN with different similarity measures (Pearson and Cosine.)

2. Fusion by CF perspective - we combined the predictions of the User-User k-NN model, and the Item-Item k-NN model.
3. k-NN Fusion by CF perspective & similarity metric -   combination of the two previous fusion schemes (total of four models in the ensemble).

For matrix factorization, we applied fusion to models which were constructed using different vector sizes of the latent factors.

4.4   Randomness Injection

All ensemble methods described so far in this section are generic in the sense that they are not limited to a specific CF prediction algorithm. Thus one of their parameters is BaseCF- the base CF prediction algorithm, which is used to construct the base models in the ensemble.

A different approach to create an ensemble is to take a base algorithm and modify it such that it will create various sub models and combine their results. A popular way to achieve this is by introducing randomness to the basic learning schema. By doing so, it is possible to run the algorithm several times, and receive a different model each time. These models can then be joined to provide a combined prediction. In this work the randomness to the CF algorithms was injected as follows:

Random k-NN - Instead of selecting the top k nearest neighbors (users or items) for the prediction rule, we randomly select any k users/items from the top 2*k nearest neighbors. We can repeat this process K times (the ensemble size) to get K different predictions, and then use a simple average on them for the final one.

The MF algorithms are naturally randomized, since in the initialization process of the learning phase, we assign small random numbers to the latent factors. If we simply repeat this process K times (the ensemble size), each time with random initial values, we will receive K different MF models. These models can then be combined into an ensemble by a simple average.

5   Modified CF Algorithms

Some ensemble methods require that the base prediction algorithm can handle datasets with reoccurring or weighted samples. Accordingly, we had to modify CF algorithms which assume that each rating appears once, and that all ratings weights are equal. The first step in our modification was to update the original CF prediction problem from Section 2, by adding a new element W to the problem formalization. W is a vector of weights whose size is equal to the number of ratings, where $w_{u,i}$ indicates the relative distribution of $r_{u,i}$. It is important to notice that when all weights are equal, all modified algorithms coincide with the original ones.

5.1   Modified k-NN Algorithms

The main modification in these algorithms is to include the rating's weights into the similarity measures. For the user-user k-NN prediction, we suggest using the modified

Pearson correlation coefficient, and the modified cosine-based similarity measures as described in Eqs.1 and 2 respectively, where S(u,v) is the set of items that both users u and v rated, r_u is the weighted average rating of user u, and w_{uvj} is the maximum between w_{uj} and w_{vj}. We use the same modified measures for item-item similarity, updating the required indices.

$$pearson(u,v) = \frac{\sum\limits_{j \in S(u,v)} w_{uvj}(r_{u,j} - r_u)(r_{v,j} - r_v)}{\sqrt{\sum\limits_{j \in S(u,v)} w_{uvj}(r_{u,j} - r_u)^2 \sum\limits_{j \in S(u,v)} w_{uvj}(r_{v,j} - r_v)^2}} \tag{1}$$

$$\cos ine(u,v) = \cos(\vec{u},\vec{v}) = \frac{\vec{u} \bullet \vec{v}}{\|\vec{u}\|_2 \times \|\vec{v}\|_2} = \frac{\sum\limits_{j \in S(u,v)} w_{uvj} r_{u,j} r_{v,j}}{\sqrt{\sum\limits_{j \in S(u,v)} w_{uvj} r_{u,j}^2} \sqrt{\sum\limits_{j \in S(u,v)} w_{uvj} r_{v,j}^2}} \tag{2}$$

5.2   Modified MF Algorithms

In this work we modified the ISMF, RISMF and BRISMF algorithms from [5] to handle weighted datasets. The modified algorithm continues to the minimize SSE (Sum of Square Errors), while applying new gradient steps as presented in Eq.3, taking into consideration the associated weight for each rating.

$$p'_{uk} = p_{uk} + \eta \cdot w_{ui} \cdot (e_{ui} \cdot q_{ki} - \lambda \cdot p_{uk}) \tag{3}$$

$$q'_{ki} = q_{ki} + \eta \cdot w_{ui} \cdot (e_{ui} \cdot p_{uk} - \lambda \cdot q_{ki})$$

In a similar way, we modify the Factorized Neighborhood Model (FNM) [7] to handle weighted datasets. We chose this model due to the high accuracy of its predications. The new gradient steps are presented in Eq. 4.

$$q_i = q_i + \eta \cdot w_{ui}(e'_{ui} \cdot p_u - \lambda \cdot q_i)$$
$$b_u = b_u + \eta \cdot w_{ui}(e'_{ui} - \lambda \cdot b_u)$$
$$b_i = b_i + \eta \cdot w_{ui}(e'_{ui} - \lambda \cdot b_i) \tag{4}$$
$$x_i = x_i + \eta \cdot w_{ui} \cdot (|R(u)|^{-\alpha} \cdot (r_{ui} - b_{ui}) \cdot sum\_error - \lambda \cdot x_i)$$
$$y_i = y_i + \eta \cdot w_{ui} \cdot (|N(u)|^{-\alpha} \cdot sum\_error - \lambda \cdot y_i)$$

6   Evaluation

6.1   Experimental Setup

The evaluation of the algorithms described in sections 4 and 5 was mainly based on the 100K MovieLens dataset. We used RMSE for measuring accuracy over 5 different random 80:20 on the dataset. We compared the following configurations: all k-NN models were evaluated by applying the modified Pearson and Cosine similarity measures and 3 k-NN sizes (5, 10, 20). The three matrix factorization algorithms from section 5.2, were evaluated using different sizes of latent factors (3, 4, 5, 10, 20, 30, 40, 50); The Factorized

Neighborhood Mode algorithm (FNM) was evaluated using different latent factor sizes: 3, 4, 5, 10, 20 and 30, all other parameters of the MF algorithms were consistent with the original work. For each configuration we evaluated its original RMSE (baseline) without any ensemble enhancement. We apply all ensemble methods from Section 4 to each configuration with different ensemble sizes and compare the results to the baseline: The ensemble size of Bagging ranged from 5 to 20 for the k-NN algorithms, and from 5 to 50 for the MF algorithms; The ensemble size of AdaBoost.RT ranged from 1 to 10; The ensemble size of Fusion was either 2 or 4 for the k-NN algorithm, and ranged between 5 and 10 for the MF algorithms; Finally, the ensemble size of Randomness Injection was set from 1 to 10. We now report various results and insights from these experiments.

6.2   Accuracy Results

Due to space restrictions, we are unable to report all possible RMSE results. We therefore limit Table 1 to the best configuration of each method. For example, from all k-NN User-User ensemble configurations using Bagging, the ensemble over k=20 produced the best results and is hence reported in the table.

We organized the base CF model according to their relative "strength", where simple/less accurate models appear on the left, and more advanced/complex/accurate appear on the right. The final row of the table indicates the improvement percentages of the best ensemble model compared to the baseline model.

Table 1. ML (100K ratings) Accuracy Results (RMSE)

CF / Ensemble	k-NN-User	k-NN-Item	ISMF	RISMF	BRISMF	FNM
Baseline	0.9535	0.9526	0.9434	0.9407	0.9268	0.9231
Bagging	0.9495 (20)	0.9464 (20)	*0.9173 (50)*	*0.9152 (50)*	*0.9170 (50)*	0.9333
AdaBoost.RT	0.9410 (10)	0.9459 (10)	0.9397 (10)	0.9415	0.9332	0.9367
Fusion	*0.9383 (4)*	*0.9383S (4)*	0.9411 (10)	0.9383 (10)	0.9241 (10)	0.9158 (10)
Random	0.9462 (10)	0.9437 (10)	0.9407 (10)	0.9381 (10)	0.9237 (10)	**0.9153 (10)**
Improvement	1.57%	1.47%	2.76%	2.66%	0.97%	0.87%

We use the following notations in the table: ensemble enhancements which improved with statistical significance the RMSE measure over the baseline accuracy are presented with the ensemble size (in parentheses). The best model in each column is displayed in bold-face font. Ensemble models of relatively weak algorithms which improve the RMSE to a level of more advanced models are displayed in italic font. We check for statistical significance using One-Way ANOVA with repeated measures (applying the Greenhouse-Geisser test) with confidence level $\alpha = 0.05$, followed by a simple paired t-test, with confidence level $\alpha = 0.05$.

Our results indicate the following: We were able to significantly improve the baseline results of every base CF model type in our work, by at least two different ensemble approaches; The improvement level was between 0.87% and 2.76%. These improvements may seem modest, but lowering the RMSE is a difficult problem, and every reduction in RMSE is difficult to achieve. The improvement level depends on the base CF models - more complex models are more difficult to improve. This agrees with the idea that ensemble should be applied to boost the performance of weak CF models, not to improve complex models. The Fusion and Random Injection ensemble methods were able to improve the accuracy of all base CF models; Bagging failed to improve FNM, and AdaBoost failed to improve RISMF, BRISMF and FNM, however, these ensemble approaches may achieve better results than other ensembles, when applied to other base CF algorithms. The performances of the suggested boosting approaches on the matrix factorization models require exploring additional boosting methods such as the Stochastic Gradient Boosting [18] or designing special ones for this task. In the spirit of the "No Free Lunch" theorem, none of the evaluated ensemble method was optimal for all given scenarios. Consequently, one should look for the (base model, ensemble) pair that achieves the best results for the dataset at hand.

6.3   The Effect of the Ensemble Size

Table 1 show that if the ensemble method improves the accuracy of the basic model, then the ensemble model that achieved the best result is the one with the highest number of members. Consequently, the strategy in this case is to use as many ensemble members as possible provided that the improvement is significant, and feasible with the amount of computation resources. Fig.3 demonstrates this idea by using Randomness injection on FNM. Adding more members to the ensemble may be practical, as the complexity of all the ensemble methods grows linearly in the number of ensemble members. Fig.3a also demonstrates that the accuracy improves with the ensemble sizes , the more members of the ensemble, the higher the accuracy, up to a certain limit, after which the improvement is marginal.(e.g.: adding the first member to the ensemble improved the accuracy by 0.43%, while the last member contributed only 0.02% improvement). This observation was stable in all the experiments of our studied ensemble methods, leading us to define the maximum ensemble sizes as described in section 6.1, as in all evaluated models the last ensemble member contributed up to 0.07% RMSE improvement which is considered insignificant.

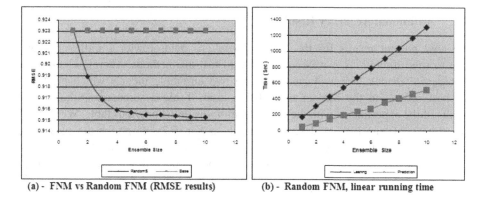

(a) - FNM vs Random FNM (RMSE results)   (b) - Random FNM, linear running time

Fig. 3. The effect of the ensemble size in Random FNM

6.4 Computational Cost and Accuracy Tradeoff

As described in sub-section 6.2 in several scenarios an ensemble of relatively weak models achieved better accuracy than a single stronger model. Fig.4 present the RMSE obtained by various methods as function of the computation cost (training time - presented in log scale). The graph shows the following results: An ensemble of the k-NN-User method achieves competitive performance with two MF methods (ISMF and RISMF) at an order of magnitude less computational cost (4 seconds instead of 24-26). An ensemble of MF methods (ISMF and RSIMF) achieves a competitive performance with a BRISMF method at a much lower computational cost (170 seconds instead of 490).

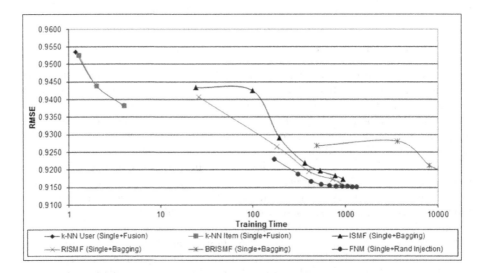

Fig. 4. Computational cost VS. RMSE

6.5 Additional Accuracy Results

We now present results for the larger MovieLens dataset with 1 million ratings. Due to time limitations it was not feasible to test all methods. Therefore, for each of base CF models, we evaluated only the two ensemble models which produced the best results on the MovieLens 100K dataset. Table 2 summarizes the RMSE results of the experiments with the same notations as in the previous table. We applied the same configurations as described in sub-section 6.2 except for the maximum ensemble size that was set to 30 in the Bagging experiments with MF algorithms. The accuracy results in this experiment are consistent with the ones in previous sections.

Table 2. ML (1M ratings) Accuracy Results (RMSE)

BaseCF	Ensemble Model	RMSE-MLB (Ensemble Size)	BaseCF	Ensemble Model	RMSE-MLB (Ensemble Size)
KNN-User	*Base (None)*	0.9302	*RISMF*	*Base (None)*	0.8712
	Fusion	**0.8972 (4)**		*Bagging*	***0.8480 (30)***
	AdaBoost.RT	0.9116 (10)		*Random*	0.8673 (10)
KNN-Item	*Base (None)*	0.9029	*BRISMF*	*Base (None)*	0.8620
	Fusion	0.8972 (4)		*Bagging*	**0.8519 (30)**
	Random	**0.8954 (10)**		*Random*	0.8570 (10)
ISMF	*Base (None)*	0.8812	*FNM*	*Base (None)*	0.8654
	Bagging	***0.8523 (30)***		*Fusion*	0.8469 (10)
	Random	0.8759 (10)		*Random*	**0.8465 (10)**

7   Conclusions

In this work we presented a novel systematic framework for applying ensemble methods to collaborative filtering models. Our framework used four popular ensemble techniques (Bagging, Boosting, Fusion and Randomness Injection) which were adapted to solve the collaborative filtering based rating prediction task. Typical collaborative filtering algorithms neither handle datasets with reoccurring samples, nor weighted samples. We thus modify the original base collaborative filtering algorithms to handle such settings.

Empirical evaluation shows an RMSE improvement by applying the suggested ensemble methods to the base CF algorithms. These improvements may increase the accuracy of relatively weak models to the level of more advanced ones. We found that in most cases it is preferable to add more base models to the ensemble, as we obtain a more accurate model compared to the combined model. Since all our ensemble methods have a linear running time and space complexity with respect to the ensemble size, it may be feasible to add more models to the ensemble as long as the improvement level is significant. These encouraging results indicate that ensemble methods can be used to enhance collaborative filtering algorithms. The boosting approach suggested in this paper is preliminary and requires further research. In the future we plan to evaluate our suggestions on other datasets and also on other problems, such as the recommendation of items.

A key issue that needs further investigation is how to find a data-driven criterion for choosing the optimal (ensemble, base model) pair for a given dataset. Other issues that need to be addressed are: application of other boosting/ensemble methods, and evaluation of additional collaborative filtering models.

References

1. Sarwar, B., Karypis, G., Konstan, J., Riedl, J.: Analysis of Recommendation Algorithms for E-Commerce. In: Proc. of the 2nd ACM Conference on E-commerce, pp. 158–167 (2000)

2. Sarwar, B., Karypis, G., Konstan, J., Riedl, J.: Item-based collaborative filtering recommendation algorithm. In: Proceedings of the 10th International Conference on World Wide Web, pp. 285–295 (2001)
3. Koren, Y.: Factorization Meets the Neighborhood: a Multifaceted Factor in the Neighbors: Scalable and Accurate Collaborative Filtering. In: Proc. 14th ACM Int. Conference on Knowledge Discovery and Data Mining (KDD 2008). ACM Press (2008)
4. Burke, R.: Hybrid recommender systems: Survey and Experiments. User Model. User-Adapt. Interact. 12(4), 331–370 (2002)
5. Takacs, Pilaszy, I., Nemeth, B., Tikk, D.: Scalable Collaborative Filtering Approaches for Large Recommender Systems. JMLR 10, 623–656 (2009)
6. Bell, R., Koren, Y.: Lessons from the Netflix Prize Challenge. SIGKDD Explorations 9 (2007)
7. Bell, R., Koren, Y.: Scalable Collaborative Filtering with Jointly Derived Neighborhood Interpolation Weights. In: IEEE International Conference on Data Mining (ICDM 2007), pp. 43–52 (2007)
8. Candillier, L., Meyer, F., Fessant, F.: Designing Specific Weighted Similarity Measures to Improve Collaborative Filtering Systems. In: IEEE International Conference on Data Mining (2008)
9. Su, X., Khoshgoftaar, T.M., Zhu, X., Greiner, R.: Imputation-boosted collaborative filtering using machine learning classifiers. In: Proceedings of the 2008 ACM Symposium on Applied Computing (2008)
10. Wu, M.: Collaborative Filtering via Ensembles of Matrix Factorizations. In: Proceedings of KDD Cup and Workshop (2007)
11. Lee, J.-S., Olafsson, S.: Two-way cooperative prediction for collaborative filtering recommendations. Expert Systems with Applications (2008)
12. Schclar, A., Meisels, A., Gershman, A., Rokach, L., Tsikinovsky, A.: Ensemble Methods for Improving the Performance of Neighborhood-based Collaborative Filtering. In: Proceedings of ACM RecSys 2009 (2009)
13. Freund, Y., Schapire, R.E.: Experiments with a New Boosting Algorithm. In: Machine Learning: Proceedings of the Thirteenth International Conference (1996)
14. Jahrer, M., Töscher, A., Legenstein, R.: Combining predictions for accurate recommender systems. In: Proc. 16th ACM SIGKDD, pp. 693–702 (2010)
15. Breiman, L.: Bagging Predictors. Machine Learning 24, 123–140 (1996)
16. Shrestha, D., Solomatine, D.: Experiments with AdaBoost.RT, an improved boosting scheme for regression. Neural Computation 18 (2006)
17. Cherkauer, K.J.: Human expert level performance on a scientific image analysis task by a system using combined artificial neural networks. In: Proc. AAAI 1996 Workshop on Integrating Multiple Learned Models for Improving and Scaling Machine Learning Algorithms, Portland, OR, pp. 15–21. AAAI Press, Menlo Park (1996)
18. Friedman, J.H.: Stochastic gradient boosting. Computational Statistics & Data Analysis 38(4), 367–378 (2002)

Combining Instance Information to Classify Bags

Veronika Cheplygina, David M.J. Tax, and Marco Loog

Pattern Recognition Laboratory, Delft University of Technology
{v.cheplygina,d.m.j.tax,m.loog}@tudelft.nl

Abstract. Multiple Instance Learning is concerned with learning from sets (bags) of feature vectors (instances), where the bags are labeled, but the instances are not. One of the ways to classify bags is using a (dis)similarity space, where each bag is represented by its dissimilarities to certain prototypes, such as bags or instances from the training set. The instance-based representation preserves the most information, but is very high-dimensional, whereas the bag-based representation has lower dimensionality, but risks throwing away important information. We show a connection between these representations and propose an alternative representation based on combining classifiers, which can potentially combine the advantages of the other methods. The performances of the ensemble classifiers are disappointing, but require further investigation. The bag-based representation preserves sufficient information to classify bags correctly and produces the best results on several datasets.

1 Introduction

Multiple-instance learning (MIL) [8] extends traditional supervised learning methods in order to learn from objects that are described by a set (*bag*) of feature vectors (*instances*), rather than a single feature vector only. For example, instead of representing an image or a document by a single feature vector, we could represent each segment or paragraph by its own feature vector. This is a more flexible representation, that can potentially preserve more information than if we were to compress all segments or paragraphs into a single feature vector.

MIL problems are often considered to be two-class problems, i.e., a bag of instances can belong either to the positive or the negative class. The bag labels are available, but the labels of the individual instances are not defined. The standard assumption here is that a bag is positive if and only if at least one instance inside the bag is positive. For example, an image labeled as "cat" would have a cat in at least one of its segments, whereas images without this label would not portray any cats at all. In this setting, it is possible to say that only one instance (the segment containing the cat) is informative.

It has been argued that there are more general kinds of MIL problems where the assumption above does not apply [22,5]. For example, for an image of the category "beach", it would be difficult to say which part of the image is informative. We would need to identify several objects (such as water and sand) to say that it is a beach, so at least a few instances in a positive bag must be informative. This reasoning can be extended even further to consider cases where simply

Z.-H. Zhou, F. Roli, and J. Kittler (Eds.): MCS 2013, LNCS 7872, pp. 13–24, 2013.

the presence of particular objects is not enough: consider how much of an image has to be covered by trees for you to call it a forest. Here, a certain fraction of instances is required for the positive class label[22], and therefore most, or even all instances can be informative.

One of the ways to classify bags in MIL problems is by representing the bags in a similarity or dissimilarity space [18]: each bag is then represented by its dissimilarities to certain prototypes. In our work[21], these prototypes are (a subset of) bags from the training set. Because a single dissimilarity is defined between two bags, information provided by the more informative instances in the bags might be overlooked. In MILES [5], an alternative representation using all the instances from the training set as the prototypes is used. A 1-norm SVM is then used to automatically select the most informative dissimilarities (and therefore instances). More investigation into the instance-based representation with other base classifiers has been done in [11].

A challenge in both settings is how to define the (dis)similarity measure between a bag and a prototype. In MILES, the similarity of a bag and a prototype instance is determined by the minimum distance between the bag's instances and the prototype instance. In our work[21,6], we define the dissimilarity of two bags as the combination (such as minimum, average or maximum) of minimum instance distances between these bags.

The way the information from different instances is combined links the instance-based and bag-based dissimilarity representations. In the former case, dissimilarities are concatenated, thus extending the dissimilarity representation, whereas in the latter they are combined into a single number by an operation such as averaging[17]. We also investigate a third alternative, i.e., combining classifiers trained on different subsets of dissimilarities. Comparing these representations can help us gain more insight into the informativeness of bags or instances as prototypes, and thus improve performances on real-life MIL problems.

2   Dissimilarity Representations

2.1   In Multiple Instance Learning

In Multiple Instance Learning, an object is represented by a bag $B_i = \{x_{ik} | k = 1, ..., n_i\} \subset \mathbb{R}^d$ of n_i feature vectors or instances. The training set $T = \{(B_i, y_i) | i = 1, ...N\}$ consists of positive ($y_i = +1$) and negative ($y_i = -1$) bags. The traditional assumption for MIL is that there are instance labels y_{ik} which relate to the bag labels as follows: a bag is positive if and only if it contains at least one positive, or *concept* instance[8]. In this case, it might be worthwhile to search for only these informative instances. Alternative formulations, where a fraction or even all instances are considered informative, have also been proposed [10].

We can represent an object, and therefore also a MIL bag B_i, by its dissimilarities to prototype objects in a representation set R[18]. In our work, R is taken to be a subset of size M of the training set T of size N ($M \leq N$). Each bag is represented as $\mathbf{d}(B_i, T) = [d(B_i, B_1), ...d(B_i, B_M)]$: a vector of M

dissimilarities. Therefore, each bag is represented by a single feature vector and the MIL problem can be viewed as a standard supervised learning problem.

MILES [5] considers a different definition of prototypes, using all the instances in the training set. The motivation is that, with just a few concept instances per bag, it is better to consider just these informative instances rather than the bag as a whole. MILES is originally a similarity-based approach, but in its dissimilarity-based counterpart, each bag would be represented as

$$\mathbf{d}(B_i, T) = [d(B_i, x_{1,1}), d(B_i, x_{1,2}), ..., d(B_i, x_{1,n_1}), ...d(B_i, x_{M,n_M})].$$

2.2   In Combining

When several dissimilarity representations for the same data are available, it can be an advantage to combine these sources of information. Assume that we are given L dissimilarity representations $D^1, D^2, \cdots, D^L$. In [17], three main ways of combining such representations are outlined:

- Concatenating the representations: $D^{ext} = [D^1 D^2 \cdots D^L]$.
- Averaging the representations: $D^{sum} = \sum_{i=1}^{L} D^i$.
- Training a base classifier on each D^i and combining the L outputs using a fixed rule (such as averaging) or a trained combiner [14,9].

3   Approach

In previous work[21,6], we have focused on defining $d(B_i, B_j)$ through the pairwise instance dissimilarities $[d(\mathbf{x}_{ik}, \mathbf{x}_{jl})]_{n_i \times n_j}$. We use the squared Euclidean distance for the instance dissimilarity, but other choices are also possible. In all the dissimilarities considered, the first step is to find, for each instance in B_i, the distance to its closest instance in B_j. Using these minimum instance distances, we can define many bag dissimilarities, for instance, by averaging these minimum distances. Assume that we are only given one prototype B_j. With the bag dissimilarity, the bag representation of B_i using prototype B_j would be:

$$D_{B_j}^{bag}(B_i) = \frac{1}{n_i} \sum_{k=1}^{n_i} \min_l d(\mathbf{x}_{ik}, \mathbf{x}_{jl}) \qquad (1)$$

In MILES, the similarity between a bag and a prototype instance is defined as the maximum similarity between the bag's instances and the prototype instance: $s(B_i, x) = \max_k \exp\left(-\frac{d(x_{ik}, x)}{\sigma^2}\right)$. In terms of distances this corresponds to the minimum instance distance between the bag and the prototype. Therefore, the instance representation of B_i using the instances of B_j would be:

$$D_{B_j}^{inst}(B_i) = [\min_l d(\mathbf{x}_{i1}, \mathbf{x}_{jl}), \min_l d(\mathbf{x}_{i2}, \mathbf{x}_{jl}), \cdots, \min_l d(\mathbf{x}_{in_i}, \mathbf{x}_{jl})] \qquad (2)$$

It is not difficult to now see that $D_{B_j}^{bag}(B_i) = \frac{1}{n_i} \sum_{k=1}^{n_i} D_{B_j}^{inst}(B_i)$. Another way to see this is that with D^{inst}, we can potentially give every prototype instance a different weight, whereas in D^{bag}, all instances from the same bag get the same weight.

Note that averaging as in (1) is not the only way to condense several dissimilarities into a single value: for instance, minimum or maximum operations could also achieve the same goal. However, these would essentially select a single instance per bag, rather than combining the information from all instances, as in (2). Therefore, we chose averaging as a combiner.

Previous results[21,5] suggest that both the bag-based and instance-based representations are (at least partly) informative: there are at least some prototypes (bags or instances) that distinguish between positive and negative bags in the dissimilarity space. We believe that comparing D^{bag} and D^{inst} directly, we can gain more insight into the structure of Multiple Instance Learning problems: how many instances are informative and what is a good (bag or instance) prototype.

Furthermore, we introduce two other representations that can help us in this understanding. In the "bag set" representation D^{BS}, a separate classifier is built on the instances of each prototype, to form M classifiers in total. In the "random set" representation D^{RS}, random sets of instances are used to build M separate classifiers. Each set of classifiers (built on bag sets or on random sets) forms an ensemble, where the individual classifier decisions are combined.

A diagram clarifying all the representations is shown in Fig.1. In terms of the initial dissimilarity matrix, D^{inst}, D^{BS} and D^{RS} are identical, but D^{inst} is used as a single input to a single classifier, whereas D^{BS} and D^{RS} have several feature subsets and classifiers associated with them. In fact, D^{RS} is just D^{inst} used together with the random subspace method [12].

These representations are also interesting in terms of speed and information trade-off. We assume that the data is available offline, so that all dissimilarity matrices can be computed beforehand. D^{inst} contains all instance information, but is very high-dimensional, which can severely slow down and/or deteriorate

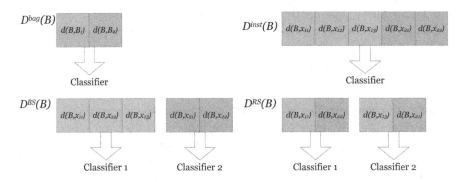

Fig. 1. Different ways for constructing dissimilarity representations of bag B using two prototype bags (green with 3 instances and blue with 2 instances). D^{bag} consists of just two dissimilarities (one for each bag), whereas D^{inst} consists of dissimilarities to all 5 instances. In D^{BS}, a separate classifier is built on each prototype's instance dissimilarities. In D^{RS}, classifiers are built on random selections of all available instances.

the performance of many classifiers. D^{bag} might lose some information, but is a more compact representation, reducing training time and the possibility of over-fitting. The ensembles D^{RS} and D^{BS} have access to all the information, although the information is now split up into subspaces. Although several classifiers have to be trained, each classifier can be very fast due to the reduced dimensionality, and the greater choice of classifiers that could be applied.

Alternatively, dimensionality reduction or rather, prototype selection techniques could be applied to D^{bag} or D^{inst} directly. This adds several more variables to the problem under investigation: which method for selection is used, and how many prototypes are selected. We do not pursue this line of investigation further, but we refer the reader to [19] for an overview of possible techniques.

4   Experiments

4.1   Artificial Data

Fig.2 shows two artificial datasets that help to gain some more understanding about the different representations. The first dataset originates from [16] and shows a classical concept in the middle of the plot. We call this the "Concept" dataset. A positive bag here consists of one such concept instance, the other $n_i - 1$ instances are from the background distribution, whereas negative bags have n_i instances from the background. In the second datasets, instances of positive and negative bags are generated by two Gaussians with the same mean, but different variance. We call this the "Distribution" dataset.

The Concept dataset has N bags with 25 instances each. Due to the dense concept, distances of the concept instances are informative: they are lower for positive bags, than for negative bags. In this case, a sparse classifier used on the $N \times 25N$ matrix D^{inst} should be able to find these informative distances. Averaging over the distances as in the $N \times N$ matrix D^{bag}, however, would dilute this important information. Indeed, from the learning curve we can see that D^{bag} performs very poorly in this case.

The Distribution dataset also has bags with 25 instances each. Here, the bag as a whole is a more discriminative source of information than a particular instance, because the distributions overlap. D^{inst} and D^{bag} would both contain the necessary information to classify the bag correctly, so the extra flexibility of D^{inst} would only result in more computation, not better classifiers. The learning curve also demonstrates that D^{bag} provides enough information for good performance.

4.2   Real-Life MIL Data

We test all representations on several MIL datasets. Because of the number of different experiments and the running times using the instance-based representation D^{inst} and, we limit ourselves to a few MIL datasets with a reasonable total number of instances. A list is shown in Table 1.

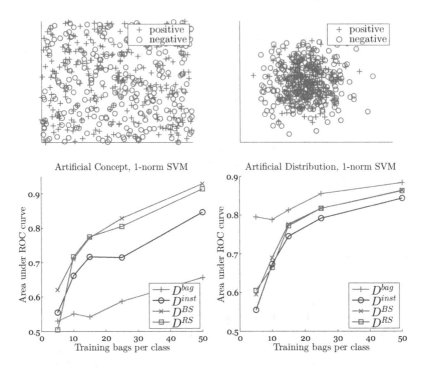

Fig. 2. Artificial datasets and corresponding learning curves. In the datasets, + and ◯ are instances of positive and negative bags respectively.

The Musk datasets[8] are molecule activity prediction problems, where bags are molecules and instances are different conformers (thus with different activity) of these molecules. Fox, Tiger and Elephant are image datasets, where the bags are images and instances are segments (of which at least some segments contain foxes, tigers or elephants). These datasets are strongly expected to have a concept, and methods that explicitly search for concept instances, have been quite successful.

In Newsgroups[24] and Web Recommendation [23], both text categorization datasets, the situation might be different. In Newsgroups, a bag is a collection of posts where a post is represented by counts of frequently-occurring words. At the first glance, it seems that this is a typical Concept-type dataset: a positive bag for the category "politics" contains 3% of posts about politics, whereas negative bags contain only posts about other topics. What is different here, is that posts about politics may have nothing in common and thus be very far apart in the feature space, unlike the concept instances in the artificial Concept dataset.

In the bird song datasets [3], a bag is a audio fragment consisting of bird songs of different species. Whenever a particular species is heard in the fragment, the bag is positive for that category. It could be expected that birds of the same species have similar songs, therefore there should be different concepts for different bird species. It is also possible that some species are heard together

Table 1. MIL Datasets. Number of positive and negative bags as well as the total, minimum, average and maximum number of instances per bag are specified.

Dataset	+bags	-bags	total	min	mean	max
Musk 1 [8]	47	45	476	2	5	40
Musk 2 [8]	39	63	6598	1	65	1044
Fox [1]	100	100	1302	2	7	13
Tiger[1]	100	100	1220	1	6	13
Elephant [1]	100	100	1391	2	7	13
Alt.atheism [24]	50	50	5443	22	54	76
Rec.motorcycles [24]	50	50	4730	22	47	73
Politics.mideast [24]	50	50	3376	15	34	55
Web recommendation 1 [23]	21	92	2212	4	30	131
Web recommendation 4 [23]	88	25	2291	4	31	200
Web recommendation 7 [23]	54	59	2400	4	32	200
Brown Creeper [3]	197	351	10232	2	19	43
Winter Wren [3]	109	439	10232	2	19	43
Pacific slope Flycatcher [3]	165	383	10232	2	19	43

more often.[1] In this case, instances which are negative for one species, could still be helpful in classifying fragments as containing that species or not.

We want to compare different data representations using the same base classifier, therefore, this classifier should be applicable to both large (D^{inst}) and small (D^{BS}, as some bags may contain just 2 or 3 instances) dimensionalities. We use the 1-norm SVM (or Liknon classifier [2]) and the Winnow classifier [15] as classifiers which are able to select a few informative dissimilarities. Furthermore, we use the logistic classifier and the support vector classifier LIBSVM [4] with a linear kernel to compare the results when no such explicit selection is taking place.

Each dataset and classifier combination is tested using the four representations D^{bag}, D^{inst}, D^{BS} and D^{RS}. For D^{RS} we let the number of classifiers is equal to the number of bags (just as in D^{BS}), the number of instances for each subspace is set to the average number of instances per bag. Both ensembles are combined by averaging the posterior probabilities of the individual classifiers. These settings are chosen as reasonable default settings for a fair comparison. We use the area under the receiver-operating-characteristic (AUC) as the evaluation measure, because this is found to be more discrimnative between classifiers[13] and more suitable for MIL problems[20]. Note that many other MIL papers use the accuracy as the evaluation measure and the results cannot be compared directly.

5   Results and Discussion

The results are shown in Tables 2 and 3. Overall, on these datasets D_{bag} performs the best, followed by D^{inst} and D^{RS}, and D^{BS} in the last place. To ease the

---

[1] We have verified this, and this is indeed true for some species, e.g. the labels of Winter Wren and Pacific-slope Flycatcher have a correlation of 0.63.

Table 2. AUC and standard error ($\times 100$) of Winnow and Liknon classifiers, 5×10 cross-validation. Bold indicates results not significantly worse than best per dataset.

	Dataset	D^{bag}	D^{inst}	D^{BS}	D^{RS}
Winnow	Musk1	**89.6 (1.6)**	**90.2 (1.4)**	83.6 (1.8)	83.1 (1.9)
	Musk2	83.6 (1.9)	84.0 (2.4)	85.6 (2.2)	**88.4 (2.0)**
	Fox	62.2 (1.8)	**65.7 (1.7)**	51.4 (1.9)	54.3 (1.8)
	Tiger	83.0 (1.6)	**85.8 (1.3)**	76.2 (2.0)	80.7 (1.9)
	Elephant	**90.1 (1.0)**	89.2 (1.0)	80.3 (1.5)	84.3 (1.4)
	alt.atheism	**70.0 (2.7)**	54.0 (2.7)	57.3 (2.8)	55.0 (2.8)
	rec.motorcycles	**73.7 (2.8)**	54.2 (2.6)	54.8 (2.3)	54.5 (2.8)
	pol.mideast	**70.8 (2.4)**	63.4 (2.3)	61.3 (2.4)	59.7 (2.2)
	Web1	79.9 (2.8)	80.7 (2.8)	75.9 (3.1)	**87.3 (2.5)**
	Web4	**76.4 (2.7)**	62.7 (3.0)	70.6 (3.3)	70.6 (2.9)
	Web7	67.8 (2.5)	**74.6 (2.7)**	66.7 (2.5)	**71.7 (2.7)**
	Brown Creeper	67.9 (1.2)	72.6 (1.4)	**85.7 (0.7)**	85.7 (0.7)
	Winter Wren	**94.0 (1.0)**	**93.7 (1.1)**	90.7 (1.5)	86.0 (1.8)
	PS Flycatcher	79.5 (1.5)	80.9 (1.1)	**86.1 (0.9)**	84.5 (1.1)
Liknon	Musk1	90.9 (1.4)	**93.8 (1.0)**	76.0 (2.4)	79.8 (2.3)
	Musk2	**91.1 (1.8)**	90.4 (1.8)	84.6 (2.1)	**90.3 (1.6)**
	Fox	**67.8 (1.7)**	66.4 (1.6)	56.4 (2.0)	61.8 (2.1)
	Tiger	**86.7 (1.5)**	87.9 (1.3)	83.4 (1.5)	84.9 (1.5)
	Elephant	**90.9 (1.0)**	90.3 (0.9)	85.2 (1.5)	88.1 (1.3)
	alt.atheism	**65.5 (2.5)**	56.1 (2.8)	59.0 (2.5)	59.2 (2.6)
	rec.motorcycles	**72.6 (2.5)**	55.0 (2.7)	51.3 (2.4)	50.0 (2.5)
	pol.mideast	**69.1 (2.6)**	**63.8 (2.5)**	50.4 (2.4)	50.2 (2.3)
	Web1	82.1 (2.6)	77.7 (2.6)	81.9 (2.7)	**88.7 (2.0)**
	Web4	**76.7 (3.1)**	55.0 (3.3)	69.8 (2.9)	**72.5 (2.8)**
	Web7	**69.5 (2.9)**	59.8 (3.2)	62.5 (2.6)	**70.4 (2.9)**
	Brown Creeper	**87.1 (0.6)**	**87.6 (0.6)**	85.9 (0.6)	86.0 (0.6)
	Winter Wren	**96.8 (0.4)**	**96.8 (0.5)**	**97.2 (0.3)**	97.0 (0.3)
	PS Flycatcher	**89.5 (0.6)**	**89.2 (0.6)**	86.4 (0.7)	84.6 (0.7)

comparison, we performed a Friedman rank test[7] on each classifier. The ranks are shown in Table 4. Although there are not always significant differences, the ordering of the ranks follows the same pattern in each case.

The fact that D^{bag} and D^{inst} perform comparably (except Newsgroups, as will be explained later) suggests that D^{bag} is able to capture the important information that D^{inst} contains. One conclusion is that real life datasets are less like the Concept, and more like the Distribution dataset from Fig.2. In other words, even the non-concept instances in positive bags may be very informative. Consider images of foxes and tigers. Because foxes and tigers live in a different habitats, the parts of the images containing trees, sand and so forth can tell us something about which animal is probably in the image. Or, as in the bird songs datasets, some birds species can be heard together often.

Although D^{inst} potentially contains more information than D^{bag}, there are few cases where it is a clear winner in terms of performance. It is possible that the low sample size of these datasets limits the full potential of D^{inst}: there are just too many features to deal with. It must be noted that D^{inst} is not completely the same as MILES[5] because there, an exponential similarity function is used which gives more importance to low distances. The effects of such a transformation (on all studied representations) are left for further investigation.

One surprising result is that for Newsgroups, only D^{bag} is able to produce some reasonable performances. One of the reasons is that positive and negative bags

Table 3. AUC and standard error ($\times100$) of support vector and logistic classifiers, 5×10 cross-validation. Bold indicates not significantly worse than best per dataset.

	Dataset	D^{bag}	D^{inst}	D^{BS}	D^{RS}
LIBSVM	Musk1	**93.3 (1.2)**	**93.7 (1.4)**	76.5 (2.4)	80.0 (2.3)
	Musk2	**93.4 (1.4)**	**93.7 (1.3)**	86.5 (2.0)	90.6 (1.7)
	Fox	**67.9 (1.5)**	**67.9 (1.5)**	52.9 (2.0)	60.6 (2.0)
	Tiger	**88.1 (1.5)**	**86.5 (1.4)**	83.2 (1.5)	84.8 (1.5)
	Elephant	**91.3 (0.8)**	89.3 (1.0)	85.6 (1.6)	87.4 (1.3)
	alt.atheism	**70.7 (2.5)**	48.8 (2.7)	49.2 (2.4)	51.1 (2.8)
	rec.motorcycles	**71.4 (2.4)**	42.9 (2.7)	39.0 (2.2)	40.4 (2.5)
	pol.mideast	**69.4 (2.3)**	50.6 (3.0)	52.4 (2.3)	51.8 (2.5)
	Web1	**86.1 (2.4)**	80.2 (2.7)	83.4 (2.1)	**89.6 (1.8)**
	Web4	**75.1 (2.8)**	60.4 (3.2)	**70.7 (3.2)**	**73.7 (2.7)**
	Web7	**71.0 (2.8)**	**68.7 (3.0)**	58.6 (2.8)	**71.6 (2.8)**
	Brown Creeper	**88.0 (0.6)**	**87.6 (0.6)**	84.6 (0.7)	84.2 (0.7)
	Winter Wren	**97.2 (0.4)**	95.7 (0.4)	96.5 (0.4)	96.2 (0.4)
	PS Flycatcher	**89.3 (0.6)**	**89.0 (0.7)**	86.6 (0.7)	84.9 (0.7)
Logistic	Musk1	**92.3 (1.4)**	**91.2 (1.6)**	80.6 (2.4)	83.8 (2.1)
	Musk2	91.4 (1.3)	86.2 (1.8)	91.5 (1.4)	**93.7 (1.1)**
	Fox	**67.1 (1.4)**	**66.4 (1.3)**	60.4 (1.9)	63.5 (1.8)
	Tiger	84.4 (1.5)	**86.3 (1.3)**	**85.0 (1.4)**	**85.4 (1.5)**
	Elephant	**89.2 (0.9)**	**88.8 (1.0)**	86.6 (1.4)	**88.8 (1.2)**
	alt.atheism	**71.3 (2.6)**	55.7 (2.6)	54.2 (2.7)	54.3 (2.7)
	rec.motorcycles	**75.3 (2.5)**	56.2 (2.6)	52.4 (2.6)	50.7 (2.7)
	pol.mideast	**69.9 (2.3)**	61.1 (2.3)	54.1 (2.4)	55.0 (2.2)
	Web1	**87.1 (2.2)**	79.0 (2.9)	78.0 (2.7)	80.4 (2.6)
	Web4	**79.4 (2.9)**	64.3 (3.1)	73.0 (3.3)	73.2 (3.6)
	Web7	69.6 (2.9)	**72.7 (2.8)**	69.3 (3.1)	**75.8 (2.9)**
	Brown Creeper	75.5 (1.0)	80.2 (0.8)	**86.9 (0.6)**	86.6 (0.6)
	Winter Wren	91.5 (0.7)	**94.8 (0.4)**	93.3 (0.4)	91.9 (0.5)
	PS Flycatcher	75.5 (0.9)	80.4 (0.8)	**84.8 (0.7)**	84.4 (0.7)

contain many instances that are very close together: similar to the background instances in the Concept dataset in Fig.2. However, a few true positive instances are very far away from all other instances, and from each other. There are so few of them, that even the sparse classifiers using D^{inst} are not able to select only the correct ones. However, they are so far away from everything, that they can still sufficiently influence the dissimilarities in D^{bag}. The asymmetry of D^{bag} also plays a role here; by transposing D^{bag}, much better results can be achieved [6]. However, this is not as straightforward for D^{inst}, so we did not pursue this possibility here.

It is interesting that D^{RS} often performs better than D^{BS}. To find out why, we examined the performances of the individual classifiers of both ensembles. Some typical results are shown in Fig.3. It is, indeed, often the case that D^{RS} produces more accurate classifiers. One reason for this could be the dimensionalities per classifier: D^{BS} can often have classifiers built on just 2 or 3 dimensional spaces (especially for Musk, Fox, Tiger and Elephant). However, sometimes D^{BS} and D^{RS} have classifiers with similar performances, but the ensemble using D^{RS} is still much better. This suggests that classifiers built on each bag separately provide more correlated information, than classifiers built on random selections of instances.

A way to improve the performance of an ensemble is to use a trained combiner which would learn which individual classifier outputs most often correspond with

Table 4. Ranks of Friedman test (best possible is 1, worst is 4). The null hypothesis H_0 is that there are no significant differences between ranks. H_0 is rejected (for significance of 5%) when the F-value of the ranks is larger than the critical value (CV). Significant differences are those larger than the critical difference (CD).

Classifier	D^{bag}	D^{inst}	D^{BS}	D^{RS}	F	CV	Reject H_0?	CD
Winnow	2.14	2.36	2.85	2.64	0.82	2.85	No	-
Liknon	1.50	2.50	3.29	2.71	6.48	2.85	Yes	1.25
LIBSVM	1.29	2.64	3.36	2.71	10.86	2.85	Yes	1.25
Logistic	2.14	2.36	3.07	2.43	1.38	2.85	No	-

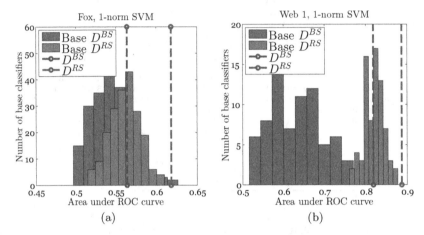

Fig. 3. Histograms of performances of individual base classifiers and the final ensembles of D^{BS} and D^{RS} for the Fox and Web Recommendation 1 datasets

the true labels of the training set. This could filter out the less accurate classifiers from the ensemble, increasing the overall performance. Looking at Fig.3, we would expect such performance improvements to be possible, especially for D^{BS}. Following [9], we have performed a few experiments with the nearest mean combiner, both on D^{BS} and D^{RS}. The results, however, were quite disappointing: for both ensembles, only minor improvements, if any, could be achieved. A possible cause for this is that nearest mean combiner was trained on normalized posterior probabilities, while the original classifier outputs might have been more informative.

We have found that for D^{BS}, there is little relation between the label of the prototype and the performance of the classifier. This is in line with the idea that positive and negative bags may not have the same type of background instances. On the other hand, we have found medium to strong correlations between dimensionality and the AUC of the individual classifiers. It might be worth investigating whether this can help us to select more informative prototypes a priori, before creating the dissimilarity matrix. Furthermore, there might be room for improvement for D^{RS}. The subspaces are allowed to be larger than the average number of instances per bag, because the dissimilarities are sampled with replacement.

6   Conclusion

We examined several dissimilarity representations for Multiple Instance Learning. These representations are based on distances of bags to prototype bags or instances. We investigated how such distances can be combined in order to create informative dissimilarities, and how this affects the dimensionality of the final representation. We considered combining such distances by averaging, by concatenating or by ensembling subspace classifiers, where each classifier is trained on a selection of the instance distances.

Averaging instance distances into a bag-based representation reduces the dimensionality and performs very well. Although the concatenated, instance-based representation is potentially the most informative, its rather high dimensionality might be harmful for performance. Lower dimensionality can also be achieved by combining subspace classifiers. However, in this case it is more difficult to achieve good performances because more variables, such as subspace size and the combining rule, are involved. It remains a question how to create and select such informative subspaces.

The bag representation produces good results, which means that averaging does not dilute the information of the individual instances. This suggests that in practice, most instances in a bag can be informative. In other words, the distributions of instances from positive and negative bags may be very different in general, and not only in terms of the presence or absence of a concept. A reasonable conclusion is that in such cases, it is better to use the bag representation, which requires less resources but still provides good performances.

References

1. Andrews, S., Hofmann, T., Tsochantaridis, I.: Multiple instance learning with generalized support vector machines. In: National Conference on Artificial Intelligence, pp. 943–944 (2002)
2. Bhattacharyya, C., Grate, L., Rizki, A., Radisky, D., Molina, F., Jordan, M., Bissell, M., Mian, I.: Simultaneous classification and relevant feature identification in high-dimensional spaces: application to molecular profiling data. Signal Processing 83(4), 729–743 (2003)
3. Briggs, F., Lakshminarayanan, B., Neal, L., Fern, X., Raich, R., Hadley, S., Hadley, A., Betts, M.: Acoustic classification of multiple simultaneous bird species: A multi-instance multi-label approach. J. Acoust. Soc. of America 131, 4640 (2012)
4. Chang, C., Lin, C.: Libsvm: a library for support vector machines. ACM Transactions on Intelligent Systems and Technology 2(3), 27 (2011)
5. Chen, Y., Bi, J., Wang, J.: Miles: Multiple-instance learning via embedded instance selection. Pattern Analysis and Machine Intelligence 28(12), 1931–1947 (2006)

6. Cheplygina, V., Tax, D., Loog, M.: Class-dependent dissimilarity measures for multiple instance learning. In: Gimel'farb, G., Hancock, E., Imiya, A., Kuijper, A., Kudo, M., Omachi, S., Windeatt, T., Yamada, K. (eds.) SSPR&SPR 2012. LNCS, vol. 7626, pp. 602–610. Springer, Heidelberg (2012)
7. Demšar, J.: Statistical comparisons of classifiers over multiple data sets. The Journal of Machine Learning Research 7, 1–30 (2006)
8. Dietterich, T., Lathrop, R., Lozano-Pérez, T.: Solving the multiple instance problem with axis-parallel rectangles. Artificial Intelligence 89(1-2), 31–71 (1997)
9. Duin, R.P.W., Tax, D.M.J.: Experiments with classifier combining rules. In: Kittler, J., Roli, F. (eds.) MCS 2000. LNCS, vol. 1857, pp. 16–29. Springer, Heidelberg (2000)
10. Foulds, J., Frank, E.: A review of multi-instance learning assumptions. Knowledge Engineering Review 25(1), 1 (2010)
11. Foulds, J., Frank, E.: Revisiting multiple-instance learning via embedded instance selection. In: Wobcke, W., Zhang, M. (eds.) AI 2008. LNCS (LNAI), vol. 5360, pp. 300–310. Springer, Heidelberg (2008)
12. Ho, T.: The random subspace method for constructing decision forests. IEEE Transactions on Pattern Analysis and Machine Intelligence 20(8), 832–844 (1998)
13. Huang, J., Ling, C.: Using auc and accuracy in evaluating learning algorithms. IEEE Transactions on Knowledge and Data Engineering 17(3), 299–310 (2005)
14. Kittler, J.: Combining classifiers: A theoretical framework. Pattern Analysis & Applications 1(1), 18–27 (1998)
15. Littlestone, N.: Learning quickly when irrelevant attributes abound: A new linear-threshold algorithm. Machine Learning 2(4), 285–318 (1988)
16. Maron, O., Lozano-Pérez, T.: A framework for multiple-instance learning. In: Advances in Neural Information Processing Systems, pp. 570–576. Morgan Kaufmann Publishers (1998)
17. Pękalska, E., Duin, R.P.W.: On combining dissimilarity representations. In: Kittler, J., Roli, F. (eds.) MCS 2001. LNCS, vol. 2096, pp. 359–368. Springer, Heidelberg (2001)
18. Pękalska, E., Duin, R.P.W.: The dissimilarity representation for pattern recognition: foundations and applications, vol. 64. World Scientific Pub. Co. Inc. (2005)
19. Pękalska, E., Duin, R.P.W., Paclík, P.: Prototype selection for dissimilarity-based classifiers. Pattern Recognition 39(2), 189–208 (2006)
20. Tax, D.M.J., Duin, R.P.W.: Learning curves for the analysis of multiple instance classifiers. In: da Vitoria Lobo, N., Kasparis, T., Roli, F., Kwok, J.T., Georgiopoulos, M., Anagnostopoulos, G.C., Loog, M. (eds.) SSPR&SPR 2008. LNCS, vol. 5342, pp. 724–733. Springer, Heidelberg (2008)
21. Tax, D.M.J., Loog, M., Duin, R.P.W., Cheplygina, V., Lee, W.-J.: Bag dissimilarities for multiple instance learning. In: Pelillo, M., Hancock, E.R. (eds.) SIMBAD 2011. LNCS, vol. 7005, pp. 222–234. Springer, Heidelberg (2011)
22. Weidmann, N., Frank, E., Pfahringer, B.: A two-level learning method for generalized multi-instance problems. In: Lavrač, N., Gamberger, D., Todorovski, L., Blockeel, H. (eds.) ECML 2003. LNCS (LNAI), vol. 2837, pp. 468–479. Springer, Heidelberg (2003)
23. Zhou, Z., Jiang, K., Li, M.: Multi-instance learning based web mining. Applied Intelligence 22(2), 135–147 (2005)
24. Zhou, Z., Sun, Y., Li, Y.: Multi-instance learning by treating instances as non-iid samples. In: Int. Conf. on Machine Learning, pp. 1249–1256. ACM (2009)

Similarity Weighted Ensembles for Relocating Models of Rare Events

Claire D'Este and Ashfaqur Rahman

Intelligent Sensing and Systems Laboratory*, CSIRO
Castray Esplanade, Hobart, Australia 7000
claire.deste@csiro.au

Abstract. Spatially distributed regions may have different influences that affect the underlying physical processes and make it inappropriate to directly relocate learned models. We may also be aiming to detect rare events for which we have examples in some regions, but not others. A novel method is presented for combining classifiers trained on regions with known sensor data and predicting rare events in new regions, specifically the closure of shellfish farms. The proposed similarity weighted ensemble method demonstrates an average 10 fold improvement in accuracy over One Class classification and 3 fold improvement over rules hand-crafted by an expert.

1 Introduction

Environmental and biological monitoring applications often require detection of previously unseen events; including harmful algal blooms [1], and tsunamis [2]. We may have records of events from some of our monitoring regions, but not others. Models may not be able to be directly relocated because of differences in the underlying physical processes at the different locations influenced by factors such as geography, meteorology, human activity, and land use. Monitoring shellfish farm contamination is an example of this problem.

Farmed shellfish contaminated with harmful microbes can cause significant risk to public health. Areas where shellfish are grown in Australia, and many other parts of the world, are monitored to ensure product does not make it to market that might cause serious illness or death when consumed. Farmers desire the shortest possible closure times, as being unable to harvest cost them around US$5,000 per day and, in some areas, closures may last for months.

Closure rules have been developed by the public health authority based on the response of thermotolerant coliforms (faecal bacteria) to changes in salinity, rainfall and river flow. The salinity, rainfall and river flow are then monitored remotely to ensure they have not exceeded thresholds. A new model is created for each growing zone because of the effect of geography on the physical processes as well as differences in distance and availability of sensors.

---

* Assisted by the Tasmanian Government administered by the Tasmanian Department of Economic Development, Tourism and the Arts and the CSIRO Food Futures Flagship.

Z.-H. Zhou, F. Roli, and J. Kittler (Eds.): MCS 2013, LNCS 7872, pp. 25–36, 2013.

The Aquaculture Decision Support (AquaDS) project has been focusing on developing a real-time decision support system for the health authority to reduce the dependence on a single expert, and provide nowcast and forecasts of possible closures over the web to allow farmers to manage proactively. The AquaDS project is developing data-driven models to predict closures from a database of manual samples linked to closure dates. The closures, however, are relatively rare with most growing zones having less than 10% manual samples taken when they were closed. Although we have demonstrated this problem can be learned with standard machine learning techniques using class balancing techniques, such as bagging and boosting [3] there are still many locations where we have insufficient, or no, examples of closures. Delivering a decision support system that predicts for only approximately 20% of growing zones will not be satisfactory. There is also a high probabilty in the future that new growing zones will open and it would be beneficial to begin providing closure predictions as soon as possible.

This paper investigates exploiting models developed for other locations that have sufficient positive examples. The hypothesis is that we can utilise each negative example to determine to what degree we can reuse models from other locations. We use the example of the shellfish farm contamination problem to explore this approach.

We are not able to measure coliform levels in real-time for use in the decision support system, however we do have coliform levels for every manual sample we use training the classification models. The proposed solution classifies farm closures with an ensemble of classifiers each weighted by their similarity in response to thermotolerant coliforms. The similarity weights are derived by creating a regression model for each zone to predict coliforms and calculating the accuracy of its ability to predict coliforms for each other zone.

2   Related Work

Chigbu et al. [4] developed a system for regulators that integrated rainfall and streamflow data, from the United States agencies National Oceanic and Atmospheric Administration (NOAA) and the National Weather Service (NWS), and compared with regulated thresholds to recommend a close/open decision. Their work did not seek to ingest feedback from decisions, make predictions and/or address the needs of the farmers themselves.

Kelsey [5] investigated the assumed relationship between rainfall/streamflow and faecal coliforms. They performed a multiple-parameter regression analysis over a much wider range of historical environmental and water sampling data in four US estuaries to identify whether closure decisions would be better based on other proxies. They point out that "These ... could be used to develop real-time predictions of bacteria concentration for use in a closure decision system. Rainfall measurements at gauges or from radar are available in near-real time through the NOAA NWS, and it may be relatively straightforward to develop real-time data sources for salinity and water temperature as well."

Biological monitoring also has the need to detect rare events or unusual conditions [6] that need to be generalised to different patients with potentially very different characteristics.

Rare event detection is generally performed using class imbalance techniques that alter the training set using oversampling of positive examples and/or undersampling of negative examples [7]. However, these still require positive examples to be available.

One class classification [8] provides the ability to learn in the absence of examples of the target class. A model is created on the negative examples and outliers are presumed to represent the target class.

Minku and Yao [9] use the output of different software development companies to predict the productivity of another company using regression trees with an accuracy weighted majority vote.

Ensemble approaches consistently demonstrate improved performance on classification tasks [10], but the use of ensembles is less explored on one rare event problems. Tax et. al. [11] showed an improvement in classifying via a one class model when an ensemble of One Class classifiers is used. However, these techniques do not allow the reuse of models for which we might have adequate examples.

3   Expert Closure Rules

As outlined in Section 1, closure rules have been developed by a human expert that consist of thresholds in salinity, rainfall or river flow that will prompt a closure. For example, the expert closure rule at Fleurty's Point is 'salinity < 30'.

The thresholds themselves are developed by graphing the environmental phenomenon (salinity, rainfall or river flow) against the thermotolerant coliform levels. The expert then determines the point at which the coliform level rises to unsafe level. The authorities use some base statistics on the number of coliforms per 100 millilitres; the median must be below 14, and 90% of samples must be under 21.

Figure 1 demonstrates this method for salinity at Fleurty's Point. We can visualise that the percentage of samples with over 21 coliforms per 100ml starts rising when salinity drops below 31 PSU. The actual salinity threshold is < 30 PSU for this zone. In comparison, Great Swanport West shows risky coliform levels rising when salinity drops under 25 PSU (Figure 2), which is consistent with the actual salinity trigger of < 26. Figure 3 shows the same method for river flow at Great Swanport West. We see the coliform levels rising dramatically when there is 5 cumecs of river flow, which is consistent with the actual flow threshold of > 5 cumecs.

The manager performs outlier detection and adjustments based on her expertise. For example, accounting for sensor drift on various platforms. She also monitors trends in rainfall, which are not captured by these threshold-based rules. This creates a system in which human interpretation of the sensor data is crucial to when the rule is applied in reality, and farms actually closed.

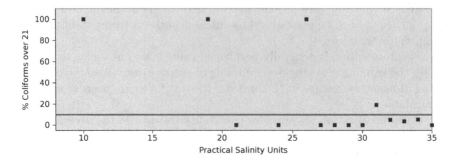

Fig. 1. Expert decision visualisation for Fleurty's Point

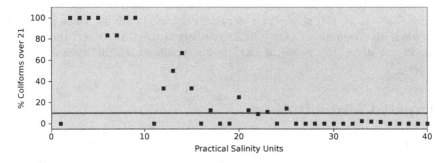

Fig. 2. Expert decision visualisation for Great Swanport West

4   Similarity Weighted Ensembles

Due to the lack of positive examples for many locations we have developed an ensemble method that makes use of locations for which we do have sufficient examples.

Our previous work, confirmed with the health authority, has shown that the coliforms themselves are not good direct predictors for closure [3] as the authorities want to close farms in anticipation of rises in the harmful bacteria. The real-time decision support system will want to warn of closures based on the causes of rainfall and river flow rather than waiting until faecal bacteria is already high. However, how coliform levels are affected by trends in rainfall and river flow in the negative example should give us an indication that two zones are similar, and provide a basis for prediction of the class of interest.

A regression model r is trained for each location to predict the coliform level, which is a continuous value representing the level of faecal bacteria present per 100 millilitres. The regression model is then evaluated for each of the other

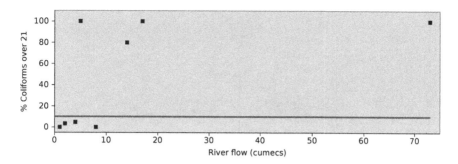

Fig. 3. Expert decision visualisation for Great Swanport West

locations on a dataset X that is different to the training and final evaluation test set. Equation 1 shows how a weight w for a classification model c for location l is derived from the accuracy of the regression model r on location l, where t represents the target value for each example x. Only negative examples are used to train the regression model to simulate locations where we only have negative examples.

$$w_{cl} = 1 - \frac{\sum\limits_{x \in X} |t_{rx} - t_{lx}|}{|X|} \tag{1}$$

The weights are then scaled between 0 and 1.

Equation 2 shows how the class probability of an example x is calculated through the sum of the weighted probabilities from the classifiers from other locations C. The classifier of the current location l is removed from C. The approach is further illustrated in Figure 4.

$$p_l(x) = \sum_{c \in C} \frac{p_c(x) \cdot w_{cl}}{\sum\limits_{c \in C} w_{cl}} \tag{2}$$

The ensemble weighting method has similarities with the Accuracy Weighted Ensembles of Wang et al. [12], whereas the means of obtaining the weights is novel.

5   Experimental Setup

The dataset is derived from 18692 manual water samples taken by the Tasmanian Shellfish Quality Assurance Program (TSQAP) between 1983 and May 2012 from 45 shellfish locations all over Tasmania. This has been combined with environmental data from the Tasmanian Department of Primary Industry, Water and the Environment (DPIPWE) and the Bureau of Meteorology (BOM). It includes 66 features:

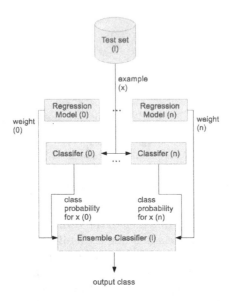

Fig. 4. Classification with Similarity Weighted Ensembles

- **1. Salinity** - salt level in the water sample in Practical Salinity Units (PSU) from manual sampling.
- **2. Coliforms** - Level of faecal bacteria present per 100 millilitres from manual water sampling.
- **3. Tide** - six discrete tide states from BOM; ranging from low rising to low falling.
- **4. Wind** - a discrete direction state from BOM, where 1-8 represent directions clockwise from North, and 9 represents a calm state.
- **5. Temperature** - air temperature in degrees celcius from BOM.
- **6. Biotoxins** - presence of harmful algae.
- **7-14. Rain at Site 1** - Rainfall in millimetres recorded for last 1 to 8 days from the closest BOM weather station.
- **15-21. Cumulative Rain at Site 1** - Cumulative rainfall in millimetres recorded for last 1 to 7 days.
- **22-36. Rain and Cumulative Rain at Site 2**
- **37-44. Flow at Site 1** river flow rate in cubic metres per second (cumecs) from the closest gauging station. This data is provided by DPIPWE.
- **45-52. Cumulative Flow at Site 1** river flow rate in cubic metres per second (cumecs).
- **53-66. Flow and Cumulative Flow at Site 2**

The sample dates have been merged with farm closure dates taken from TSQAP annual reports to create an output class, which is the *Open/Closed* decision made by the TSQAP manager. This is a much more complete data set than that used in [3], which had only five inputs; salinity, coliforms, location, rainfall and flow.

For the purposes of this work we describe an example with a *Closed* state a positive example and *Open* state a negative example.

There are 10 growing zones out of 45 that have greater than 9% of examples with a *Closed* state. In order to evaluate the proposed Similarity Weighted Ensemble method we focus on these 10 zones, so that we can determine actual versus predicted accuracy. We could perform these tests on the zones for which we do not have sufficient *Closed* examples, but it would be impossible to evaluate the prediction quality. Table 1 describes the class percentages for each of the selected zones.

Table 1. Percentages of each class for growing zones studied

Zone	Closed%	Open%	Instances	Zone	Closed%	Open%	Instances
Fleurtys Point	9.1	90.1	484	Hastings	15.8	84.2	941
Montagu	23.5	76.5	337	Moulting	11.7	88.3	367
Big Bay B	32.5	67.5	209	Big Bay C	22.2	77.8	266
Big Bay E	21.1	78.9	665	Duck Bay	35.5	64.5	900
Blackman	11.7	88.3	729	Deep Bay	15.4	84.6	415

5.1 Machine Learning Methods

Unless otherwise stated the machine learning methods were implemented using the Weka toolbox [13] with cross-validated parameter selection. The Weka classifiers used were:

Method	Abbreviation	Weka Classifier
Random Forest	RF	RandomForest
Decision Tree	DT	J48
Nearest Neighbour	KNN	IBk
Support Vector Machine	SVM	SMO
One Class	One Class	OneClassClassifier
Expert Rules	Expert	*Custom*

All ensembles used AdaBoost [14] to perform class balancing; as this proved simple and effective in previous experiments [3]. All experiments use 10 cross-fold validation.

Experimentation on the data set suggest that nearest neighbour was the most effective regression method so the Weka *IBk* library was chosen for determining

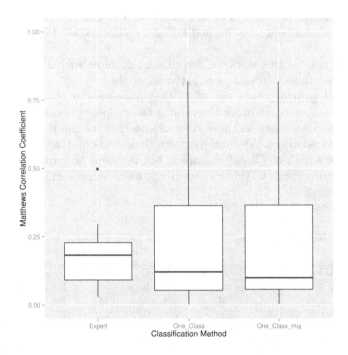

Fig. 5. Matthews Correlation Coefficient for all zones for One Class, One Class ensembles with majority voting (maj), and the Expert closure rules

weights. A new *Vote* meta-classifier was developed to implement the Similarity Weighted combination rule described in Section 3. The expert closure rules have also been modeled in a Weka Classifier so that we can also include them in ensembles.

The One Class Classifier uses the default settings of a REPT tree with Bagging. This method combines class probability and density estimation as described in [15]. The training set includes only negative examples, simulating a zone for which we only have *Open* states, and is tested on both a set with both *Open* and *Closed* states.

The learned models are compared using Matthews Correlation Coefficient, which is an effective method for evaluating a two-class classifier even when the classes are imbalanced [16]. Equation 3 describes the measure that is based on the confusion matrix where *tp* denotes the true positive count, and *fn* the false negative count.

$$mcc = \frac{tp \cdot tn - fp \cdot fn}{\sqrt{(tp + fp)(tp + fn)(tn + fp)(tn + fn)}} \tag{3}$$

6 Results

Figures 5 and 6 show the results of the shellfish farm closure dataset on all locations. Table 2 shows in detail three of the growing zones on the expert rules, different machine learning techniques and the simlarity weighted ensembles versus a majority voting scheme.

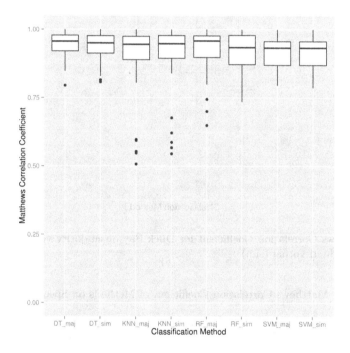

Fig. 6. Matthews Correlation Coefficient for all zones on majority voting (maj) and Similarity Weighted voting (sim)

Figure 7 shows the range of accuracy of the two voting schemes in the growing zone Duck Bay specifically.

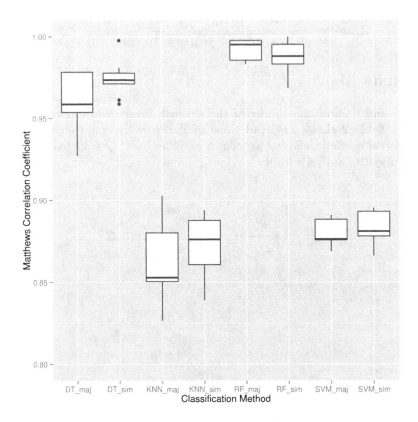

Fig. 7. Matthews Correlation Coefficient for Duck Bay on majority voting (maj) and Similarity Weighted voting (sim)

Table 2. Matthews Correlation Coefficient of Methods on Specific Zones

Method	Vote	Big Bay	Fleurty's	Moulting
Expert Rules	None	0.3	0.1	0.2
Expert Rules	Majority	0.3	0.2	0.2
One Class	None	-0.5 ($\pm$ 0.03)	-0.002 ($\pm$ 0.06)	0
One Class	Majority	-0.05 ($\pm$ 0.06)	0.05 ($\pm$ 0.07)	0.2 ($\pm$ 0.02)
Random Forest	Majority	0.82 ($\pm$ 0.07)	0.76 ($\pm$ 0.05)	0.89 ($\pm$ 0.07)
Random Forest	Similarity	0.85 ($\pm$ 0.06)	0.78 ($\pm$ 0.05)	0.85 ($\pm$ 0.07)
Decision Tree	Majority	0.92 ($\pm$ 0.07)	0.87 ($\pm$ 0.02)	0.85 ($\pm$ 0.03)
Decision Tree	Similarity	0.95 ($\pm$ 0.02)	0.9 ($\pm$ 0.03)	0.86 ($\pm$ 0.03)
KNN	Majority	0.93 ($\pm$ 0.02)	0.88 ($\pm$ 0.06)	0.56 ($\pm$ 0.07)
KNN	Similarity	0.95 ($\pm$ 0.02)	0.91 ($\pm$ 0.03)	0.62 ($\pm$ 0.08)
SVM	Majority	0.93 ($\pm$ 0)	0.87 ($\pm$ 0)	0.83 ($\pm$ 0.007)
SVM	Similarity	0.93 ($\pm$ 0)	0.87 ($\pm$ 0)	0.83 ($\pm$ 0.007)

7 Discussion

The manager performs outlier detection and adjustments based on her expertise. For example, accounting for sensor drift on various platforms. Figures 1 to 3 show significant numbers of examples where the threshold is exceeded, but the coliform levels do not rise. Because of these factors the closure rules do not predict closures well, as seen in Figure 5. There is also little benefit in combining the expert rules into an ensemble. Therefore it does not appear to be enough when a new growing zone is added to use an ensemble of closure rules until we have enough coliform examples. The creation of expert rules also requires manual samples with high coliform levels to be available, which may not occur for an unknown period of time. Reusing the data-driven models from other locations will allow predictions of closures to be made much sooner.

Using an ensemble approach produces significantly more accurate classifiers than using One Class classification with the medians of the Matthews Correlation Coefficient up to ten times higher at Fleurty's Bay with each machine learning method (Table 2). A small improvement is gained from an ensemble of One Class classifiers, however combining classifiers trained for other locations is much more successful.

The similarity weighting approach does indicate an improvement over a majority voting scheme for most zones. This is visible in Figure 7 where we can see an improved accuracy range for all methods except for Random Forest, although in Table 2 there is an improvement in all these zones. The biggest improvements are seen in the Nearest Neighbour methods. The average performance over all zones is similar between majority voting and similarity weighting.

The results suggest that the similarity technique may be preferable for some growing zones; potentially those that are different from the norm. Further work will be required on identifying the characteristics of these zones; possibly through an initial clustering stage.

8 Conclusion

This study can be applied to a range of rare event detection problems that are affected by different geographical factors. The data-driven approach was consistently shown to be significantly more accurate than the expert crafted rules.

In addition to standard techniques we have introduced a novel method that makes use of the physical processes learned from manual sampling programs and models from other locations. This showed an average 10 fold improvement over One Class classification for an ensemble of models from other locations, and an average improvement in accuracy of 2% when we weight by the similarity of the locations for many zones.

We have also determined that traditional ensemble classifiers are a significant improvement over one-class classification, which suggests that they would be useful in circumstances in which we do not have a observations of a relevant physical process to weight by.

References

1. Muttil, N., Chau, K.: Machine-learning paradigms for selecting ecologically significant input variables. Journal Engineering Applications of Artificial Intelligence 20(6), 735–744 (2007)
2. Bernard, E., Meinig, C.: History and future of deep-ocean tsunami measurements. In: OCEANS 2011, pp. 1–7. IEEE (2011)
3. D'Este, C., Rahman, A., Turnbull, A.: Predicting shellfish farm closures with class balancing methods. In: Thielscher, M., Zhang, D. (eds.) AI 2012. LNCS, vol. 7691, pp. 39–48. Springer, Heidelberg (2012)
4. Chigbu, P., Strange, T., Gordon, S., Jester, K., Baham, J., Young, J., Hughes, R., Remata, R., Martinolich, K., Hilbert, K., Mott, D., Watts, M., McIntosh, M.: Development of decision support tools for aquaculture: the pond experience. Journal of Shellfish Research 25(3), 1091–1099 (2006)
5. Kelsey, R., Scott, G., Porter, D., Siewicki, T., Edwards, D.: Improvements to shellfish harvest area closure decision making using gis, remote sensing and predictive models. Estuaries and Coasts 33, 712–722 (2010)
6. Choe, W., Ersoy, O., Bina, M.: Neural network schemes for detecting rare events in human genomic dna. Bioinformatics 16(12), 1062–1072 (2000)
7. Chawla, N., Bowyer, K., Hall, L., Kegelmeyer, W.: SMOTE: Synthetic minority over-sampling technique. Journal of Artificial Intelligence Research 16, 341–378 (2002)
8. Tax, D.: One-class classification. PhD thesis, Delft University of Technology (2001)
9. Minku, L.L., Yao, X.: Using unreliable data for creating more reliable online learners. In: The 2012 International Joint Conference on Neural Networks (IJCNN), pp. 1–8. IEEE (2012)
10. Opitz, D., Maclin, R.: Popular ensemble methods: An empirical study. Journal of Artificial Intelligence Research 11, 169–198 (1999)
11. Tax, D.M.J., Duin, R.P.W.: Combining one-class classifiers. In: Kittler, J., Roli, F. (eds.) MCS 2001. LNCS, vol. 2096, pp. 299–308. Springer, Heidelberg (2001)
12. Wang, H., Fan, W., Yu, P., Han, J.: Mining concept-drifting data streams using ensemble classifiers. In: Proceedings of the Ninth ACM SIGKDD International Conference on Knowledge Discovery and Data Mining, pp. 226–235. ACM (2003)
13. Hall, M., Frank, E., Holmes, G., Pfahringer, B., Reutemann, P., Witten, I.: The Weka data mining software: An update. SIGKDD Explorations 11(1) (2009)
14. Freund, Y., Schapire, R.: Decision-theoretic generalization of on-line learning and an application to boosting. Journal of Computer and System Sciences 55(1), 119–139 (1997)
15. Hempstalk, K., Frank, E., Witten, I.H.: One-class classification by combining density and class probability estimation. In: Daelemans, W., Goethals, B., Morik, K. (eds.) ECML PKDD 2008, Part I. LNCS (LNAI), vol. 5211, pp. 505–519. Springer, Heidelberg (2008)
16. Baldi, P., Brunak, S., Chauvin, Y., Andersen, C., Nielsen, H.: Assessing the accuracy of prediction algorithms for classification: an overview. Bioinformatics 16(5), 412–424 (2000)

Diversity in Classifier Ensembles: Fertile Concept or Dead End?

Luca Didaci, Giorgio Fumera, and Fabio Roli

Dept. of Electrical and Electronic Engineering, University of Cagliari
Piazza d'Armi, 09123 Cagliari, Italy
{luca.didaci,fumera,roli}@diee.unica.it
http://prag.diee.unica.it/en

Abstract. Diversity is deemed a crucial concept in the field of multiple classifier systems, although no exact definition has been found so far. Existing diversity measures exhibit some issues, both from the theoretical viewpoint, and from the practical viewpoint of ensemble construction. We propose to address some of these issues through the derivation of decompositions of classification error, analogue to the well-known bias-variance-covariance and ambiguity decompositions of regression error. We then discuss whether the resulting decompositions can provide a more clear definition of diversity, and whether they can be exploited more effectively for the practical purpose of ensemble construction.

Keywords: Diversity, Bias-variance-covariance decomposition, Ambiguity decomposition.

1 Introduction

The concept of "diversity" is deemed among the most important in the field of multiple classifier systems (MCSs), both theoretically, as a way to understand how MCSs work, and as a practical tool for constructing effective classifier ensembles [21,46]. However, its exact understanding and definition is still a relevant open issue. For instance, quoting from [46] (Sect. 5.1): "It is no doubt that understanding diversity is the holygrail in the field of ensemble learning".

Besides the obvious observation that combining identical classifiers is useless, the concept of diversity has roots in theoretical arguments (e.g., [36,22,15,9]), also inspired by other domains like software engineering [24,28]. In particular, it has been influenced by the *bias-variance-covariance* (BVC) [37] and the *ambiguity* [18,4] decompositions of the error of regressor ensembles. Moreover, wide empirical evidence motivates the potential usefulness of combining *non-identical* classifiers. This lead to the widely accepted idea that: (i) there exists a property of MCSs that can be defined as "diversity", can be quantitatively defined and thus measured, and is related to ensemble accuracy (together with the accuracy of individual classifiers); (ii) such a property can be practically exploited to construct an effective ensemble of classifiers.

Z.-H. Zhou, F. Roli, and J. Kittler (Eds.): MCS 2013, LNCS 7872, pp. 37–48, 2013.
© Springer-Verlag Berlin Heidelberg 2013

The concept of diversity has been investigated in the MCS literature from different aspects: several diversity measures have been defined (e.g., [17,26,8,13,32,20,1,40,23]); several methods for MCS construction have been proposed, explicitly using diversity measures (e.g., [41,13,1,2,16,40,23]); existing diversity measures have been analysed to understand whether and how much they are "correlated" with ensemble accuracy [32,20,35]; several authors analysed the concept of diversity itself [31,19,4,5,7].

However, although all the existing measures reflect intuitive notions of diversity, none of them has been derived from an exact decomposition of the ensemble error, contrary to the ambiguity decomposition; none of them exhibits a clear trade-off with the average error of ensemble members, in determining the ensemble error [20,35]; and effective techniques for MCS construction like bagging and boosting do not make explicit use of diversity measures. These issues led some authors even to question the practical usefulness of measuring diversity in MCSs: "[...] the question of the participation of diversity measures in designing classifier ensembles is still open. Directly calculating the accuracy for the chosen combination method makes more sense than calculating the diversity and trying to predict the accuracy. Even if the measure of diversity is easier to calculate than some combination methods, the ambiguous relationship between diversity and accuracy discourages optimising the diversity" [32] (sect. 7); "The quest for defining and using diversity might be heading toward a dead end or might result in powerful new ensemble-building methodologies" [21] (sect. 10.5); "It is not yet known whether diversity is really a driving force, or actually a trap since it might be just another appearance of accuracy" [46] (sect. 5.1).

On the basis of the above premises, in this paper we address the issue of diversity with the following goals: (i) deriving *exact* decompositions of the ensemble error for any combining rule and any number of classes; and exploiting them to understand (ii) whether they can provide a more clear understanding of diversity, and (iii) whether they can be exploited for ensemble construction, more effectively than existing measures. After an overview on BVC and ambiguity decompositions for regression problems in Sect. 2, in Sect. 3 we address issue (i) above by deriving the analogue of these decompositions for the ensemble classification error. In particular, we consider the Kohavi-Wolpert bias-variance decomposition [17] to derive a BVC-like decomposition, while our ambiguity-like decomposition generalises the one of [7]. We then address issues (ii) and (iii) above in Sect. 4. We finally suggest some directions for future work in Sect. 5.

2   Background: Decompositions of Regression Error

In regression problems, an unknown function has to be estimated using a set d of n samples of its input-output pairs, $(\mathbf{x}, y) \in \mathbb{R}^m \times \mathbb{R}$.[1] Assume that a learning algorithm is used, which produces the estimator $f(\mathbf{x}; d)$ when trained on d. To simplify notation, in the following we will write f in place of $f(\mathbf{x}; d)$.

---

[1] Throughout the paper we will use uppercase letters to denote random variables, and the corresponding lowercase letters to denote specific values.

The expectation of the mean squared error (MSE) of f on a given input $\mathbf{x}$, taken over random training sets D of size n and over $P[Y|\mathbf{x}]$, can be written in terms of the well-known bias-variance (BV) decomposition [12]:

$$E_{D,Y|\mathbf{x}}\left[(f - Y)^2\right] = bias_f^2 + var_f + noise , \tag{1}$$

where *noise* equals the variance of Y given $\mathbf{x}$, and is independent on D, while

$$bias_f^2 = (E_D[f] - E[Y|\mathbf{x}])^2 , \quad var_f = E_D\left[(f - E_D[f])^2\right] . \tag{2}$$

It is known that, usually, bias can be reduced only at the expense of a higher variance, and vice versa, and that an effective variance reduction technique consists of linearly combining an ensemble of N different regressors:

$$f_{\text{ens}}(\mathbf{x}; d) = \frac{1}{N} \sum_{i=1}^{N} f_i(\mathbf{x}; d) . \tag{3}$$

The BV decomposition of f_{ens} can be rewritten in the form of a bias-variance-covariance (BVC) decomposition [37,4]. Let us define

$$\overline{bias} = \tfrac{1}{N} \sum_j bias_{f_j} , \quad \overline{var} = \tfrac{1}{N} \sum_j var_{f_j} ,$$
$$\overline{cov} = \tfrac{1}{N(N-1)} \sum_{i,j \neq i} E_D\left[(f_i - E_D[f_i])(f_j - E_D[f_j])\right] . \tag{4}$$

It then follows that:

$$E_D[(f_{\text{ens}} - E[Y|\mathbf{x}])^2] = \overline{bias}^2 + \frac{1}{N}\overline{var} + (1 - \frac{1}{N})\overline{cov} . \tag{5}$$

This highlights that the variance reduction effect strongly depends on the amount of correlation between the outputs of individual regressors: the lower the correlation (i.e., the lower the term $\overline{cov}$, which can also be negative), the higher the reduction of variance.

The MSE of f_{ens} can also be written equivalently in terms of the *ambiguity* decomposition, which, for a given $(\mathbf{x}, y)$ and d, is given by [18,4]:

$$(y - f_{\text{ens}})^2 = \frac{1}{N} \sum_{i=1}^{N}(y - f_i)^2 - \frac{1}{N} \sum_{i=1}^{N}(f_i - f_{\text{ens}})^2 . \tag{6}$$

Differently from the BVC decomposition, the ambiguity decomposition highlights a trade-off between the average accuracy of individual regressors, and their deviation from the ensemble output. The latter term was called "ambiguity" (hence the name of the decomposition),[2] and can be easily interpreted in terms of diversity between individual regressors. Therefore, this provides a clear, formal definition of "diversity" for regression problems [4].

---

[2] The ambiguity term is related to the correlation among individual regressors. The beneficial effects of negative correlation had already been pointed out in [29].

From the practical viewpoint of ensemble construction, the ambiguity decomposition was successfully exploited by the Negative Correlation Learning (NCL) method [25].[3] NCL is a *parallel*, gradient-descent learning algorithm, whose objective function is given by the linear combination of the MSE of individual regressors (the first term of the ambiguity decomposition), minus a term proportional to the corresponding ambiguity. Instead of independently training the individual regressors first, and then computing the coefficients of their linear combination, NCL pursues both goals simultaneously. This may allow it to attain a better trade-off between accuracy and diversity.

In principle, the ambiguity decomposition could also be exploited in the context of an overproduce and choose strategy, for selecting the best subset of regressors s' out of a given, larger set s. The members of s can be first independently trained (by minimising their individual MSE); then, the subset s' exhibiting the highest ambiguity should be selected. However, all works on regressor ensemble selection we are aware of did not use this approach, but relied on the direct estimate of the ensemble MSE [43,30,14,39,27,34,38], with the only exception of [11]. However, in [11] diversity measures inspired by the ones defined for classification problems were used, instead of the clean ambiguity term. The above overproduce and choose strategy, implemented by maximising some diversity measure, is used by several classifier ensemble construction techniques, instead. We will further discuss this point in Sect. 4.

3   Decompositions of the Ensemble Classification Error

Several BV decompositions of classification error have been proposed, e.g., [3,17,10], and have been used to empirically investigate the variance (and sometimes bias) reduction effect of classifier combination techniques. However, no decomposition analogue to BVC (i.e., explicitly including the outputs of individual classifiers) has been derived yet. This is not straightforward, for instance because the concept of covariance is undefined for categorical outputs (class labels), as pointed out in [4]. Similarly, no ambiguity-like decomposition (i.e., including the average error of individual classifiers) has been derived for MCSs, with the only exception of the one of [7] for two-class problems. Accordingly, existing diversity measures have not been derived from exact decompositions of classification error. However, we point out that they have been empirically and theoretically analysed by investigating whether and how their trade-off with the average error of individual classifiers is related to the ensemble error. In other words, they have been (often implicitly) considered as the equivalent of the ambiguity term in the corresponding decomposition (6).

In the following we show how an analogue of the BVC decomposition can be derived, as well as the analogue of the ambiguity decomposition, which generalises the one of [7] to any number of classes.

---

[3] NCL was actually defined "heuristically" in [25], with no reference to the ambiguity decomposition. The strong relationship between NCL and the ambiguity decomposition was pointed out and thoroughly analysed in [5].

3.1   A Bias-Variance-Covariance Decomposition for Classifier Ensembles

We consider the Kohavi-Wolpert BV decomposition of classification error (0/1-loss), for a L-class problem [17]. We denote class labels by $y_1, \ldots, y_L$. To further simplify the notation, we define: $P[y_i] = P[Y = y_i | \mathbf{x}]$, and $\hat{P}[y_i] = P_D[f(\mathbf{x}; D) = y_i]$. The loss of a classifier $f(\mathbf{x}; d)$ on a given sample $(\mathbf{x}, y)$, which we denote by $e(\mathbf{x}, y; d)$, equals $I[f(\mathbf{x}; d) \neq y]$, where $I[a] = 1$ (0), if $a = $ True (False). Bias and variance are defined in [17] as follows:

$$
bias_f = \frac{1}{2} \sum_{y_i} \left(P[y_i] - \hat{P}[y_i] \right)^2, \quad var_f = \frac{1}{2} \left(1 - \sum_{y_i} \hat{P}[y_i]^2 \right) . \tag{7}
$$

It follows that [17]:

$$
E_{D,Y|\mathbf{x}}[e(\mathbf{x}, Y; D)] = bias_f + var_f + noise , \tag{8}
$$

where $noise = \frac{1}{2} \left(1 - \sum_{y_i} P[y_i]^2 \right)$.

We now rewrite the above bias and variance terms for a MCS $\{f_1, \ldots, f_N\}$. We denote by f_{ens} the ensemble output, with no restriction on the combining rule. Adding and subtracting to the expressions of $bias_{f_{ens}}$ and $var_{f_{ens}}$ the two terms indicated below, after some manipulations we obtain:

$$
\begin{aligned}
bias_{f_{ens}} &= \frac{1}{2} \sum_{y_i} \left(P[y_i] - \frac{1}{\sqrt{N}} \sum_j \hat{P}_j[y_i] + \frac{1}{\sqrt{N}} \sum_j \hat{P}_j[y_i] - \hat{P}_{ens}[y_i] \right)^2 \\
&= \overline{bias} + b, \\
var_{f_{ens}} &= \frac{1}{2} \left(1 - \frac{1}{N^2} \sum_{j,y_i} \hat{P}_j^2[y_i] + \frac{1}{N^2} \sum_{j,y_i} \hat{P}_j^2[y_i] - \sum_{y_i} \hat{P}_{ens}[y_i]^2 \right) \\
&= \frac{1}{N} \overline{var} + v ,
\end{aligned}
\tag{9}
$$

where $\overline{bias} = \frac{1}{N} \sum_j bias_{f_j}$, $\overline{var} = \frac{1}{N} \sum_j var_{f_j}$, and the terms b and v are given in the online appendix of this paper[4] (they are not reported here due to lack of space). This easily leads us to the analogue of the BVC decomposition (5):

$$
E_{D,Y|\mathbf{x}} e_{ens}(\mathbf{x}, Y; D) = \overline{bias} + \frac{1}{N} \overline{var} + b + v + noise . \tag{10}
$$

The term $b + v$ corresponds to the covariance term of (5), and its interpretation is under analysis at the time of submitting the camera-ready of this paper.

3.2   Ambiguity-Like Decomposition for Classifier Ensembles

The only decomposition of the classification error (0/1-loss) of an ensemble, analogue to the ambiguity decomposition, has been derived in [7], for two-class problems. Denoting the class labels by the values $\{-1, +1\}$, the loss of a classifier

---

[4] http://prag.diee.unica.it/pra/bib/didaciMCS2013

on a sample $(\mathbf{x}, y)$ can be expressed as $e_f(\mathbf{x}, y; d) = \frac{1}{2}(1 - y \times f)$. Denoting by $\overline{e}(\mathbf{x}, y; d)$ the average loss of an ensemble of N classifiers, it follows that [7]:

$$e_{\mathrm{ens}}(\mathbf{x}, y; d) = \overline{e}(\mathbf{x}, y; d) - y \times f_{\mathrm{ens}} \times \frac{1}{N} \sum_{j=1}^{N} \delta_j(\mathbf{x}, y; d) , \qquad (11)$$

where $\delta_j(\mathbf{x}, y; d) = \frac{1}{2}(1 - f_j \times f_{\mathrm{ens}})$. This term is a measure of the disagreement between classifier f_j and the ensemble. The decomposition (11) appears thus very similar to the ambiguity decomposition (6). However, the second term in the right-hand side (RHS) of (11) also includes the true class label y, contrary to the ambiguity term in (6). The interpretation of decomposition (11) is very clear: it shows that a lower average accuracy of individual classifiers can be compensated by a higher disagreement with the ensemble, as far as the ensemble remains correct. This latter condition is due to the presence of the y term in the RHS of (11). We point out that decomposition (11) is valid for any combining rule, although in [7] only majority voting was considered.

Here we show that a more general decomposition can be obtained, for any number of classes. To this aim, we can exploit the BVC-like decomposition (10). We denote the expected average misclassification probability of individual classifiers on a point $\mathbf{x}$, $E_{D,Y|\mathbf{x}}\left[\frac{1}{N}\sum_j e_j(\mathbf{x}, y; d)\right]$, by $\overline{e}(\mathbf{x})$. It is easy to see that $\overline{e}(\mathbf{x}) = \overline{bias} + \overline{var} + noise$. Rewriting (10) by adding and subtracting the term $\frac{N-1}{N}\overline{var}$, after some manipulations we obtain:

$$\begin{aligned} e_{\mathrm{ens}}(\mathbf{x}) &= \overline{e}(\mathbf{x}) - \sum_{y_i} P[y_i]\frac{1}{N}\sum_j \left(\hat{P}_{\mathrm{ens}}[y_i] - \hat{P}_j[y_i]\right) \\ &= \overline{e}(\mathbf{x}) - \frac{1}{N}\sum_j \left(P_{D,Y|\mathbf{x}}[f_{\mathrm{ens}} = Y|\mathbf{x}] - P_{D,Y|\mathbf{x}}[f_j = Y|\mathbf{x}]\right) . \end{aligned} \qquad (12)$$

The same result can also be obtained by directly computing $E_{D,Y|\mathbf{x}}[e_{ens}(\mathbf{X}, Y; D) - \overline{e}(\mathbf{X}, Y; D)]$, which is the approach followed in [7]. Obviously, for $L = 2$, the expectation of (11) with respect to $D, Y|\mathbf{x}$ equals (12).

We can further rewrite decomposition (12) in the case of a fixed training set d, i.e., by taking the expectation of $e_{\mathrm{ens}}(\mathbf{x}, y; d)$ with respect of $P[Y|\mathbf{x}]$ only:

$$e_{\mathrm{ens}}(\mathbf{x}; d) = \overline{e}(\mathbf{x}; d) - \frac{1}{N}\sum_j \left(P_{Y|\mathbf{x}}[f_{\mathrm{ens}} = Y|\mathbf{x}] - P_{Y|\mathbf{x}}[f_j = Y|\mathbf{x}]\right) . \qquad (13)$$

In particular, for a single sample $(\mathbf{x}, y)$ and a single training set d, we obtain the generalisation of (11) for $L > 2$:

$$e_{f_{\mathrm{ens}}}(\mathbf{x}, y; d) = \overline{e}(\mathbf{x}, y; d) - \frac{1}{N}\sum_j \left(I[f_{\mathrm{ens}} = y] - I[f_j = y]\right) . \qquad (14)$$

Expressions (12)–(14) are thus three different versions of a general ambiguity-like decomposition of the ensemble error, that is valid for any number of classes and any combining rule.[5] Comparing (12) to the BVC-like decomposition (10), to understand the correspondence between their terms (as done in [4] for regression error), is the subject of our ongoing work.

---

[5] This decomposition can also be easily extended to any loss function.

4 Discussion

Let us recall the second and third issues mentioned in the introduction. Can the second term of decomposition (12)–(14) be interpreted as a diversity measure? Can it be practically exploited for ensemble construction? In particular, is it more effective than existing diversity measures, and than the direct estimate of the ensemble error, e.g., in terms of estimation reliability, computational complexity, or the possibility of estimating it using unlabelled samples only? We address these issues in the following.

4.1 Interpretation of the Ambiguity-Like Decomposition

For two-class problems, the second term in the RHS of decomposition (11) can be interpreted as a diversity measure, in terms of the disagreement between the individual classifiers and the ensemble [7], similarly to the ambiguity term in (6). In the general case when $L > 2$, it is easy to see that a similar interpretation can be given for the second term in the RHS of decomposition (14). However, for $L > 2$ the disagreement is not expressed in terms of the class labels, but in terms of the *correctness* of such choices (they coincide only when $L = 2$).

In [7] the decomposition (11) was further analysed by considering the case of zero Bayes error (i.e., when y is a deterministic function of $\mathbf{x}$), and by taking the expectation of (11) over $P[\mathbf{X}]$, which gives the error probability of the ensemble.[6] Taking into account that $y \times f_{\mathrm{ens}} = +1$ (-1) when the ensemble is correct (wrong), and denoting by $\mathbf{x}+$ and $\mathbf{x}-$ the corresponding regions in feature space, one obtains [7]:

$$e_{\mathrm{ens}}(d) = \bar{e}(d) - \int_{\mathbf{x}+} \frac{1}{N} \sum_{j=1}^{N} \delta_j(\mathbf{x}; d)\mathrm{d}\mathbf{x} + \int_{\mathbf{x}-} \frac{1}{N} \sum_{j=1}^{N} \delta_j(\mathbf{x}; d)\mathrm{d}\mathbf{x} . \quad (15)$$

This highlights that increasing the disagreement is beneficial on samples where the ensemble is correct, while it is detrimental on samples where the ensemble is wrong. Accordingly, the corresponding diversity components were named respectively "good" and "bad" diversity in [7].

It is now easy to see that the same interpretation can be given when $L > 2$, from decomposition (14), provided that "disagreement" is intended as explained above. On samples where the ensemble is correct, increasing the disagreement is beneficial, i.e., the highest number of individual classifiers should *misclassify* such samples, *independently* on the specific class labels they choose. Increasing the disagreement is detrimental on samples misclassified by the ensemble, instead: this means as well that the highest number of individual classifiers should *misclassify* such samples. Accordingly, the concept of good and bad diversity can be extended to $L > 2$, by considering the above definition of disagreement.

---

[6] This analysis can be easily extended to the case of non-zero Bayes error.

4.2   Practical Exploitation of Diversity Measures

Here we discuss the practical usefulness of diversity measures, including the diversity term of the ambiguity-like decomposition (14), for ensemble construction.

We first point out that the diversity term in (14) depends on the specific combining rule. This is a consequence of the fact that the error of a given ensemble depends on the combining rule, and that the first term in the RHS of (14) is the average error of ensemble members. However, existing diversity measures do not depend on the combining rule. Actually, although they are not explicitly tailored to a specific rule, most of them seem related to majority voting [35]. Even interesting measures recently proposed, using information theory, do not take into account the combining rule [6,44]. The pros and cons of using a single diversity measure for all combining rules, and of using different measures tailored to specific rules, have been discussed in [21] (sect. 10.5): "The problem is that the 'clean' diversity measure might be of little use due to its weak relationship with the ensemble accuracy [...]. On the other hand, the more we involve the ensemble performance into defining diversity, the more we are running onto the risk of trying to replace a simple calculation of the ensemble error by a clumsy estimate that we call diversity.'

On the other hand, as pointed out in Sect. 3, existing measures are usually considered as the equivalent of the ambiguity term in regression. Indeed, they have often been analysed by investigating whether and how their trade-off with the average error of individual classifiers is related to the ensemble error, but no clear correlation has been found [32,20,35]. This raised some doubts about the usefulness of existing measures for ensemble construction. Some authors even argued that a direct estimation of ensemble accuracy can be more effective. For instance, see the quote from [32], reported in Sect. 1; and: "In our opinion, the existing diversity measures are [...] not [sufficient] for [selecting base classifiers]" [35] (sect. 4). Such doubts are strengthened by the following fact: overproduce and choose methods for ensemble construction, that make explicit use of diversity measures, did not provide evidence that such an approach is more effective than directly estimating ensemble accuracy [41,13,1,2,16,40]. In particular, besides [41,40], where such a comparison has not been made, in [13,1,2] the use of diversity measures did not provide any significant accuracy improvement, and in [16] the direct estimation of classifier accuracy turned out to significantly outperform the use of diversity measures.

Consider now the *exact* decomposition of the ensemble error (14), or the equivalent (for two-class problems) decomposition (11). Can their diversity terms be exploited more effectively in the context of overproduce and choose methods? At least at a first glance, the answer seems negative. The reason is that to compute these diversity terms (on a given set of samples, e.g., a validation set) one needs to know both the ensemble output and the correct class label of each sample. However, this also allows one to directly estimate the ensemble accuracy. We point out that a similar issue arises about the use of the ambiguity decomposition for regression problems as well, as mentioned in Sect. 2. However, even though computing the ambiguity term in (6) is not computationally cheaper than directly computing the ensemble MSE, the former does not involve the correct output y, which allows one to compute it using also *unlabelled* samples. It would be interesting to investigate whether this can be actually advantageous. It is worth noting that the use of unlabelled samples to promote diversity in MCSs has been suggested in [42].

Consider now the use of diversity measures for ensemble construction strategies analogue to NCL, i.e., for directly constructing a MCS without overproducing first, and then selecting a subset of classifiers. In this context, it is pertinent to note that well-known MCS construction techniques like bagging, random forests, random subspace, and AdaBoost, are effective even though they do not explicitly use any diversity measure (see, e.g., [21], chapter 10). By the way, they are all tailored to majority voting (or weighted voting, in the case of AdaBoost) [35]. On the one hand, it is commonly believed that such techniques "can be interpreted as building diverse base classifiers implicitly" [35]. This fact has also inspired the idea of investigating what objective function, and thus, what diversity measure, is implicitly optimised by such techniques.[7] On the other hand, the above discussion about existing measures and about the diversity terms in (14) and (11), strengthens the doubt that they are not more useful in practice than directly estimating ensemble accuracy. Indeed, they seem only "descriptive", i.e., they formalise the intuition that (at least for the majority voting rule) an effective ensemble is made up of classifiers that are accurate "enough" on different regions of the feature space, such that (ideally) a majority of them correctly classifies each sample. This is exactly the goal that techniques like bagging pursue, using different strategies, without explicitly relying on diversity measures. To our knowledge, the only method analogue to NCL proposed so far (besides the direct use of NCL with base classifiers like neural networks) is the one of [45]. It simultaneously trains a set of two-class linear classifiers, and computes the weights of their linear combination, using a SVM-like learning algorithm. The objective function aims at jointly maximising individual accuracy and diversity. Diversity is measured as the average pairwise disagreement between individual classifiers. This method exhibited comparative performance with bagging and AdaBoost. On the other hand, the considered diversity measure does not coincide with the ambiguity term (11). A further investigation of this method is thus interesting.

To sum up, existing diversity measures are at most an approximation of the "real" diversity term, in the context of exact decompositions of the ensemble error like (14) and (11), in which the first term is given by the average error of ensemble members. On the other hand, the practical usefulness of diversity measures, even exact ones, remains questionable. In the next section we will indicate possible research directions to address this issue.

5   Suggestions for Future Research

On the basis of the above results and discussion, we conclude this paper by suggesting some research directions, aimed at better understanding whether

---

[7] Zhi-Hua Zhou, MCS 2010 panel discussion: `http://www.diee.unica.it/mcs/mcs2010/panel discussion.html`

the explicit use of diversity measures can be useful in practice, for ensemble construction.

1. The BVC-like decomposition (10), and in particular the term corresponding to covariance, deserves further analysis. A comparison with the ambiguity-like decomposition (12) is interesting, to understand the correspondence between their terms, as done in [4] for the BVC and ambiguity decompositions of regression error.
2. The ambiguity-like decomposition (12) should be extended to loss functions different than 0/1. It should also be further analysed with respect to specific combining rules, different from majority voting (which has been considered in [7]). In particular, it would be interesting to investigate whether the average error of individual classifiers is the most suitable as the first term of such a decomposition, for any combining rule.
3. The effectiveness of explicitly using diversity measures in ensemble construction methods with the overproduce and choose strategy, should be thoroughly compared with the direct estimation of ensemble accuracy. This should be done also for regression problems, where the ambiguity term seems in principle more advantageous than the corresponding one for classification problems.
4. It is also interesting to compare the diversity terms derived from exact decompositions of the ensemble error like (14) and (11), with existing diversity measures. This can help understanding which of these measures is a better approximation to the "real" diversity, also with respect to a specific combining rule. This could even suggest new diversity measures that better approximate the "real" one, and are also of practical use.

Acknowledgments. The authors thank the reviewers for their constructive comments and criticisms to the first version of this paper. This work has been partly supported by the project CRP-18293 funded by Regione Autonoma della Sardegna, L.R. 7/2007, Bando 2009.

References

1. Banfield, R.E., Hall, L.O., Bowyer, K.W., Kegelmeyer, W.P.: A new ensemble diversity measure applied to thinning ensembles. In: Windeatt, T., Roli, F. (eds.) MCS 2003. LNCS, vol. 2709, pp. 306–316. Springer, Heidelberg (2003)
2. Banfield, R.E., Hall, L.O., Bowyer, K.W., Kegelmeyer, W.P.: Ensemble diversity measures and their application to thinning. Information Fusion 6, 49–62 (2005)
3. Breiman, L.: Bias, variance, and arcing classifiers. Technical Report 460, Statistics Department, University of California, Berkeley, CA (1996)
4. Brown, G., Wyatt, J.L., Harris, R., Yao, X.: Diversity creation methods: a survey and categorisation. Information Fusion 6, 5–20 (2005)
5. Brown, G., Wyatt, J.L., Tino, P.: Managing diversity in regression ensembles. Journal of Machine Learning Research 6, 1621–1650 (2005)

6. Brown, G.: An information theoretic perspective on multiple classifier systems. In: Benediktsson, J.A., Kittler, J., Roli, F. (eds.) MCS 2009. LNCS, vol. 5519, pp. 344–353. Springer, Heidelberg (2009)

7. Brown, G., Kuncheva, L.I.: "Good" and "Bad" diversity in majority vote ensembles. In: El Gayar, N., Kittler, J., Roli, F. (eds.) MCS 2010. LNCS, vol. 5997, pp. 124–133. Springer, Heidelberg (2010)

8. Dietterich, T.G.: An experimental comparison of three methods for constructing ensembles of decision trees: Bagging, boosting, and randomization. Machine Learning 40, 139–157 (2000)

9. Dietterich, T.G.: Ensemble methods in machine learning. In: Kittler, J., Roli, F. (eds.) MCS 2000. LNCS, vol. 1857, pp. 1–15. Springer, Heidelberg (2000)

10. Domingos, P.: A unified bias-variance decomposition for zero-one and squared loss. In: 7th Int. Conf. on Artificial Intelligence, pp. 564–569 (2000)

11. Dutta, H.: Measuring Diversity in Regression Ensembles. In: 4th Indian Int. Conf. on Artificial Intelligence, pp. 2220–2236 (2009)

12. Geman, S., Bienenstock, E., Doursat, R.: Neural networks and the bias/variance dilemma. Neural Computation 4, 1–58 (1992)

13. Giacinto, G., Roli, F.: Design of effective neural network ensembles for image classification purposes. Image and Vision Computing 19, 699–707 (2001)

14. Hernandez-Lobato, D., Martinez-Munoz, G., Suarez, A.: Pruning in ordered regression bagging ensembles. In: Int. Joint Conf. Neural Net., pp. 1266–1273 (2006)

15. Kittler, J., Hatef, M., Duin, R.P.W., Matas, J.: On combining classifiers. IEEE Transactions on Pattern Analysis and Machine Intelligence 20, 226–239 (1998)

16. Ko, A.H.-R., Sabourin, R., DeSouza Britto Jr., A.: Compound diversity functions for ensemble selection. Int. J. Patt. Rec. Artificial Intelligence 23, 659–686 (2009)

17. Kohavi, R., Wolpert, D.H.: Bias plus variance decomposition for zero-one loss functions. In: 13th Int. Conf. Mac. Learn., pp. 275–283. Morgan Kaufmann (1996)

18. Krogh, A., Vedelsby, J.: Neural network ensembles, cross validation, and active learning. In: Adv. in Neural Inf. Proc. Systems, vol. 7, pp. 231–238. MIT Press (1995)

19. Kuncheva, L.I.: That elusive diversity in classifier ensembles. In: Perales, F.J., Campilho, A.C., Pérez, N., Sanfeliu, A. (eds.) IbPRIA 2003. LNCS, vol. 2652, pp. 1126–1138. Springer, Heidelberg (2003)

20. Kuncheva, L.I., Whitaker, C.J.: Measures of diversity in classifier ensembles and their relationship with the ensemble accuracy. Mac. Learn. 51, 181–207 (2003)

21. Kuncheva, L.I.: Combining Pattern Classifiers: Methods and Algorithms. John Wiley & Sons, Hoboken (2004)

22. Lam, L., Suen, C.Y.: Application of majority voting to pattern recognition: an analysis of its behavior and performance. IEEE Transactions on Systems, Man, and Cybernetics - Part C 27, 553–568 (1997)

23. Li, N., Yu, Y., Zhou, Z.-H.: Diversity Regularized Ensemble Pruning. In: Flach, P.A., De Bie, T., Cristianini, N. (eds.) ECML PKDD 2012, Part I. LNCS, vol. 7523, pp. 330–345. Springer, Heidelberg (2012)

24. Littlewood, B., Miller, D.R.: Conceptual modeling of coincident failures in multiversion software. IEEE Transactions on Software Engineering 15, 1596–1614 (1989)

25. Liu, Y.: Negative Correlation Learning and Evolutionary Neural Network Ensembles. PhD thesis, University College, The University of New South Wales, Australian Defence Force Academy, Canberra, Australia (1998)

26. Margineantu, D.D., Dietterich, T.G.: Pruning adaptive boosting. In: 14th Int. Conf. on Machine Learning, pp. 211–218 (1997)

27. Partalas, I., Tsoumakas, G., Hatzikos, E.V., Vlahavas, I.P.: Greedy regression ensemble selection: Theory and an application to water quality prediction. Information Sciences 178, 3867–3879 (2008)
28. Partridge, D., Krzanowski, W.J.: Software diversity: practical statistics for its measurement and exploitation. Information & Software Technology 39, 707–717 (1997)
29. Perrone, M.P., Cooper, L.N.: When networks disagree: Ensemble methods for neural networks. In: Mammone, R.J. (ed.) Artificial Neural Networks for Spech and Vision, pp. 126–142. Chapman & Hall, New York (1993)
30. Rooney, N., Patterson, D., Nugent, C.: Reduced ensemble size stacking. In: 16th Int. Conf. on Tools with Artificial Intelligence, pp. 266–271 (2004)
31. Sharkey, A.J.C., Sharkey, N.E.: Combining diverse neural nets. The Knowledge Engineering Review 12, 231–247 (1997)
32. Shipp, C.A., Kuncheva, L.I.: Relationships between combination methods and measures of diversity in combining classifiers. Information Fusion 3, 135–148 (2002)
33. Sirlantzis, K., Hoque, S., Fairhurst, M.C.: Diversity in multiple classifier ensembles based on binary feature quantisation with application to face recognition. Applied Soft Computing 8, 437–445 (2008)
34. Sun, Q., Pfahringer, B.: Bagging Ensemble Selection for Regression. In: Thielscher, M., Zhang, D. (eds.) AI 2012. LNCS, vol. 7691, pp. 695–706. Springer, Heidelberg (2012)
35. Tang, E.K., Suganthan, P.N., Yao, X.: An analysis of diversity measures. Machine Learning 65, 247–271 (2006)
36. Tumer, K., Ghosh, J.: Analysis of decision boundaries in linearly combined neural classifiers. Pattern Recognition 29, 341–348 (1996)
37. Ueda, N., Nakano, R.: Generalization error of ensemble estimators. In: Int. Conf. on Neural Networks, pp. 90–95 (1996)
38. Wang, D., Alhamdoosh, M.: Evolutionary extreme learning machine ensembles with size control. Neurocomputing 102, 98–110 (2013)
39. Yu, Y., Zhou, Z.-H., Ting, K.M.: Cocktail Ensemble for Regression. In: 7th Int. Conf. Data Mining, pp. 721–726. IEEE Computer Society (2007)
40. Yu, Y., Li, Y.-F., Zhou, Z.-H.: Diversity regularized machine. In: 22nd Int. Joint Conf. on Artificial Intelligence, pp. 1603–1608 (2011)
41. Zenobi, G., Cunningham, P.: Using Diversity in Preparing Ensembles of Classifiers Based on Different Feature Subsets to Minimize Generalization Error. In: Flach, P.A., De Raedt, L. (eds.) ECML 2001. LNCS (LNAI), vol. 2167, pp. 576–587. Springer, Heidelberg (2001)
42. Zhang, M.-L., Zhou, Z.-H.: Exploiting unlabeled data to enhance ensemble diversity. Data Min. Knowl. Disc. 26, 98–129 (2013)
43. Zhou, Z.-H., Wu, J., Tang, W.: Ensembling neural networks: many could be better than all. Artificial Intelligence 137, 239–263 (2002)
44. Zhou, Z.-H., Li, N.: Multi-information Ensemble Diversity. In: El Gayar, N., Kittler, J., Roli, F. (eds.) MCS 2010. LNCS, vol. 5997, pp. 134–144. Springer, Heidelberg (2010)
45. Yu, Y., Li, Y.-F., Zhou, Z.-H.: Diversity regularized machine. In: Proc. 22nd Int. Joint Conf. on Artificial Intelligence, pp. 1603–1608 (2011)
46. Zhou, Z.-H.: Introduction to Ensemble Methods. CRC Press (2012)

Gender Classification Using Mixture of Experts from Low Resolution Facial Images

Yomna Safaa El-Din[1], Mohamed N. Moustafa[2], and Hani Mahdi[1]

[1] Department of Computer and Systems Engineering,
Ain Shams University,
Cairo, Egypt
{yomna.safaa-eldin,hani.mahdi}@eng.asu.edu.eg
[2] Department of Computer Science and Engineering,
American University in Cairo,
New Cairo, Egypt
moustafa@ieee.org, m.moustafa@aucegypt.edu

Abstract. In this study, we propose a novel two-stages mixture of experts scheme estimating gender from facial images. The first stage combines a couple of complementary gender classifiers with a third arbiter in case of decision discrepancy. Experimentally, we have verified the common thinking that one appearance-based (Haar-features cascade) classifier with another shape-based (landmarks positions metrology with SVM) classifier form a complementary couple. Subsequently, the second stage in our scheme is a Bayesian framework that is activated only when the arbiter cannot take a confident decision. We demonstrate that the proposed scheme is capable of classifying gender reliably from faces as small as 16x16 thumbnails on benchmark databases, achieving 95% gender recognition on FERET database, and 91.5% on the Labeled Faces in the Wild dataset.

Keywords: gender classification, committee machines, Bayes, resolution.

1 Introduction

Many human activities and machine applications depend on accurate gender recognition. It can be used as a prior step to face recognition and verification [1,2]. Gender discrimination also helps in the indexing and retrieval of images and videos [3].

Automatic gender classification has been widely investigated in literature. Most studies have used 2D face images for classification, which can be done using either appearance-based or shape-based methods. Appearance-based approaches use the cropped, resized, and illumination normalized texture of the face (or portions of it) as a classification attribute, while the shape-based approaches extract a set of discriminative face shape features and uses them for the classification process.

Z.-H. Zhou, F. Roli, and J. Kittler (Eds.): MCS 2013, LNCS 7872, pp. 49–60, 2013.
© Springer-Verlag Berlin Heidelberg 2013

It is expected that the classification accuracy can be improved by combining more than one classifier [4], especially when each method relies on different input features extracted from the face.

In this work we compare several state-of-the art classification techniques on different features and image resolution. Then we propose a Mixture of Experts (MoE) technique based on Naive Bayes theorem for merging some of these well known classifiers to achieve boosted performance.

The rest of the paper is organized as follows; Section 2 reviews previous related work, Section 3 describes the individual classification methods used along with their input feature types, and Section 4 introduces our proposed method for merging these classifiers. Section 5 states the databases we used, then explains and discusses the experiments and results. In the final section, we conclude our work.

2   Related Work

Mäkinen et al. [4] presented an overview on the topic of gender classification from face images. They experimented on FERET database [5] and WWW [4] (another dataset containing images they randomly collected from the web). They compared six state-of-the-art gender classification algorithms, none of which is shape-based. They used different face normalizations and alignments, and they introduced combined results of these classifiers.

Another comparative study was presented by Calfa et al. in [6], giving special attention to linear techniques and their relations, due to their simplicity and low computational requirements. Their work proves that, with a linear feature selection, Linear Discriminant Analysis on the linearly selected set of features achieves results comparable to the best gender classifiers based on Support Vector Machines with Radial Basis Function kernel (SVM+RBF) [7] and Boosting.

Shan [8] investigated gender classification on real-life faces acquired in unconstrained conditions. Boosted Local Binary Patterns (LBP) [9] were used with SVM, where LBP was employed to describe faces, then Adaboost was used to select the discriminative LBP features, followed by an SVM for classification. The author reported results on the Labeled Faces in the Wild (LFW) database [10].

Cao et al. [11] presented a shape-based approach where they used topological information extracted from facial landmarks to perform gender classification. The authors compared their technique to Local Binary Patters which is an appearance-based classifier, and showed a slightly lower performance that is due to the simplicity and small amount of information encoded in the metrology features. I our work we focus on combining shape information with the face appearance for boosted accuracy.

3   Individual Experts and Features

In this section, we shed some light on the classifiers we used as independent experts in our merger, along with the features supplied for each classifier.

3.1 Features

We compared four different types of features; three of which are appearance-based and one is shape-based.

Appearance-Based Approach. We used three types of features that can be extracted from the appearance of a face in the image; which are normalized pixel values (to be in the range $[0-1]$), Principal component analysis (PCA) [12], and Haar-like features [13].

Shape-Based Approach. We used the positions of 76 facial landmarks, that were automatically located on the face then their coordinates values were shifted to have the nose-tip at the center. The positions are then normalized by scaling them so that all faces have a constant inter-eyes distance.

3.2 Individual Gender Classification Methods

To perform classification, we chose to use Support Vector Machines (SVM) [7] which are well known for their accuracy and speed. However, for the Haar-like features, due to their high-dimensional vector, we used Adaboost [14] to select the most discriminant features.

For SVM, we specifically use Least-Square SVM (LS-SVM) [15] which are the least squares versions of SVM in which the solution is found by solving a set of linear equations instead of the convex quadratic programming problem for classical SVMs. We use SVM with Radial Basis Function (RBF) Kernel.

4 Proposed Mixture of Experts

We present here the details of our approach for merging several individual classification methods in order to achieve higher gender recognition accuracy.

4.1 Mixture of Experts

A Mixture of Experts (MoE) is a form of dynamic committee machines, where the outputs of the constituent experts (classifiers) are non-linearly combined by some form of gating system to produce an overall output that is superior to that of any single expert alone. In MoE, the input signal is also directly involved in actuating the integration mechanism as shown in Fig. 1.

4.2 Proposed Basic Mixture of Experts with Bayesian Combiner

The Naive Bayes classifier is based on the Bayes' theorem;

$$p(C|F_1, \ldots, F_n) = \frac{p(C)p(F_1, \ldots, F_n|C)}{p(F_1, \ldots, F_n)} , \qquad (1)$$

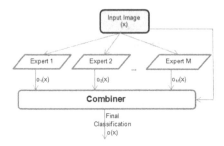

Fig. 1. Structure of a general Mixture of Experts network

where F_1 to F_n are the input features, and C is the class of these features. The denominator of this fraction can be neglected as it does not depend on the class C; then the theorem can be stated as;

$$posterior \propto (prior \times likelihood) \,. \tag{2}$$

We adopt this Naive Bayes approach for the merger stage of our mixture machine using the scores as inputs, and C representing the chosen expert. In other words, for each image I and M different experts, the combiner will return the decision of one chosen expert, depending on the set of scores returned by the M experts. The scores are independent of each other given a certain expert, hence the Naive concept.

Training. Each expert is trained using a training subset of the database images. We then train the Bayesian merger using another subset of images. This is done by calculating the prior of each of the contributing individual methods and the likelihood of their outputs' scores as follows:

For each expert, we run its previously-trained classifier on this subset and obtain the following:

- *Prior* This is the classifier's achieved accuracy on this data subset. At merge time we normalize it with the other experts' priors, so that $\sum_{m=1}^{M} prior(m) = 1$; where $prior(m)$ is the prior of the m's classifier and M is the number of experts to be merged.

- *Likelihood* We use the scores returned by the classifier from this subset, then split the score range into N intervals. For each interval, we calculate the percentage of images that were correctly classified (true percent) and those that were wrong (false percent). Likelihood of this interval is the true percent minus the false percent, which might result in a negative value if there are more falsely classified samples in an interval than the correct ones. In this case, we shift all the values to have a $min = 1$, and finally, the values are normalized to $[0 - 1]$. For our machine, we choose $N = 20$.

Merging. When a new face is to be classified, each expert, m, is run on the input image returning a binary output o_m along with its score c_m. The prior of each expert alone is retrieved, and these priors are normalized. Then we retrieve the likelihoods of the returned scores, using the trained Bayesian merger. The posterior of each method is then calculated by:

$$posterior(m, c_m) = pr(m) \times likelihood(m, c_m) . \qquad (3)$$

The final output (class) is then taken to be that of the method with the highest posterior.

4.3   Proposed 2-Stages MoE

The experts contributing in the MoE should be chosen such that they classify more images differently, which will happen if each expert relies on different cues for its decision.

The effect of increasing the number of experts used in the machine will be discussed in Section 5.4, and it can be expected that using more experts might lead to higher accuracy.

However, we introduce an enhancement to the basic Bayesian MoE proposed in the previous subsection, which allows us to achieve these high correct rates using only two main experts, by adding another stage to this basic MoE, prior to the Bayesian stage.

Using two experts only, will cause the merge machine to be invoked only if each decides the image to belong to a different class.

Training Stage(1). Stage(1) is a classifier trained on the same subset of images used to train the Bayesian MoE (which is now Stage(2)).

Each trained expert is used to classify this subset and return scores for its decisions. The scores for each image are concatenated to its normalized pixel values to form a single vector used to train the classifier of Stage(1). A threshold score $Conf_{Thr}$, is calculated as the average score of the correctly classified faces in this subset.

Merging. If both experts disagree on which class an image belongs to, then their scores are fed to Stage(1) along with the image itself, to return a decision and a score. The decision of Stage(1) supports one of the experts to be the final decision only if its confidence is above the calculated threshold score Sc_{Thr}; otherwise, its decision is discarded and Stage(2) is used to resolve the conflict based on the experts' posteriors as done in the basic Bayesian MoE. The use of the image itself in the decision of Stage(1) is what separates our MoE from a standard dynamic committee machine.

The structure of the proposed 2-Stages MoE is depicted in Fig 2.

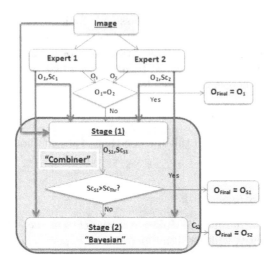

Fig. 2. Structure of the proposed 2-Stages Mixture of Experts network. O is output class with score Sc. Sc_{Thr} is the threshold score of Stage (1).

5   Experiments and Results

5.1   Datasets and Experimental Setup

Our experiments were performed on three face image databases; two of which are well known and publicly available: the FERET image database [5], the LFW database [10]. We have used also a dataset we refer to as MixDB, containing images that were privately collected including several ethnicities; Caucasians, Asians and also some from African descent. We used frontal and near-frontal face images.

While FERET DB contains studio-setting constrained images; LFW offers a unique collection of faces captured from the web, which represents a variation of expressions and lightings. From the FERET database we used one image per subject from the frontal *fa* gallery. For the LFW we selected the frontal faces and formed a set containing 4500 males and 2340 females, having at most two images of the same subject.

For each DB, we created three subsets;
- Training set: used for training individual classifiers;
- Extra-Training set: used for training the combiner stages; and, finally,
- Testing set: used to evaluate the classification performance.

Five-folds cross validation is used for our experiments, where the images are divided into 5 folds, keeping the same ratio between male and female faces; duplicate images of the same subject are placed in the same fold. One fold is used for Testing, two for Training and the remaining two for Extra-Training; this process is repeated five times and the average is reported. The number of faces used for each database is shown in Table 1.

Table 1. Number of faces used for each database (Male/Female)

	FERET	MixDB	LFW
All	600/405	1185/1096	4500/2340
1 Fold	120/81	237/219	900/468

For implementation, images preprocessing, training and testing, we used MAT-LAB. In all images, the eyes positions were manually located, however in practice, the eyes can be located automatically using active appearance model (AAM). Each image was then rotated so that both eyes lie on a horizontal line, and the face area was extracted to be a square with dimensions relative to the inter-eyes distance. Colored images were transformed to grayscale, then the lighting was enhanced using MATLAB's built-in function *imadjust* which increases the contrast by remapping the intensity values to fill the entire range of $[0 - 255]$.

5.2 Individual Experts

The experts we use in our mixture machine, adopt two classifiers; SVM and Adaboost, each with different features extracted from the image.

The notations we use, are:

- SVM[Norm]: Normalized image pixels,
- SVM[PCA]: Dimensionally reduced vector using PCA,
- SVM[LM]: Landmarks positions, and
- Ada[Haar]: Haar-features.

For SVM[PCA], we varied the number of principal components (PCs) used from 50 to 300, then tested SVM's classification on different image sizes, from 16×16 to 40×40 with step 8 pixels per side. We obtained best classification results using 150 PCs regardless of the initial images' size. So, on the following experiments we will use 150 PCs for SVM[PCA].

For SVM[LM], 76 landmarks' positions are located automatically using Stasm [16] which is an extended version of Active Shape Model. The coordinates of the landmarks are manipulated as explained in 3.1. Subsequently, SVM is trained on these manipulated coordinates; a vector of size $nLM \times 2$, where nLM is the number of landmarks detected.

We carried out two experiments; the first one regards the choice of the resolution of face images to be used. Then the second experiment tests the performance of our proposed Mixture of Experts.

5.3 Experiment (1): Studying the Effect of Image Resolution on Individual Classifiers

The purpose of this experiment is to investigate the effect of changing the input face image size on the accuracy of the individual classifiers.

Fig. 3 compares the weighted average performance of the experts listed in 5.2, except for SVM[LM] which is independent of the image size. Images are resized using bi-cubic interpolation. For Haar, due to the very high dimension of the Haar-like feature vector, we stopped at 24×24 images, in which case the Haar-features vector's dimension is 136,656.

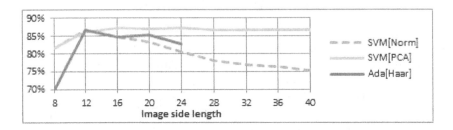

Fig. 3. Classifiers' responses over different image resolution

From Fig. 3, the following observations are made:

- Increasing the image resolution for the used classifiers does not improve the achieved classification accuracy; but even degrades it significantly when using the image pixels values.

- Using SVM: For a certain higher resolution image; e.g. $40 \times 40 = 1600$, reducing the dimension yields better accuracy; which can be done by two ways; either reducing the image size using simple down-sampling; i.e. SVM[Norm] on lower resolution images, or using PCA; SVM[PCA]. Obviously PCA gave higher performance.

- SVM[PCA]: Its performance is much better than using simple pixel values, and as the number of PCs remain constant (150 in this case), the initial image size does not affect the performance.

- Computational time: Reducing the images' resolution significantly reduces training time for some experts. For example, on the FERET dataset, Adaboost took 7 minutes for training using Haar features on 400 face images of size 16×16. When the size increased to 24×24, the training time jumped to 1 and a half hour.

For the coming experiment we will use low resolution; 16×16 images, since it achieves not only less computational time, but also better accuracy.

Table 2 presents the classification rates of each individual expert on 16×16 images.

5.4   Experiment (2): Proposed MoE

We used our proposed 2-Stages MoE to merge pairs of expert on 16×16 images. Experts contributing in the machine must each be using different features for classification as explained in Section 4, so we specifically chose to merge the shape-based expert, SVM[LM], with each of the other appearance based experts.

For Stage(1), we use SVM as an aiding expert trained with normalized pixel values of the face images concatenated with the score values returned by the two

Table 2. Classification Results of Individual Experts on 16×16 face images, sorted in a Descending Order

Expert	FERET	MixDB	LFW
Ada[Haar]	**93.71%**	**86.89%**	**88.41%**
SVM[Norm]	92.91%	85.58%	88.36%
SVM[PCA]	91.11%	82.68%	87.41%
SVM[LM]	80.19%	77.48%	78.76%

experts. The result of Stage(1) supports one of the main contributing experts' decision only if they both disagree, and the SVM's returned confidence is above the threshold score Sc_{Thr}.

Table 3. Results of the proposed Bayesian Mixture of Experts on 16×16 images, when one of the experts is shape-based. These results are the weighted average of all databases.

Experts	Best Expert	Basic Bayesian MoE	2-Stages MoE
(1) SVM[LM]+Ada[Haar]	88.59%	90.01%	**91.54%**
(2) SVM[LM]+SVM[PCA]	86.71%	88.41%	**90.51%**
(3) SVM[LM]+SVM[Norm]	88.19%	89.66%	-
(4) SVM[LM]+Ada[Haar]+SVM[Norm]	88.59%	**90.57%**	-
(5) SVM[LM]+Ada[Haar]+SVM[PCA]	88.59%	90.17%	-
(6) SVM[LM]+SVM[PCA]+SVM[Norm]	88.19%	89.74%	-
(7) SVM[LM]+Ada[Haar]+SVM[PCA]+SVM[Norm]	88.59%	90.53%	-

Table 3 presents the best achieved weighted average results using our proposed MoE on 16×16 images, which shows an improvement over the best contributing expert by up to 3%. Best classification rate is achieved using our proposed 2-Stages MoE to merge the shape-based SVM[LM] with the appearance-based Ada[Haar].

By comparing row(1) with row(4), and row(2) with row(6), it is observed that the accuracy of the 2-Stages MoE is about 1% higher than that of the Basic Bayesian MoE, when using 3 experts, two of which are the same experts used in the 2-Stages MoE, and the third expert is SVM[Norm]. Using SVM[Norm] as an expert in this 3-experts Basic MoE is the closest to Stage(1) of the 2-Stages MoE, yet trained on normalized image pixels only (without the confidence values of the other two experts). From this we can say that using the aiding expert (here SVM[Norm]) as a part of the combiner stages in the 2-Stages MoE is better than including it from the beginning as a contributing expert while using Bayesian combiner only for the merge.

Fig. 4 compares the performance of the basic MoE with varying number of contributing experts, with the 2-Stages MoE. From this figure it is seen that increasing the number of experts for the basic MoE beyond three does not improve the accuracy, yet using the proposed 2-Stages MoE does.

Fig. 4. Best mixture results on 16×16 images. These results are the weighted average of all databases.

Table 4. Best 5-folds Results on each Database

	Best Individual Expert	2-Stages MoE	Other Methods		Classification by humans
FERET	93.71% Ada[Haar]	**95.10%±1.2%**	- OpenCV: - Calfa el at. [6]:	90.31% 93.95%±2.6%	85.50%
FERET [4]	91.59%	**95.33%**	- Mäkinen et al. [4]: 92.86%		
MixDB	86.89% Ada[Haar]	**90.12%±3.4%**	- OpenCV:	81.67%	77.80%
LFW	88.41% Ada[Haar]	**91.49%±1.1%**	- OpenCV: - Shan [8]:	81.58% 94.81%±1.1%	86.67%

5.5   Best Results on each Database

Table 4 presents the best achieved results on each database alone using our proposed 2-Stages MoE for merging SVM[LM] and Ada[Haar]. We compare these results to OpenCV's gender classification [17] on our subsets, and to published results on the same databases. For FERET, Calfa et al. [6] used the frontal set as we did, and reported results using 5-folds cross validation.

Mäkinen et al. [4] achieved best results using combination by voting of six classification methods, on frontal images with hair. Their subset contained 760 face images divided equally between male and female, from which they used 80% for training and 20% for testing. We report results of our method on their set, splitting the 80% that are used for training equally between the 'Training' and 'Extra-Training' subsets.

Shan [8] tested SVM on boosted LBP, on the LFW database; a set containing 4500 males and 2943 female, approximately the same number of images we used. They performed 5-folds cross validation for their reported results as we did. We think Shan's results are better on LFW because of using the boosted LBP which are complicated features compared to the simple features we used for our mixture. However, they have not reported results on any other database.

We also report the average rate of classification done by a small group of 5 people; 3 male and 2 female, on a subset of the 16×16 adjusted face images that was randomly selected from the datasets, 100M/100F from FERET and the same for MixDB, while 150M/150F from LFW. The poor classification rate proves that even though humans can easily classify gender from high resolution face images, it is much harder to classify from very small images like the ones we used.

6   Conclusions and Future Work

In this research, we presented a new approach for merging several gender classification methods using Naive Bayes. We used a Mixture of Experts formed of two stages used to combine results of two experts; the first stage uses the input image along with the scores of the experts for classification, while the second stage implements a Naive Bayes approach for the merge. Our experiments showed that the multilevel MoE composed of only two contributing experts achieves comparable results and sometimes better than using basic MoE which supports more than two experts.

We tested on both appearance-based and shape-based data, and the best merge result obtained are when these were combined together, as shown in Table 3, where the best results are obtained when combining SVM on landmarks positions, with Adaboost on Haar-like features.

The effect of varying the resolution of images on the classification accuracy was studied, from which we proved one of the aspects of the 'curse of the dimensionality' problem, showing that higher resolution images are not necessary for better performance. We demonstrated in Section 5.3 that the time taken by the classification process can be reduced a lot without loosing much accuracy using low resolution face images.

For future work we plan to use different features extracted from the face image like SIFT, SURF, or boosted LBP along with other classifiers.

References

1. Kumar, N., Berg, A.C., Belhumeur, P.N., Nayar, S.K.: Attribute and Simile Classifiers for Face Verification. In: IEEE International Conference on Computer Vision, ICCV (2009)
2. Ross Beveridge, J., Givens, G.H., Jonathon Phillips, P., Draper, B.A.: Factors that influence algorithm performance in the Face Recognition Grand Challenge. Computer Vision and Image Understanding, 750–762 (2009)

3. Kumar, N., Belhumeur, P.N., Nayar, S.K.: FaceTracer: A Search Engine for Large Collections of Images with Faces. In: Forsyth, D., Torr, P., Zisserman, A. (eds.) ECCV 2008, Part IV. LNCS, vol. 5305, pp. 340–353. Springer, Heidelberg (2008)
4. Mäkinen, E., Raisamo, R.: An experimental comparison of gender classification methods. Pattern Recognition Letters 29(10), 1544–1556 (2008)
5. Phillips, P.J., Moon, H., Rizvi, S.A., Rauss, P.J.: The FERET Evaluation Methodology for Face-Recognition Algorithms. IEEE Transactions on Pattern Analysis and Machine Intelligence 22, 1090–1104 (2000)
6. Bekios-Calfa, J., Buenaposada, J.M., Baumela, L.: Revisiting Linear Discriminant Techniques in Gender Recognition. IEEE Transactions on Pattern Analysis and Machine Intelligence 33(4), 858–864 (2011)
7. Moghaddam, B., Yang, M.-H.: Learning Gender with Support Faces. IEEE Transactions on Pattern Analysis and Machine Intelligence 24(5), 707–711 (2002)
8. Shan, C.: Learning local binary patterns for gender classification on real-world face images. Pattern Recognition Letters 33, 431–437 (2012)
9. Ojala, T., Pietikainen, M.: Multiresolution gray-scale and rotation invariant texture classification with local binary patterns. IEEE Transactions on Pattern Analysis and Machine Intelligence 24(7), 971–987 (2002)
10. Huang, G.B., Ramesh, M., Nick, J., Berg, T., Learned Miller, E.: Labeled Faces in the Wild: A Database for Studying Face Recognition in Unconstrained Environments, Technical report, University of Massachusetts, Amherst (2007)
11. Cao, D., Chen, C., Piccirilli, M., Adjeroh, D., Bourlai, T., Ross, A.: Can Facial Metrology Predict Gender? In: Proc. of International Joint Conference on Biometrics (IJCB), Washington, DC, USA (October 2011)
12. Turk, M., Pentland, A.: Eigenfaces for recognition. Journal of Cognitive Neuroscience 3(1), 71–86 (1991)
13. Viola, P., Jones, M.: Rapid Object Detection Using a Boosted Cascade of Simple Features. In: IEEE Computer Society Conference on Computer Vision and Pattern Recognition, vol. 1, pp. 511–518 (2001)
14. Freund, Y., Schapire, R.E.: A Decision-Theoretic Generalization of On-Line Learning and an Application to Boosting. Journal of Computer and System Sciences 55(1), 119–139 (1997)
15. Suykens, J.A.K., Vandewalle, J.: Least Squares Support Vector Machine Classifiers. Neural Process. Lett. 9(3), 293–300 (1999)
16. Milborrow, S., Nicolls, F.: Locating Facial Features with an Extended Active Shape Model. In: Forsyth, D., Torr, P., Zisserman, A. (eds.) ECCV 2008, Part IV. LNCS, vol. 5305, pp. 504–513. Springer, Heidelberg (2008), http://www.milbo.users.sonic.net/stasm
17. OpenCV Gender Classification, http://docs.opencv.org/modules/contrib/doc/facerec/index.html

Multi-view Multi-class Classification for Identification of Pathogenic Bacterial Strains

Evgeni Tsivtsivadze[1], Tom Heskes[2], and Armand Paauw[1]

[1] MSB Group, The Netherlands Organization for Applied Scientific Research,
Zeist, The Netherlands
firstname.lastname@tno.nl
[2] Institute for Computing and Information Sciences,
Radboud University, The Netherlands
firstname.lastname@science.ru.nl

Abstract. In various learning problems data can be available in different representations, often referred to as views. We propose multi-class classification method that is particularly suitable for multi-view learning setting. The algorithm uses co-regularization and error-correcting techniques to leverage information from multiple views and in our empirical evaluation notably outperforms several state-of-the-art classification methods on publicly available datasets. Furthermore, we apply the proposed algorithm for identification of the pathogenic bacterial strains from the recently collected biomedical dataset. Our algorithm gives a low classification error rate of 5%, allows rapid identification of the pathogenic microorganisms, and can aid effective response to an infectious disease outbreak.

1 Introduction

Frequently the problem at hand requires considering a classification task involving more than two classes, namely when the label y is chosen from the set $\mathcal{Y} = \{1, \ldots, \kappa\}$, where $\kappa > 2$. A number of methods have been proposed to deal with this problem and they can be divided into two main categories: i) methods that reduce the multi-class problem into simpler binary classification tasks and combine the obtained results afterwards (e.g. [1,2,3]) and ii) genuine multi-class classification algorithms (e.g. [4,5,3]) that learn a single function for discriminating between the multiple classes. In [3], authors provide a detailed overview of the methods that are frequently used for multi-class classification. They refer to the algorithms that learn a single function for discriminating between different classes as a "single machine" approach. For instance, in the one-versus-all method (e.g. [3]) the aim is to create a binary classification problem such that examples $y = l_1$ belong to the positive class and all other examples having class labels $l_{2,\ldots,\kappa}$ belong to the negative class. Another way to deal with the multi-class learning problem is described in [2], where all possible pairs of classes $l_1, l_2 \in \mathcal{Y}$ are considered. This means that $\binom{\kappa}{2}$ hypotheses have to be generated and combined. The method is referred to as the all pairs approach.

A more general suggestion on how to treat multi-class classification problem was proposed in [1] and later extended in [6]. It is known as the error-correcting

Z.-H. Zhou, F. Roli, and J. Kittler (Eds.): MCS 2013, LNCS 7872, pp. 61–72, 2013.
© Springer-Verlag Berlin Heidelberg 2013

output codes (ECOC) approach. The key idea is to construct the coding matrix $C \in \{-1, +1\}^{\kappa \times p}$, where p is some positive integer, such that the rows of the matrix have good error correcting properties (e.g. a large Hamming distance). The binary learning algorithm is then run once for each column of the output matrix, whose rows correspond to the encodings of the appropriate y labels induced by the coding matrix C. For example, this setting is a generalization of one-versus-all scheme and it can be represented with $\kappa \times \kappa$ coding matrix having all diagonal elements equal to 1 and all other elements equal to -1. Given a new example $\mathbf{x}'$ from input space $\mathcal{X}$, we can predict the corresponding label y' by finding the row of the coding matrix that is "closest" to $\mathbf{f} = (f_1(\mathbf{x}') \ldots f_p(\mathbf{x}'))$, where $f_s(\mathbf{x}'), s = 1, \ldots, p$ are the prediction functions constructed for each column of the output matrix.

Another group of algorithms that can be considered as genuine multi-class classifiers is described in e.g. [7,4,5]. These algorithms learn a single prediction function that can properly discriminate between different classes. Some of the algorithms (e.g. [4,5]) use so-called joint feature maps/views $\Phi(\mathbf{x}, y)$ on the data from $\mathcal{X}$ and labels $\mathcal{Y}$ to learn the prediction function. In this case the predicted class is the one that maximizes the output of the learnt function for the new data point. We note that the initial problem considered in [4] is structured output prediction and the presented formulation allows considering multi-class classification as a special case. In fact, using multiple views for learning has been shown to be beneficial for the predictive performance of the algorithm in many tasks beyond multi-class classification (e.g. [8]). Multi-class classification algorithms that construct a single prediction function appear to have good generalization performance and are usually computationally more efficient compared to approaches that need to train several binary classifiers to solve the problem. On the other hand, methods such as ECOC have the attractive property of error correction, they are extensively used in practice and continuously improved (see, for example, a recent work on decreasing number of binary classifiers needed for class prediction [9]).

This work aims to combine the benefits of the methods described above and presents a novel multi-class classification algorithm that is particularly suitable for multi-view learning setting. We extends co-regularization [10] framework to be applicable to multi-class problems as well as describe a loss-based decoding approach for estimating class labels when using multiple views/feature representations. In our empirical evaluation on publicly available datasets from the UCI, Statlog, and other repositories proposed algorithm outperforms several state-of-the-art multi-class classification methods. Furthermore, we apply our algorithm to the task of identification of pathogenic bacterial species from the dataset containing MS spectra of highly infectious Brucella microorganism [11]. The genus Brucella contains infectious species that have been found to cause infections in a wide variety of mammals. Most Brucella species have a narrow host range. Infection in humans arises from direct or indirect contact with infected animals or through consumption of contaminated meat or dairy products. We demonstrate that our algorithm gives a low classification error rate of 5% and allows rapid identification of the pathogenic strains. Proposed algorithm can be helpful for timely and effective response to an infectious disease outbreak, regardless of whether the outbreak is natural or deliberate.

2   Preliminaries

In the following subsections we briefly describe our framework for constructing our multi-view multi-class classification algorithm and outline related research directions. We are interested in the selection of suitable prediction function $f \in \mathcal{H}$. Following [12], we consider the reproducing kernel Hilbert space (RKHS) determined by the input space $\mathcal{X}$ and the positive definite kernel function $k : \mathcal{X} \times \mathcal{X} \to \mathbb{R}$. Using the RKHS $\mathcal{H}$ as our hypothesis space, we consider the optimization problem

$$\min_{f \in \mathcal{H}} J(f) = c(f, \mathcal{D}) + \lambda \|f\|_{\mathcal{H}}^2 \tag{1}$$

where $c(\cdot, \cdot)$ is the loss measuring the error of the prediction function f on the training set $\mathcal{D} = (X, Y)$, $\| \cdot \|_{\mathcal{H}}$ denotes the norm in $\mathcal{H}$, and $\lambda \in \mathbb{R}^+$ is a regularization parameter controlling the tradeoff between the error on the training set and the complexity of the hypothesis. We note that by specializing the loss in the above formulation we can obtain support vector machines or regularized least-squares (RLS) [13] as well as other closely related algorithms such as proximal vector machines and ridge regression.

Next, we describe a straightforward extension of the RLS algorithm to handle multiple outputs. Suppose instead of having a single column matrix for the outputs, we now have an $n \times p$-matrix, where p is the number of outputs. Slightly overloading our notation, let the output matrix be denoted as $Y \in \mathbb{R}^{n \times p}$. In the context of multi-class classification using ECOC the rows of Y would be the same as those of the coding matrix C. We use the dataset $\mathcal{D} = (X, Y)$ originating from a set $\{(\mathbf{x}_i, \mathbf{y}_i)\}_{i=1}^n$ of data points, where $X = (\mathbf{x}_1, \dots, \mathbf{x}_n)^t \in \mathcal{X}^n$, $Y = (\mathbf{y}_1, \dots, \mathbf{y}_n)^t \in \mathbb{R}^{n \times p}$. We write the minimization problem as

$$\min_{\mathbf{f} \in \mathcal{H}} J(\mathbf{f}, \mathcal{D}) = \sum_{i=1}^p \sum_{j=1}^n (y_{ji} - f_i(\mathbf{x}_j))^2 + \lambda \|f_i\|_{\mathcal{H}}^2, \tag{2}$$

thus, the problem at hand boils down to solving p independent regression tasks. We note that using a square loss function leads to an efficient multi-output regression solution, namely we obtain predictions for each output by inverting the kernel matrix only once, therefore, complexity of the algorithm is hardly increased compared to a standard single output problem. On the other hand when using methods similar to SVMs for prediction of multiple outputs, the complexity of solving a single task is multiplied by the number of outputs.

3   Co-regularized Multi-class Classification

Co-regularization (e.g. [10,14]) is naturally applicable in situations where more than one feature representation of the same object exists. Formally, consider

M RKHSs $\mathcal{H}_1, \ldots, \mathcal{H}_M$ along with their corresponding kernel functions $k^{(v)}$: $\mathcal{X} \times \mathcal{X} \to \mathbb{R}, 1 \leq v \leq M$. In the classification task, we search for a vector $\mathbf{f} = (\mathbf{f}_1, \ldots, \mathbf{f}_M) \in \mathcal{H}_1 \times \ldots \times \mathcal{H}_M$ of prediction functions which minimizes

$$J(\mathbf{f}, \mathcal{D}) = \sum_{v=1}^{M} c(\mathbf{f}^{(v)}, \mathcal{D}) + \lambda \sum_{v=1}^{M} \|\mathbf{f}^{(v)}\|_{\mathcal{H}_v}^2 + \nu \sum_{v,u=1}^{M} \tilde{c}(\mathbf{f}^{(v)}, \mathbf{f}^{(u)}, \mathcal{D}), \quad (3)$$

where $\lambda, \nu \in \mathbb{R}^+$ are regularization parameters and $\tilde{c}$ is a loss function measuring the disagreement between the prediction functions of the views. A classical example is a web-document classification task where the document can be represented by the word features or link features it contains, thus, creating two distinct views of the same data point [15]. The setting described above is suitable for multi-view approaches, however, in many cases it is not trivial to decide which set of features is most appropriate for co-regularization (e.g. when the views are too similar the approach will not work well in practice, as the co-regularization term would not contribute to the learning).

In case of multi-class classification, the encoding represents a natural choice for constructing a different view of the data. In some sense, the encoding can be considered as a structured output that uniquely describes the label. We suggest that by taking into account correlations among such structured outputs in addition to the inputs can be beneficial for the performance of the learning algorithm. We will construct the views on input-output representations and use them for training of our algorithm. We will show that when dealing with multi-class classification problems co-regularization among the views constructed from the input-output representations leads to improved classification performance. Our key contribution is an efficient algorithm for multi-class classification that retains benefits of an error-correcting approach and can learn from multiple views based on expressive input-output feature representations.

One problem associated with using multi-view representation of the inputs and outputs for multi-class classification arises when estimating the class label of the new data point. Due to the kernel function used to construct the feature space, the encoding of the label of the new data point has to be provided to the algorithm. We solve above problem by making use of the fact that in multi-class classification tasks the set of possible class labels that can be assigned to the new data point is known. The strategy is to compute predictions by considering all possible labels. Once predictions are available loss-based decoding [6] can be used to find the true class label.

Below we propose a simple loss-based decoding approach for estimating the class label when using multiple feature representations. More formally, applying the representer theorem [12] in this context of the co-regularization problem described in (3) shows that the minimizers $\mathbf{f}^{(v)} \in \mathcal{H}^{(v)}$ for $v = 1, \ldots, M$ have the form $\mathbf{f}^{(v)}(\mathbf{x}', \mathbf{y}') = \sum_{i=1}^{n} \mathbf{a}_i^{(v)} k^{(v)}((\mathbf{x}', \mathbf{y}'), (\mathbf{x}_i, \mathbf{y}_i))$, where $\mathbf{x}'$ is an unseen example, $\mathbf{y}' \in \{C_{1,\cdot}, \ldots C_{\kappa,\cdot}\}$ is the encoding of the label, and $\mathbf{a}_1^{(v)}, \ldots, \mathbf{a}_n^{(v)} \in \mathbb{R}^p$ are the coefficients. We take the average over all views $\mathbf{f}^*(\mathbf{x}', \mathbf{y}') = \frac{\sum_{v=1}^{M} \mathbf{f}^{(v)}(\mathbf{x}', \mathbf{y}')}{M}$, and define the loss based decoding function that calculates the distance between the prediction and the rows of the coding matrix $d_L(C_{i,\cdot}, \mathbf{f}^*(\mathbf{x}', \mathbf{y}')) = \sum_{j=1}^{p}(C_{i,j} - f_j^*(\mathbf{x}', \mathbf{y}'))^2$. Finally, we select the label to be assigned to the new data point $\mathbf{x}'$ by choosing the class i^* with the smallest distance: $i^* = \operatorname{argmin}_i d_L(C_{i,\cdot}, \mathbf{f}^*(\mathbf{x}', \mathbf{y}_i'))$. In our empirical evaluation we have tried several strategies for selecting the class label for a new data point, however, assigning the label based on smallest distance usually leads to the best results. We note that constructing views on inputs-outputs has been shown to be beneficial (e.g. [8]) in structured output prediction problems. We suggest that our approach can be adapted for such tasks, however, in this work we primarily aim to address multi-class classification problem.

4   Computational Issues

We have mentioned that using square or hinge loss functions leads to two closely related algorithms and argued that our choice of the square loss leads to considerable computational benefits when addressing the problem of multiple output prediction. Thus, using square loss and matrix notations we can reformulate (3) as

$$J(\mathbf{A}) = \sum_{v=1}^{M} \operatorname{tr}\left(Y - K^{(v)}A^{(v)}\right)^t \left(Y - K^{(v)}A^{(v)}\right) + \lambda \sum_{v=1}^{M} \operatorname{tr}A^{(v)^t} K^{(v)} A^{(v)} +$$

$$\nu \sum_{v,u=1}^{M} \operatorname{tr}\left(K^{(v)}A^{(v)} - K^{(u)}A^{(u)}\right)^t \left(K^{(v)}A^{(v)} - K^{(u)}A^{(u)}\right),$$

where $A^{(v)} = (\mathbf{a}_1^{(v)}, \dots, \mathbf{a}_n^{(v)})^t \in \mathbb{R}^{n \times p}$ and $\mathbf{A} = (A_1^t, \dots, A_M^t)^t \in \mathbb{R}^{Mn \times p}$. The matrix $K^{(v)} \in \mathbb{R}^{n \times n}$ has entries of the form $\left[K^{(v)}\right]_{i,j} = k^{(v)}((\mathbf{x}_i, \mathbf{y}_i), (\mathbf{x}_j, \mathbf{y}_j))$.

The function $k((\mathbf{x}, \mathbf{y}), (\mathbf{x}', \mathbf{y}'))$ is referred to as a joint kernel. The main idea behind the joint kernel is to describe the similarity between input-output pairs by mapping pairs into a joint space [8]. A joint kernel can encode more than just information about inputs or outputs independent of each other: it can also encode known dependencies between inputs and outputs. For example, several variations of joint kernels have been proposed in [8]. One way to define a joint kernel k is to multiply kernels constructed on input data $k_{\mathcal{X}} : \mathcal{X} \times \mathcal{X} \to \mathbb{R}$ with kernel constructed on outputs $k_{\mathcal{Y}} : \mathcal{Y} \times \mathcal{Y} \to \mathbb{R}$:

$$k((\mathbf{x}, \mathbf{y}), (\mathbf{x}', \mathbf{y}')) = k_{\mathcal{X}}(\mathbf{x}, \mathbf{x}') \cdot k_{\mathcal{Y}}(\mathbf{y}, \mathbf{y}'). \tag{4}$$

This feature space corresponds to the tensor product of the features space on $\mathcal{X}$ with that on $\mathcal{Y}$. The kernel we use in the experiments is the standard Gaussian kernel:

$$k((\mathbf{x}, \mathbf{y}), (\mathbf{x}', \mathbf{y}')) = \exp\left[\frac{-||(\mathbf{x}, \mathbf{y}) - (\mathbf{x}', \mathbf{y}')||^2}{2\sigma^2}\right].$$

Note that this kernel can be decomposed into a tensor product as in (4) with Gaussian kernels both on inputs and outputs.

4.1   Derivation of the Algorithm

Consider following optimization problem $\min_{A \in \mathbb{R}^{Mn \times p}} J(\mathbf{A})$. Given this formulation of our optimization problem, we can follow the framework described in [16] to find a closed form for the solution by taking the partial derivative of $J(\mathbf{A})$ with respect to $A^{(v)}$

$$\frac{d}{dA^{(v)}} J(\mathbf{A}) = -2K^{(v)t}(Y - K^{(v)}A^{(v)}) + 2\lambda K^{(v)}A^{(v)}$$

$$-4\nu \sum_{u=1, u \neq v}^{M} K^{(v)t}(K^{(u)}A^{(u)} - K^{(v)}A^{(v)}).$$

By defining $G_\nu^{(v)} = 2\nu(M-1)K^{(v)t}K^{(v)}$, $G_\lambda^{(v)} = \lambda K^{(v)}$ and $G^{(v)} = K^{(v)t}K^{(v)}$, we can rewrite the above term as

$$\frac{d}{dA^{(v)}} J(\mathbf{A}) = 2(G^{(v)} + G_\nu^{(v)} + G_\lambda^{(v)})A^{(v)} - 2K^{(v)t}Y - 4\nu \sum_{u=1, u \neq v}^{M} K^{(v)t}K^{(u)}A^{(u)}.$$

At the optimum we have $\frac{d}{dA^{(v)}} J(\mathbf{A}) = 0$ for all views, thus we get the exact solution by solving

$$
\begin{pmatrix}
\bar{G}^{(1)} & -2\nu K^{(1)t}K^{(2)} \cdots \\
-2\nu K^{(2)t}K^{(1)} & \bar{G}^{(2)} \quad \cdots \\
\vdots & \vdots \quad \ddots
\end{pmatrix}
\begin{pmatrix}
A^{(1)} \\
A^{(2)} \\
\vdots
\end{pmatrix}
=
\begin{pmatrix}
K^{(1)t}Y \\
K^{(2)t}Y \\
\vdots
\end{pmatrix}
$$

with respect to $A^{(1)}, \ldots, A^{(M)}$, where $\bar{G}^{(v)} = G^{(v)} + G_\nu^{(v)} + G_\lambda^{(v)}$. The left-hand side matrix is positive definite and therefore invertible. By defining

$$
B = \begin{pmatrix} G^{(1)} & 0 & \cdots \\ 0 & G^{(2)} & \cdots \\ \vdots & \vdots & \ddots \end{pmatrix} \quad
D = \begin{pmatrix} G_\lambda^{(1)} & 0 & \cdots \\ 0 & G_\lambda^{(2)} & \cdots \\ \vdots & \vdots & \ddots \end{pmatrix} \quad
E = \begin{pmatrix} K^{(1)t}Y \\ K^{(2)t}Y \\ \vdots \end{pmatrix}
$$

$$
C = \begin{pmatrix} G_\nu^{(1)} & -2\nu K^{(1)t}K^{(2)} \cdots \\ -2\nu K^{(2)t}K^{(1)} & G_\nu^{(2)} \quad \cdots \\ \vdots & \vdots \quad \ddots \end{pmatrix}
$$

we can formulate the solution of the system as follows:

$$\mathbf{A} = (B + C + D)^{-1}E. \tag{5}$$

Furthermore, we suggest an approach that allows searching for the optimal parameter without increasing the computational cost (see online supplementary material). The computational complexity of constructing the vector E is

Table 1. Classification errors in % on resampled datasets of the RANDOM baseline, MSVMLIN, MSVM, MRLS, and MVRLS algorithms. Bold numbers indicate the method with the lowest classification error on the particular dataset.

DATASET	RANDOM	MSVMLIN	MSVM	MRLS	MVRLS
COVERTYPE	85.7	46.2	45.6	46.1	**44.1**
LETTER	96.1	68.2	67.7	67.9	**67.1**
MNIST	90	41.6	41.1	41.0	**39.5**
NEWS20	95	82.3	81.3	81.7	**78.6**
POKER	90	55.5	55.1	55.2	**54.5**
SVMGUIDE4	83.3	56.6	56.2	56.7	**55.7**
USPS	90	32.3	31.8	32.5	**30.4**
VOWEL	90.9	60.2	59.9	59.9	**59.4**

$\mathcal{O}(Mn^2p)$. Further, the matrices B and D can be constructed in $\mathcal{O}(Mn^3)$, and the matrix C in $\mathcal{O}(M^2n^3)$ time. The resulting matrix $(B + C + D) \in \mathbb{R}^{Mn \times Mn}$ can be inverted in $\mathcal{O}(M^3n^3)$. Thus, for fixed parameters $\lambda, \nu \in \mathbb{R}^+$ the solution of the optimization problem can be found in $\mathcal{O}(M^3n^3 + Mn^2p)$ time (we demonstrate how to further decrease computation complexity of the algorithm in the online supplementary material).

5 Empirical Evaluation

In the following section we refer to the multi-output regularized least-squares algorithm for multi-class classification as MRLS. The shorthand notation for our multi-view multi-class classification algorithm is MVRLS. Further, we denote a multi-class SVM[1] using a linear and Gaussian kernel with shorthand notation MSVMLIN and MSVM, respectively.

Various datasets are used to evaluate performance of the multi-class classifiers. Following previous works, we test the methods on the benchmark datasets from the UCI, Statlog, and other repositories that are made publicly available at the LIBSVM webpage[2] in the format directly usable by MSVM and our algorithms. We note that the attributes in the datasets are linearly scaled to [-1,1].

For the MVRLS algorithm we construct two views: one using a joint kernel over inputs and outputs, another one purely based on inputs. We use training data points and their encodings for constructing the views. For estimating the class of the unseen data point using these two views we follow the procedure described in Section 3. Similar to [16] we set the width of $k(\mathbf{x}, \mathbf{x}') = \exp\left[-(\mathbf{x} - \mathbf{x}')^2/2\sigma^2\right]$ to $\sigma^2 = 1/n \sum_{i,j=1}^n (\mathbf{x}_i - \mathbf{x}_j)^2$. For the joint kernel the width parameter is chosen identically. We have checked on several datasets whether the computed width is in a good agreement with the one estimated using 10-fold cross-validation.

---

[1] http://svmlight.joachims.org/svm_multiclass.html
[2] http://www.csie.ntu.edu.tw/~cjlin/libsvmtools/datasets/multiclass.html

In all cases we have found the computed value to be near optimal. For selecting the regularization parameters for the MVRLS as well as for the other learning algorithms we use a 10-fold cross-validation procedure. Estimating optimal regularization parameters can be computationally prohibitive when conducting cross-validation. For the MVRLS we use the procedure described in Section 4 to efficiently search for the parameters. Namely, for each ν on a grid, we apply the fast parameter selection procedure to obtain results for all λ's with a single run. A similar approach can be applied to find the optimal regularization parameter for the MRLS algorithm.

5.1   Experimental Setup

Following [6] we construct a dense encoding for the multi-class classification problems. The encoding has $10 \log_2(\kappa)$ columns where each entry is chosen to be -1 or 1 with equal probability. We generate 10000 random matrices and select the one having the largest Hamming distance among the rows. We also conduct several experiments using sparse random codes [6] (e.g. in addition to -1 and 1, entries containing 0s are allowed), as well as with one-vs-all encoding. Using dense encoding leads to slightly better performance in these trials and we choose it for comparing the algorithms in our final experiments. Somewhat similar observations were reported in [3], where it is also noted that when parameters for underlying binary classifiers are tuned correctly, the one-vs-all scheme becomes quite competitive to error-correcting approaches.

We consider an experimental setup similar to [17], where the classification performance of the binary classifiers is compared on large UCI datasets. From every dataset we randomly sample 100 examples. We ensure that examples belonging to different classes are present in the randomly selected subset. Two thirds of these examples are used for training, while one third is reserved for testing. On the training set we use 10-fold cross-validation to estimate regularization parameters. We select parameters that on average lead to the best performance over 10-folds. With these parameters fixed, we retrain the algorithm on the training set (two thirds) and test it on the test set (one third). Finally, we repeat the complete experiment 50 times and average the results. The obtained results for each dataset are reported in the Table 1.

We note that although the MSVM algorithm is efficient with a linear kernel, its runtime when using a nonlinear kernel is large[3]. Thus, one of the reasons for adapting the experimental setup proposed in [17] is the possibility to compare our algorithm to the non-linear MSVM method. It can clearly be seen that MVRLS consistently has good classification performance. We use the Friedman test [18] with significance level ($p < 0.01$) on the results obtained from 50 independent runs of the algorithm and demonstrate that statistical significance of the obtained results. To further study the performance differences between the algorithms we use post-hoc Nemenyi test. Only in one case (vowel dataset)

---

[3] See http://svmlight.joachims.org/svm_multiclass.html for details.

where MVRLS performs better than the other methods the obtained differences are not significant.

6   Identification of Pathogenic Bacterial Strains

We apply proposed multi-view multi-class classification algorithm to recently collected biomedical dataset described in [11]. The dataset obtained via matrix-assisted laser desorption/ionization time-of-flight mass spectrometry consists of spectra of 170 Brucella isolates. The genus Brucella contains highly infectious species that have been found to cause infections in a wide variety of mammals. Infection in humans arises from direct or indirect contact with infected animals or through consumption of contaminated meat or dairy products. Also, diagnostic laboratory workers are also at risk; 2% of all cases of brucellosis are laboratory acquired. Brucella species have a low infectious dose and are capable of transmission via aerosols, and the treatment of infections is lengthy with a risk of complications. The Brucella species primarily considered to be pathogenic for humans are B. melitensis, B. suis, B. abortus, and sporadically B. canis, B. suis biovars. The collected data requires preprocessing such as re-sampling, de-noising, baseline correction, normalization and finally peak detection and qualification before it can be used for training the classifier. We follow [19] to extract relevant features for each data point. We evaluate performance of the proposed algorithm on multi-class classification problem with 5 classes containing pathogenic Brucella species (B. melitensis, B. suis, B. abortus, B. canis, and B. suis biovars) and one additional class containing the rest of non-pathogenic microorganisms. We also test classification performance of our method with lower number of classes representing pathogenic species. A motivation for such experiment is an observation that B. canis and B. suis biovars have not been documented to be human pathogens and can be merged into separate class. In all classification experiments we follow the same parameter estimation and view construction procedure as described in Section 5. We randomly split complete dataset into training set (two thirds) and test it on the test set (one third). We ensure that examples belonging to different classes are represented in both subsets. Once the parameters are estimated via 10-fold cross-validation on the training set, we retrain the algorithm on the training set and test it on the test set. The obtained classification performance on the test set is depicted in the Figure 1. Similarly to the experiments on the benchmark datasets it can observed that MVRLS consistently outperforms other algorithms. The performance differences are statistically significant according to the Friedman test [18] ($p < 0.01$). The main result is shown in the subfigure (a). Experiment 1 corresponds to the multi-class classification problem with 5 classes of pathogenic and 1 class of non-pathogenic species. Experiment 2 corresponds to the classification problem with B. canis and B. suis biovars species merged into a single class. The rest of the experiments correspond to different reductions of 5 pathogenic classes into lower number of classes. With the exception of experiment 1, the subfigures (a), (b), and (c) depict results on 5-class classification and the subfigures (d), (e), and (f) depict results on to 4-class classification tasks.

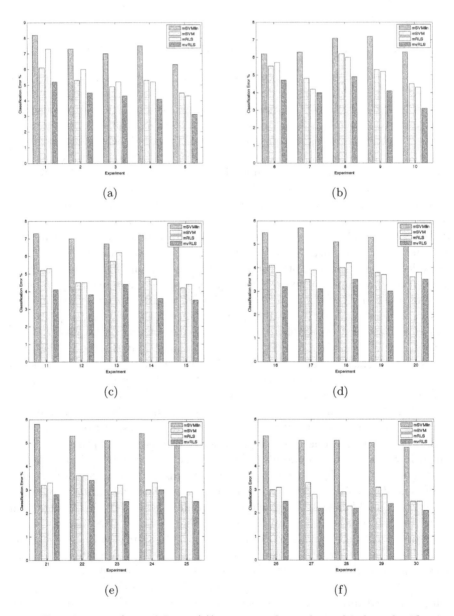

Fig. 1. Experiment 1 (see subfigure (a)) corresponds to the multi-class classification problem with 5 classes of pathogenic and 1 class of non-pathogenic species. The rest of the experiments correspond to different reductions of 5 pathogenic classes into lower number of classes. With the exception of experiment 1, the subfigures (a), (b), and (c) depict results on 5-class classification and the subfigures (d), (e), and (f) depict results on to 4-class classification tasks.

7   Conclusions and Discussion

This work concerns a novel multi-class classification algorithm that is particularly suitable for multi-view learning setting. We formulate the algorithm by extending co-regularization [10] framework to be applicable to multi-class problems as well as suggest new loss-based decoding approach for estimating class labels when using multiple views/feature representations. In our empirical evaluation on benchmark datasets the proposed algorithm outperforms several state-of-the-art multi-class classification methods. We apply our algorithm to the task of identification of pathogenic bacterial species of highly infectious Brucella microorganism. The results show low classification error rate of 5% on this task, which suggests that our method can be helpful for timely and effective response to an infectious disease outbreak.

Our approach can be extended in various situations. For instance, the algorithm can be adapted for the task of structured output prediction. It also allows efficient construction of different views for the co-regularization problem (see online supplementary material) and leads to notable speed up when dealing with large-scale learning tasks.

Acknowledgements. This work was financially supported by the Dutch Ministry of Defense, grant number V1036. We also acknowledge support from the Netherlands Organization for Scientific Research (NWO), in particular Vici grant (639.023.604).

References

1. Dietterich, T.G., Bakiri, G.: Solving multiclass learning problems via error-correcting output codes. Journal of Artificial Intelligence Res. 2, 263–286 (1995)
2. Hastie, T., Tibshirani, R.: Classification by pairwise coupling. In: Proceedings of the Neural Information Processing Systems, pp. 507–513. MIT Press, Cambridge (1998)
3. Rifkin, R., Klautau, A.: In defense of one-vs-all classification. Journal of Machine Learning Research 5, 101–141 (2004)
4. Tsochantaridis, I., Hofmann, T., Joachims, T., Altun, Y.: Support vector machine learning for interdependent and structured output spaces. In: Proceedings of the International Conference on Machine Learning, p. 104. ACM (2004)
5. Zien, A., Ong, C.S.: Multiclass multiple kernel learning. In: Proceedings of the International Conference on Machine Learning, pp. 1191–1198. ACM, New York (2007)
6. Allwein, E.L., Schapire, R.E., Singer, Y.: Reducing multiclass to binary: a unifying approach for margin classifiers. Journal of Machine Learning Research 1, 113–141 (2001)
7. Crammer, K., Singer, Y.: On the algorithmic implementation of multiclass kernel-based vector machines. Journal of Machine Learning Research 2, 265–292 (2002)
8. Weston, J., Schölkopf, B., Bousquet, O.: Joint kernel maps. In: Cabestany, J., Prieto, A.G., Sandoval, F. (eds.) IWANN 2005. LNCS, vol. 3512, pp. 176–191. Springer, Heidelberg (2005)

9. Park, S.-H., Fürnkranz, J.: Efficient decoding of ternary error-correcting output codes for multiclass classification. In: Buntine, W., Grobelnik, M., Mladenić, D., Shawe-Taylor, J. (eds.) ECML PKDD 2009, Part II. LNCS (LNAI), vol. 5782, pp. 189–204. Springer, Heidelberg (2009)

10. Sindhwani, V., Niyogi, P., Belkin, M.: A co-regularization approach to semi-supervised learning with multiple views. In: Proceedings of ICML Workshop on Learning with Multiple Views (2005)

11. Lista, F., Reubsaet, F., De Santis, R., Parchen, R., de Jong, A., Kieboom, J., van der Laaken, A., Voskamp-Visser, I., Fillo, S., Jansen, H.J., Van der Plas, J., Paauw, A.: Reliable identification at the species level of brucella isolates with maldi-tof-ms. BMC Microbiology 11(1), 267 (2011)

12. Schölkopf, B., Herbrich, R., Smola, A.J.: A generalized representer theorem. In: Helmbold, D.P., Williamson, B. (eds.) COLT/EuroCOLT 2001. LNCS (LNAI), vol. 2111, pp. 416–426. Springer, Heidelberg (2001)

13. Rifkin, R., Yeo, G., Poggio, T.: Regularized least-squares classification. In: Advances in Learning Theory: Methods, Model and Applications, pp. 131–154. IOS Press, Amsterdam (2003)

14. Tsivtsivadze, E., Pahikkala, T., Boberg, J., Salakoski, T., Heskes, T.: Co-regularized least-squares for label ranking. In: Hüllermeier, E., Fürnkranz, J. (eds.) Preference Learning, pp. 107–123 (2010)

15. Blum, A., Mitchell, T.: Combining labeled and unlabeled data with co-training. In: Proceedings of the Eleventh Annual Conference on Computational Learning Theory, pp. 92–100. ACM, New York (1998)

16. Brefeld, U., Gärtner, T., Scheffer, T., Wrobel, S.: Efficient co-regularised least squares regression. In: Proceedings of the International Conference on Machine Learning, pp. 137–144. ACM, New York (2006)

17. Brefeld, U., Scheffer, T.: Auc maximizing support vector learning. In: Proceedings of ICML Workshop on ROC Analysis in Machine Learning (2005)

18. Demšar, J.: Statistical comparisons of classifiers over multiple data sets. Journal of Machine Learning Research 7, 1–30 (2006)

19. Liu, Q., Sung, A.H., Qiao, M., Chen, Z., Yang, J.Y., Yang, M.Q.Q., Huang, X., Deng, Y.: Comparison of feature selection and classification for MALDI-MS data. BMC Genomics 10(suppl. 1) (2009)

Feature Level Multiple Model Fusion Using Multilinear Subspace Analysis with Incomplete Training Set and Its Application to Face Image Analysis

Zhen-Hua Feng[1,2], Josef Kittler[2], William Christmas[2], and Xiao-Jun Wu[1]

[1] School of IoT Engineering, Jiangnan University, Wuxi, 214122, China
`xiaojun_wu_jnu@163.com`
[2] Centre for Vision, Speech and Signal Processing, University of Surrey, Guildford, GU2 7XH, United Kingdom
`{Z.Feng,J.Kittler,W.Christmas}@surrey.ac.uk`

Abstract. In practical applications of pattern recognition and computer vision, the performance of many approaches can be improved by using multiple models. In this paper, we develop a common theoretical framework for multiple model fusion at the feature level using multilinear subspace analysis (also known as tensor algebra). One disadvantage of the multilinear approach is that it is hard to obtain enough training observations for tensor decomposition algorithms. To overcome this difficulty, we adopted the M^2SA algorithm to reconstruct the missing entries of the incomplete training tensor. Furthermore, we apply the proposed framework to the problem of face image analysis using Active Appearance Model (AAM) to validate its performance. Evaluations of AAM using the proposed framework are conducted on Multi-PIE face database with promising results.

1 Introduction

Observations in the real world are often affected by many factors which lead to wide variations in object appearance. Typical examples are gender, pose, age and expression variations of the human face. The difficulties posed by these factors limit the performance of many existing object recognition approaches. Thus, modelling these factors is very important for image understanding and analysis, which are the ultimate goals of computer vision.

To counteract the difficulty posed by different variations, more sophisticated object modelling approaches have been proposed in recent years, such as the view-based [1] [2], bilinear-based [3] [4] and tensor-based approaches [5] [6]. The core idea of these methods is to try to decouple the original space into different subspaces and obtain a set of state-specific models which can represent their corresponding state-specific observations well. This is a common way to solve the difficulty as stated above. These models trained using subsets parametrized by specific factors can perform much better than a generic model trained from a pool of observations with many factors [7]. Implicitly, these approaches are based on different multiple model frameworks. In practical applications, we always choose one of these models or fuse some of them into a new model. Actually, both the view-based and bilinear methods can be viewed as special

Z.-H. Zhou, F. Roli, and J. Kittler (Eds.): MCS 2013, LNCS 7872, pp. 73–84, 2013.

cases of the multilinear methods. Using multilinear structure has lots of advantages in high dimensional data analysis because it offers a natural description of real-world observations.

In this paper, we develop a unified and compact theoretical framework for feature level multiple model fusion by using multilinear algebra. In practical applications, however, it is normally hard to obtain enough training samples for classical tensor decomposition algorithms, such as the Higher Order Singular Value Decomposition (HOSVD) [8]. For example, for an object with 5 different factors and each factor including 10 different variations, the total number of the required training samples is 10^5. To cope with this problem, the M^2SA [9] algorithm is implemented with our multiple model fusion system, which can reconstruct the missing entries by using a weighted scheme. The proposed framework is applied to face image analysis using Active Appearance Model(AAM)[10] to validate it and to assess its performance.

The rest of this paper is organized as follows: Section 2 introduces the basic theory of tensor algebra and our multiple model fusion framework. Section 3 addresses the M^2SA algorithm to cope with the missing entries. The application of the proposed framework to the AAM is discussed in Section 4. The experimental results obtained on the Multi-PIE face database [11] are presented in Section 5 and conclusions are summarized in the last section.

2   Multiple Model Fusion

In this paper, scalars, vectors, matrices and higher-order tensors are denoted by lower-case and upper-case letters ($a, A, b, B, \cdots$), bold lower-case letters ($\mathbf{a}, \mathbf{b}, \cdots$), bold upper-case letters ($\mathbf{A}, \mathbf{B}, \cdots$) and calligraphic upper-case letters ($\mathcal{A}, \mathcal{B}, \cdots$) respectively.

The adopted model fusion framework is carried out at the feature level of observations by using multilinear algebra. The multilinear algebra, also known as tensor algebra, is an extension of 1D vector and 2D matrix in linear algebra, which are actually 1st-order and 2nd-order tensors respectively. Normally, the term 'higher-order tensors' stands for Nth-order tensors when $N \geq 3$.

Suppose we have an observation training set $\mathbf{X}$ parametrised by M different factors and each factor has I_m ($m = 1 \cdots M$) different variations. We can divide this training set into subsets $\mathbf{X} = \{\mathbf{X}^{1,1,\cdots,1}, \cdots, \mathbf{X}^{i_1,i_2,\cdots i_M}, \cdots, \mathbf{X}^{I_1 I_2 \cdots I_M}\}$. For these observations in each subset, we can extract their features and obtain the feature level description for the corresponding subset:

$$\mathbf{Y}^{i_1,i_2,\cdots,i_M} = \{\mathbf{y}_1^{i_1,i_2,\cdots,i_M}, \cdots, \mathbf{y}_L^{i_1,i_2,\cdots,i_M}\}, \tag{1}$$

where L is the number of features, $\mathbf{y}_l \in \mathbb{R}^{D_l \times 1}$($l = 1 \cdots L$) are feature vectors and D_l is the dimensionality of these feature vectors. These features could be the shape, texture, Haar-like feature, HoG, LBP, SIFT and so on. We can choose one or some of these features for (1). The feature selection might affect the performance of the model seriously. Thus, we should choose suitable feature selection methods to achieve the best performance of the models in a given application.

Normally, given a specific model, we can train a set of those models using different state-specific training subsets:

$$\{\mathbf{M}^{i_1,i_2,\cdots,i_M} : \mathbf{Y}^{i_1,i_2,\cdots,i_M}\}, i_m = 1 \cdots I_M \tag{2}$$

where $\mathbf{M}^{i_1,i_2,\cdots,i_M}$ is the state-specific model trained from the corresponding subset $\mathbf{Y}^{i_1,i_2,\cdots,i_M}$. Then, one of these models or a fused model obtained by combining some of these models chosen by a state estimation approach can be applied to the test set with corresponding variations. This multiple model system can improve the performance greatly. However, it is not convenient to describe and analyse this incompact framework. Here, we introduce the multilinear approach to the feature level multiple model fusion framework to obtain a unified and compact structure. Fig. 1 shows an example of the proposed feature level multiple model fusion approach based on a 3rd-order tensor.

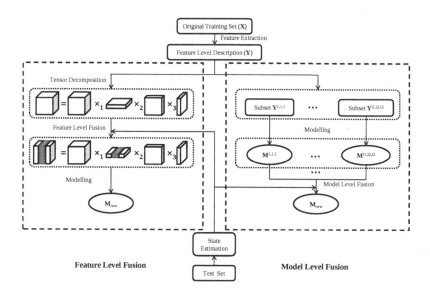

Fig. 1. Comparison between the model level and feature level multiple model fusion frameworks

In multilinear approach, each feature level training set can be rewritten as a higher-order tensor $\mathcal{Y}_l \in \mathbb{R}^{I_1 \times I_2 \times \cdots \times I_M \times I_{M+1}} (l = 1 \cdots L)$ and $I_{M+1} = D_l$. Set $N = M + 1$, the $(M + 1)$th-order tensor can be simplified to a Nth-order tensor $\mathcal{Y}_l \in \mathbb{R}^{I_1 \times I_2 \times \cdots \times I_N}$. Then, we apply tensor decomposition algorithms to these training tensors $\mathcal{Y}_l(l = 1 \cdots L)$. The most popular two types of tensor decomposition algorithms are CANDECOMP/PARAFAC (CP) based tensor decomposition and Tucker-based tensor decomposition [12]. Both of them are generalizations of SVD and PCA from a 2D matrix to a higher-order tensor. In this paper, we choose the Tucker-based tensor because its decomposition algorithm and structure are much more convenient than that of the CP-based tensor.

The feature level training tensor $\mathcal{Y}$ can be decomposed by Tucker-based HOSVD [8] decomposition algorithm (also known as the 'Tucker1' algorithm):

$$\mathcal{Y} = \mathcal{Z} \times_1 \mathbf{U}_1 \times_2 \mathbf{U}_2 \cdots \times_N \mathbf{U}_N \qquad (3)$$

where $\mathcal{Z} \in \mathbb{R}^{I_1 \times I_2 \times \cdots \times I_N}$ is the core tensor with the same dimensionality of the input tensor $\mathcal{Y}$, which stands for the interaction between the orthonormal mode matrices $\mathbf{U}_n \in \mathbb{R}^{I_n \times I_n} (n = 1 \cdots N)$. The mode-n matrix $\mathbf{U}_n$ in (3) is the left singular matrix obtained by applying SVD to mode-n flatten matrix $\mathbf{Y}_{(n)}$ of tensor $\mathcal{Y}$ [12] and the core tensor $\mathcal{Z}$ is computed by:

$$\mathcal{Z} = \mathcal{Y} \times_1 \mathbf{U}_1^T \times_2 \mathbf{U}_2^T \cdots \times_N \mathbf{U}_N^T. \tag{4}$$

When we apply the multiple model system to a specific application, we must choose one of these models or fuse them into a new model to adapt to the practical environment. Here, we do the selection or fusion process at the feature level rather than the model level. Note that, the performance of a model might be affected only by one or some of the factors. Thus, we only need to fuse the training tensor on these corresponding factor modes rather than all the modes. Suppose the model is influenced by the nth and $n+1$th factors seriously, we can obtain the fused model at the feature level by:

$$\{\mathbf{M}_{new} : \mathcal{Z} \times_1 \mathbf{U}_1 \cdots \times_n \alpha_n^T \mathbf{U}_n \times_{n+1} \alpha_{n+1}^T \mathbf{U}_{n+1} \cdots \times_N \mathbf{U}_N\}$$

$$s.t. \sum_{k=1}^{I_n} \alpha_n(k) = 1, \tag{5}$$

where $\mathbf{M}_{new}$ is the fused new model and α_n is the state parameters standing for the degree of the membership to different variations of the nth factor. We can also set $\alpha_n(k) \in \{0, 1\}$ to achieve a model selection framework. The elements in state vector α can be obtained by using different classifiers or multiple classifier systems, such as the variation estimation scheme in [6].

3   Coping with Incomplete Training Set

The feature level multiple model fusion framework using tensor algebra provides a compact structure for the multiple model system. However, this tensor-based framework can only tackle with the problems when all the entries of the input tensors are available. In practical applications, it is hard to obtain such an complete tensor. To overcome this difficulty, we introduce some state-of-the-arts tensor completion methods in this section.

One possible way to solve this problem is by using the naive mean of available entries. But this method can only be used when a small number of training samples are missing. The performance of the fused model goes down rapidly as the number of missing entries is increasing. Another way is to reconstruct these missing entries by using some prediction algorithms. Suppose we have an incomplete training tensor $\mathcal{Y} \in \mathbb{R}^{I_1 \times I_2 \times \cdots \times I_N}$. To reconstruct the missing entries for this incomplete tensor, we first define a corresponding non-negative index (weight) tensor $\mathcal{I}$ of the same size as $\mathcal{Y}$:

$$\mathcal{I}^{i_1 i_2 \cdots i_N} = \begin{cases} 1, & \text{when } \mathcal{Y}^{i_1 i_2 \cdots i_N} \text{ is available,} \\ 0, & \text{when } \mathcal{Y}^{i_1 i_2 \cdots i_N} \text{ is missing.} \end{cases} \tag{6}$$

The goal of the incomplete tensor decomposition is to minimize:

$$\|\mathcal{I} * (\mathcal{Y} - \hat{\mathcal{Y}})\|, \tag{7}$$

where $'*'$ is Hadamard product (or element-wise product), and $\hat{\mathcal{Y}}$ is the reconstructed tensor. To solve this objective function, Acar *et al.* proposed an approach named CP Weighted OPTimization (CP-WOPT) [13] by using a gradient descent optimization approach. However, the CP-WOPT is based on CP tensor and the gradient descent optimization algorithms influence the reconstruction accuracy seriously. An alternative method is the Tucker-based M^2SA algorithm [9]. Although the objective function of M^2SA is the same as that of CP-WOPT, it does not need to compute the gradient using some optimization algorithms in each iteration. Thus the time cost is greatly reduced by using M^2SA and the reconstruction performance is independent to the optimization algorithms.

The M^2SA is an iterative progress based on N-Mode tensor dimensionality reduction algorithm. The N-Mode tensor dimensionality reduction algorithm aims to find a lower $rank - (R_1, \cdots, R_N)$ approximation for an input tensor. The mode-n rank of tensor $\mathcal{Y}$ is defined as $R_n = rank(\mathbf{Y}_{(n)})$, where $\mathbf{Y}_{(n)}$ is the mode-n flatten matrix of $\mathcal{Y}$ at the nth mode. A pseudo code of the N-mode dimensionality reduction algorithm is given in Algorithm 1.

Algorithm 1. N-Mode Dimensionality Reduction

1.Pre-iteration

Set the lower rank $R_n < I_n$ for $n = 1, 2, \cdots, N$; apply HOSVD to $\mathcal{Y}$; truncate each mode matrix $\mathbf{U}_n$ to R_n columns and obtain the initial mode matrices $\mathbf{U}_1^0, \mathbf{U}_2^0, \cdots \mathbf{U}_N^0$;

2. Iterate for $k = 1, 2, \cdots$:

2.1 Set $\tilde{\mathcal{U}}_n^k = \mathcal{Y} \times_1 (\mathbf{U}_1^k)^T \cdots \times_{n-1} (\mathbf{U}_{n-1}^k)^T \times_{n+1} (\mathbf{U}_{n+1}^{k-1})^T \cdots \times_N (\mathbf{U}_N^{k-1})^T$;

2.2 Obtain $\tilde{\mathbf{U}}_n^k$ by unfolding $\tilde{\mathcal{U}}_n^k$ along the nth mode;

2.3 Orthonormalise the columns of $\tilde{\mathbf{U}}_n^k$ and truncate to R_n columns to obtain $\mathbf{U}_n^k$;

   Untill $\left\| \mathbf{U}_n^{k\,T} \cdot \mathbf{U}_n^{k-1} \right\|^2 > (1 - \varepsilon) R_n$, for $n = 1, 2, \ldots, N$;

3. Compute the core tensor by $\hat{\mathcal{Z}} = \tilde{\mathcal{U}}_N \times_N \hat{\mathbf{U}}_N^T$ and the rank-reduced approximation $\hat{\mathcal{Y}} = \hat{\mathcal{Z}} \times_1 \hat{\mathbf{U}}_1 \times_2 \hat{\mathbf{U}}_2 \cdots \times_N \hat{\mathbf{U}}_N$.

Algorithm 2. M^2SA

---

1. Fill the missing elements in training tensor $\mathcal{Y}$ with the average value of all the available elements with some corresponding contributory factors to obtain the initialization of the training tensor $\mathcal{Y}^0$;

2. Apply Algorithm 1 to $\mathcal{Y}^0$ to get the initial rank-reduced approximation $\hat{\mathcal{Y}}^0 = \hat{\mathcal{Z}}^0 \times_1 \hat{\mathbf{U}}_1^0 \times_2 \hat{\mathbf{U}}_2^0 \cdots \times_N \hat{\mathbf{U}}_N^0$;

3. Iterate for $k = 1, 2, \cdots$:

3.1 Update training tensor by $\mathcal{Y}^k = \mathcal{Y}. \times \mathcal{I} + \hat{\mathcal{Y}}^{k-1}. \times (\sim \mathcal{I})$;

3.2 Apply Algorithm 1 to $\mathcal{Y}^k$ to get the new rank-reduced approximation $\hat{\mathcal{Y}}^k = \hat{\mathcal{Z}}^k \times_1 \hat{\mathbf{U}}_1^k \times_2 \hat{\mathbf{U}}_2^k \cdots \times_N \hat{\mathbf{U}}_N^k$;

   Until $\left\| (\mathcal{Y}^k - \hat{\mathcal{Y}}^k). \times \mathcal{I} \right\| < \varepsilon$ or $k > Max\_Loop$;

4. Compute the rank-reduced approximation $\hat{\mathcal{Y}} = \hat{\mathcal{Z}} \times_1 \hat{\mathbf{U}}_1 \times_2 \hat{\mathbf{U}}_2 \cdots \times_N \hat{\mathbf{U}}_N$.

Unfortunately, the Algorithm 1 cannot be used for an incomplete tensor. To adapt this to a sparse tensor with missing entries, the M^2SA algorithm has been proposed. The M^2SA algorithm uses a weighted scheme to achieve the best reconstruction performance for the available entries as the reduction of the dimensionality of the input sparse tensor, which is summarized in Algorithm 2. The prediction of missing entries by using M^2SA completes the training tensor. Thus, we can apply the classical HOSVD to this reconstructed training tensor.

4 Application to Face Image Analysis

To validate the proposed multiple model fusion framework, we apply it to human face image analysis, which is an important problem in pattern recognition and computer vision. We develop the proposed multiple model fusion framework using 2D morphable models in this section.

It has been reported that the 2D morphable models are powerful tools for face image analysis, such as the well-known active shape model (ASM) [14], AAM [10] and constrained local model (CLM) [15]. AAM is one of the most popular 2D morpbable models due to its capability of modelling both shape and global texture for human faces [16]. Typically, an AAM is fitted to input images to achieve automatic face annotation or to attempt face recognition. However, the AAM is very sensitive to pose, expression and illumination variations, which seriously limits its applicability. In this section, we perform the tensor-based multiple model fusion framework to the AAM to overcome the fitting difficulty posed by view, expression and illumination variations when we have an incomplete training samples.

4.1 Feature Selection

The classical AAM is trained from a set of labelled images. We choose the classical shape and appearance (global texture) information as the features used in our multiple model framework. The shape is manually landmarked in the training phase and the appearance relates to shape-free surface obtained by using a piecewise affine warp from the original shape to the mean shape. Fig. 2 shows an example of the normalized shape and appearance features. Suppose the training set contains I_{id} identities with I_{pe} pose, I_{exp} expression and I_{ill} illumination variations. In practical applications, it is hard to obtain such a big training set which contains all these variations. Given an incomplete training set, we can extract the shape and appearance features and subsequently obtain the incomplete shape training tensor $S \in \mathbf{R}^{I_{id} \times I_{pe} \times I_{ill} \times I_{exp} \times I_s}$ and appearance training tensor $A \in \mathbf{R}^{I_{id} \times I_{pe} \times I_{ill} \times I_{exp} \times I_t}$, where I_s and I_t are dimensions of the shape and global texture feature vectors.

For the incomplete training shape and appearance tensors, the use of M^2SA implies constructing:

$$S = Z_S \times_1 \mathbf{V}_{id} \times_2 \mathbf{V}_{pe} \times_3 \mathbf{V}_{ill} \times_4 \mathbf{V}_{exp} \times_5 \mathbf{V}_s, \tag{8}$$

$$A = Z_A \times_1 \mathbf{W}_{id} \times_2 \mathbf{W}_{pe} \times_3 \mathbf{W}_{ill} \times_4 \mathbf{W}_{exp} \times_5 \mathbf{W}_t, \tag{9}$$

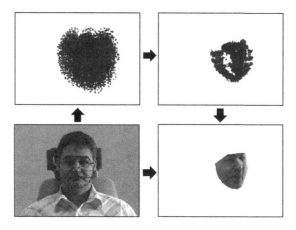

Fig. 2. Extracted shape and appearance features

where: $\mathcal{Z}_S$ and $\mathcal{Z}_A$ are shape and appearance core tensors; $\mathbf{V}_{id}, \mathbf{V}_{pe}, \mathbf{V}_{ill}, \mathbf{V}_{exp}, \mathbf{V}_s$ are mode matrices of the shape tensor for identity, pose, illumination, expression and the coordinates of the landmarks in shape respectively; $\mathbf{W}_t$ is the mode matrix of the global texture tensor based on the number of pixels in the mean shape.

Using the multiple model fusion framework we can obtain a specific AAM model which can be fitted to the input images with corresponding variations much better than a generic AAM model.

4.2   Feature Level Model Fusion

Given an input test image, we first predict the pose, expression and illumination conditions to obtain the fused shape and appearance models of AAM. The prediction algorithm can be performed either on pixel level or feature level. For model selection, this is a typical classification problem which identifies the single membership states of the input images. We can use some classical algorithms to obtain the state estimation results, such as the SVM, neural network, the discrete tensor-based estimation in [6] and so on. In principle, we could also identify the degree of membership of each input image in various states of variations to define mixing parameters for the multiple models (in contrast to model selection).

Once we obtain the state estimation results, we can train the fused shape and appearance models using the training approach in [17]:

$$\{\mathbf{M}_S : \mathcal{Z}_S \times_2 \boldsymbol{\alpha}_{pe}^T \mathbf{V}_{pe} \times_4 \boldsymbol{\alpha}_{exp}^T \mathbf{V}_{exp} \times_5 \boldsymbol{\alpha}_s^T \mathbf{V}_s\}, \tag{10}$$

$$\{\mathbf{M}_A : \mathcal{Z}_A \times_2 \boldsymbol{\alpha}_{pe}^T \mathbf{W}_{pe} \times_3 \boldsymbol{\alpha}_{ill}^T \mathbf{W}_{ill} \times_4 \boldsymbol{\alpha}_{exp}^T \mathbf{W}_{exp} \times_5 \boldsymbol{\alpha}_t^T \mathbf{W}_t\}, \tag{11}$$

where αs are the model mixing coefficient defined in (5).

It has been observed that the AAM fitting algorithms work well when the initial appearance can cover most part of the face in the input image [18]. Thus, we assume that the face region has been detected by a face detection algorithm with a sufficient accuracy to provide initialization for the AAM fitting. The estimation algorithm we adopted for predicting αs in (10) is the discrete estimation algorithm in [6]. At last, the corresponding fused shape and appearance models are used for AAM fitting by the inverse compositional algorithm in [16] to obtain the shape and global texture information of the face in the input image.

5   Experimental Results

5.1   Database and Experimental Environments

We evaluated the proposed tensor based multiple model fusion framework by applying it to face image analysis using AAM on the Multi-PIE [11] face database. The Multi-PIE face database has more than 750,000 facial images (640*480) captured from 377 people across 15 different poses, with 19 different illumination conditions and a range of different expressions across 4 sessions. It is a laborious work to landmark all the images in the Multi-PIE face database for model training and test. Although the total number of the identities is 377, only 129 identities are captured in all sessions with wide variations. From these 129 identities, we choose a subset containing 40 identities with 4 different poses (01_0, 04_1, 05_1 and 09_1), 3 different expressions (neutral from session 1, smile from session 3 and scream from session 4) and 4 different illuminations (00, 01, 07 and 13) as our training and test sets. Fig. 3 shows these pose, expression and illumination variations in our subset.

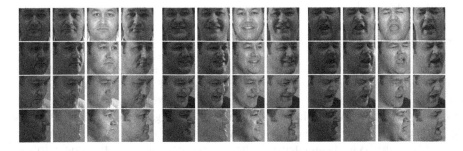

Fig. 3. Pose, illumination and expression variations of one identity from our experimental subset

The experiments were conducted on Dell PowerEdge C6145 servers with 4×AMD Opteron 6262 Processors (64 cores), 512 GB RAM and programmed by Matlab 2012a 64-bit using Tensor Toolbox 2.5 from Sandia National Laboratories [19].

5.2   Performance of AAM Using the Proposed Framework with Incomplete Training Set

In our experiments, we randomly choose 20 identities as the training set and the others as the test set. To make the evaluation meaningful, we adopt the repeated cross-validation scheme in our experiments. In each loop, we randomly remove 5% − 95% entries from the complete training shape and texture tensors to generate the incomplete

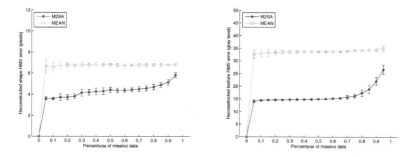

Fig. 4. Comparison of the reconstructed shape and texture errors by the M^2SA and naive mean methods

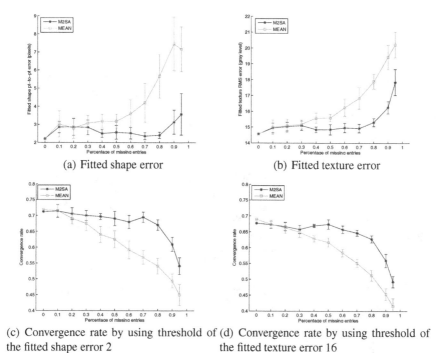

(a) Fitted shape error   (b) Fitted texture error

(c) Convergence rate by using threshold of   (d) Convergence rate by using threshold of
the fitted shape error 2   the fitted texture error 16

Fig. 5. Comparison of the AAM fitting performance by using the proposed framework

training tensors. Both the training and test subsets were landmarked manually to obtain the shape and global texture features for modelling, and the ground truth for evaluation. We took 52 landmarks for the shape feature in our experimental results. Thus, the size of the shape and texture training tensors are $20 \times 4 \times 3 \times 4 \times 104$ and $20 \times 4 \times 3 \times 4 \times 4018$ respectively. All the images have been resized to 320*240.

We first test the reconstruction performance of the M^2SA algorithm in terms of the RMS errors between the reconstructed missing entries and the ground truth data. To make a comparison, we substitute the missing entries using the mean of the available entries:

$$y_m^{i_{id},i_{pe},i_{exp},i_{ill},i} = \frac{\sum_{\mathbf{J}} y_a^{((j_{id}=i_{id})\vee(j_{pe}=i_{pe})\vee(j_{exp}=i_{exp})\vee(j_{ill}=i_{ill}))\wedge(j=i)}}{sum(\mathcal{I})}, \qquad (12)$$

where the lower right subscripts $'m'$ and $'a'$ stand for missing entries and available entries respectively; the upper right subscripts stand for the positions of the value in tensor $\mathcal{Y}$; $sum(\mathcal{I})$ gives the number of the available entries and $\mathcal{I}$ is the index tensor (6). Fig. 4 presents the reconstruction performance of M²SA in terms of shape and appearance reconstruction RMS errors. The M²SA can obtain a better reconstruction results both for the shape and appearance features used for AAM modelling. The obtained reconstructed shape and appearance features are used to cope with the problem of incomplete data in the training set in the subsequent steps.

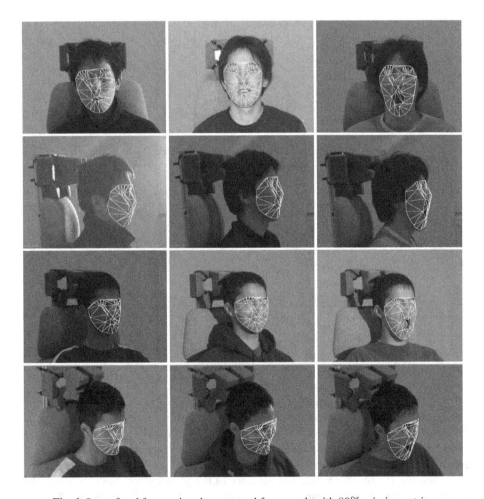

Fig. 6. Some fitted faces using the proposed framework with 80% missing entries

To evaluate the AAM fitting performance with the proposed framework, we first measured the fitting performance in terms of the pt-to-pt error between the fitted shape and the ground truth shape. Then we warped the global texture from the fitted shape to the mean shape and measured the RMS error between the warped texture and the ground truth texture. Also, we measured the convergence rates of the AAM by using threshold value of fitted shape error 2 and threshold value of fitted texture error 16. Fig. 5 shows the trends of the AAM fitting performance using the proposed multiple model fusion framework as the percentage of missing entries is increasing. It is obvious that the use of the M^2SA algorithm can maintain the AAM fitting performance even when the majority of training entries are missing. Fig. 6 shows some typical fitted results produced by the fused AAM model using the proposed framework with 80% missing entries. The proposed multiple model fusion framework can overcome the pose, expression and illuminations variations even in some extreme cases.

6 Conclusions

In this paper, we developed a unified theoretical framework for feature level multiple model fusion by using multilinear algebra. Furthermore, we applied the M^2SA algorithm to extend the proposed framework to incomplete training data. We then applied the proposed framework to face image analysis using AAM and evaluated the system performance on the Multi-PIE face database.

The experimental results obtained on the Multi-PIE face database validate the robustness of the proposed multiple model fusion framework on face image analysis in the presence of with pose, expression and illumination variations. The use of M^2SA algorithm improved the performance of our model fusion system in the case of an incomplete training set. The results show that our model can maintain good performance even when up to 80% training samples are missing.

Acknowledgement. The authors gratefully acknowledge supports of the EU-FP7 Biometrics Evaluation and Testing (BEAT) project under Grants 284989, National Natural Science Foundation of China under Grants 60973094, 61103128 and the Key Grant Project of Chinese Ministry of Education under Grants 311024.

References

1. Cootes, T., Wheeler, G., Walker, K., Taylor, C.: View-based active appearance models. Image and Vision Computing 20(9), 657–664 (2002)
2. Zhu, X., Ramanan, D.: Face detection, pose estimation, and landmark localization in the wild. In: 2012 IEEE Conference on Computer Vision and Pattern Recognition (CVPR), pp. 2879–2886. IEEE (2012)
3. Gonzalez-Mora, J., De la Torre, F., Murthi, R., Guil, N., Zapata, E.L.: Bilinear active appearance models. In: Proc. IEEE 11th Int. Conf. Computer Vision, ICCV 2007, pp. 1–8 (2007)
4. Lee, J., Moghaddam, B., Pfister, H., Machiraju, R.: A bilinear illumination model for robust face recognition. In: Tenth IEEE International Conference on Computer Vision, ICCV 2005, vol. 2, pp. 1177–1184. IEEE (2005)

5. Vasilescu, M., Terzopoulos, D.: Multilinear subspace analysis of image ensembles. In: Proceedings of Computer Vision and Pattern Recognition, vol. 2, p. II-93. IEEE (2003)
6. Lee, H.S., Kim, D.: Tensor-based AAM with continuous variation estimation: Application to variation-robust face recognition. IEEE Transactions on Pattern Analysis and Machine Intelligence 31(6), 1102–1116 (2009)
7. Gross, R., Matthews, I., Baker, S.: Generic vs. person specific active appearance models. Image and Vision Computing 23(12), 1080–1093 (2005)
8. De Lathauwer, L., De Moor, B., Vandewalle, J.: A multilinear singular value decomposition. SIAM Journal on Matrix Analysis and Applications 21(4), 1253–1278 (2000)
9. Geng, X., Smith-Miles, K., Zhou, Z., Wang, L.: Face image modeling by multilinear subspace analysis with missing values. IEEE Transactions on Systems, Man, and Cybernetics, Part B: Cybernetics 41(3), 881–892 (2011)
10. Cootes, T., Edwards, G., Taylor, C.: Active appearance models. IEEE Transactions on Pattern Analysis and Machine Intelligence 23(6), 681–685 (2001)
11. Gross, R., Matthews, I., Cohn, J., Kanade, T., Baker, S.: Multi-PIE. Image and Vision Computing 28(5), 807–813 (2010)
12. Kolda, T., Bader, B.: Tensor decompositions and applications. SIAM Review 51(3), 455–500 (2009)
13. Acar, E., Dunlavy, D., Kolda, T., Mørup, M.: Scalable tensor factorizations for incomplete data. Chemometrics and Intelligent Laboratory Systems 106(1), 41–56 (2011)
14. Cootes, T., Taylor, C., Cooper, D., Graham, J., et al.: Active shape models-their training and application. Computer Vision and Image Understanding 61(1), 38–59 (1995)
15. Cristinacce, D., Cootes, T.: Feature detection and tracking with constrained local models. In: Proc. British Machine Vision Conference, vol. 3, pp. 929–938 (2006)
16. Matthews, I., Baker, S.: Active appearance models revisited. International Journal of Computer Vision 60(2), 135–164 (2004)
17. Feng, Z.H., Kittler, J., Christmas, W., Wu, X.J., Pfeiffer, S.: Automatic face annotation by multilinear AAM with missing values. In: Proc. 21st International Conference on Pattern Recognition, ICPR (2012)
18. Edwards, G., Cootes, T., Taylor, C.: Advances in active appearance models. In: The Proceedings of the Seventh IEEE International Conference on Computer Vision, vol. 1, pp. 137–142 (1999)
19. Bader, B.W., Kolda, T.G., et al.: Matlab tensor toolbox version 2.5 (January 2012)

Kalman Filter Based Classifier Fusion for Affective State Recognition

Michael Glodek[1], Stephan Reuter[2], Martin Schels[1], Klaus Dietmayer[2], and Friedhelm Schwenker[1]

[1] Institute of Neural Information Processing, University of Ulm, 89075 Ulm, Germany
{michael.glodek,martin.schels,friedhelm.schwenker}@uni-ulm.de
[2] Institute of Measurement, Control and Microtechnology, University of Ulm, 89075 Ulm, Germany
{stephan.reuter,klaus.dietmayer}@uni-ulm.de

Abstract. The combination of classifier decisions is a common approach to improve classification performance [1–3]. However, non-stationary fusion of decisions is still a research topic which draws only marginal attention, although more and more classifier systems are deployed in real-time applications. Within this work, we study Kalman filters [4] as a combiner for temporally ordered classifier decisions. The Kalman filter is a linear dynamical system based on a Markov model. It is capable of combining a variable number of measurements (decisions), and can also deal with sensor failures in a unified framework. The Kalman filter is analyzed in the setting of multi-modal emotion recognition using data from the audio/visual emotional challenge 2011 [5, 6]. It is shown that the Kalman filter is well-suited for real-time non-stationary classifier fusion. Combining the available sequential uni- and multi-modal decisions does not only result in a consistent continuous stream of decisions, but also leads to significant improvements compared to the input decision performance.

1 Introduction

Typically, fusion approaches aim at combining classifier decisions in a time-invariant manner using classical combiners, as specified by Kuncheva [2], i.e. average, maximum, product etc. or fuzzy integrals, decision templates, Dempster-Shafer combination [7, 2, 3]. Most of these methods require fuzzy or probabilistic classifier outputs. Furthermore, adaptive fusion approaches such as the associative linear memory, pseudo-inverse, naive Bayes decision fusion have been examined [8]. However, these stationary, time-independent approaches are insufficient when dealing with a stream of classifier outputs obtained by observing an ongoing event in real-time. Intuitively, the temporally linked inputs to the fusion architecture are faced by extending standard combiners, which results in methods such as moving average, or computing the product of all inputs within one window. Jeon and Landgrebe [9] propose two fusion approaches for multi-temporal classifiers, namely the *likelihood decision fusion rule* and the *weighted majority decision fusion rule*. Both approaches make use of temporal data to

Z.-H. Zhou, F. Roli, and J. Kittler (Eds.): MCS 2013, LNCS 7872, pp. 85–94, 2013.
© Springer-Verlag Berlin Heidelberg 2013

find a global decision. Although the method is proposed for fusing temporally structured data, no Markov assumption is made. Glodek et al. [10] addresses the combination of temporally structured data using Markov fusion networks (MFN), which are related to the Markov Random Field in image processing to repair distorted images. The MFN estimates the most likely sequence of classifier outcomes by enforcing them (1) to be close to the given available classifier outcomes and (2) to maintain a similarity over time. Because of the later property, the MFN can handle missing classifier outcomes in a native way (e.g. due to silence in the audio channel).

Kalman Filters [4], which are broadly used in the area of object tracking, are well-suited for the task of classifier fusion, since they are based on a Markov chain. The goal is to reduce the noise of measurements by taking multiple measurements and the latest estimate into account. Within this work, we will adopt the Kalman Filter to use classifier outcomes as input measurements. Missing classifier outcomes (e.g. because of a missing signal) are addressed by increasing the uncertainty of the estimate using the observation noise of the model.

Recently, the recognition of human affective states became an important field of research [11–14]. Many configurations of features and classifiers using different modalities have been studied [15–18]. However, the interplay of different modalities, which is an inherent part of the recognition task, has been examined only marginally. The audio/visual emotion challenge (AVEC) is designed to investigate spatio-temporal combination of multiple classifiers [6, 5]. Within this work, we use the challenge data to study the Kalman filter for classifier fusion.

The rest of this paper is organized as follows: Section 2 introduces the proposed Kalman filter for classifier fusion. The method is then evaluated in Section 3 using the AVEC 2011 data set. Section 4 summarizes the results of the new approach and draws a conclusion on the achieved outcome.

2   Classifier Fusion Using Kalman Filter

The Kalman Filter [4] is a popular algorithm in the field of navigation and object tracking [19, 20]. A prominent feature is that all measurements are assumed to be uncertain. Furthermore, the model also handles missing measurements by an additional increase of the uncertainty.

Classifier fusion is realized by taking classifier outcomes (ranging in the interval of $[0, 1]$) as measurements. Classifier outcomes may be unavailable in case a sensor fails or does not perceive any signal. The estimation is decomposed into two steps, namely the prediction and the update step. While the prediction step calculates an estimated scalar $\widehat{x}_t$ at the time of the next measurement, the update step (also known as correction step) combines this estimate with the latest measurements z_{mt} where $t \in \{1, \ldots, T\}$ denotes the time steps and $m \in \{1, \ldots, M\}$ the index of the classifier. The prediction step's mean estimate $\widehat{x}_t$ is obtained based on the previous estimate x_{t-1} and a so-called control u. Both quantities are weighted linearly by f and b:

$$\widehat{x}_t = f \cdot x_{t-1} + b \cdot u \qquad \text{with } f, b \geq 0. \tag{1}$$

The control u can be used to bias the prediction to a certain value (e.g. the least informative classifier combination 0.5 in case of a two-class problem and with predictions ranging between $[0, 1]$). However, we decided to omit the last term. As a result, our model presumes that the mean of the current estimate is identical to the previous one. Due to the restriction of the state space in this application to the values $[0, 1]$, the usage of popular process models like dead reckoning, which propagates the state using the last state and its first derivation with respect to the time, may not be used. Consequently, a non-linear version of the dead reckoning model would be necessary to keep the state restrictions. The covariance of the prediction is given by $\widehat{p}_t$ and obtained by combining the *a posteriori* covariance with an additional covariance q_m which models the process noise:

$$\widehat{p}_t = f \cdot p_t \cdot f + q_m. \tag{2}$$

The successive update step is performed for every classifier m and requires three intermediate results, namely the residuum y, the innovation variance s and the Kalman gain k:

$$y = z_{mt} - h \cdot \widehat{x}_t \tag{3}$$
$$s = h \cdot \widehat{p}_t \cdot h + r_m \tag{4}$$
$$k = \widehat{p}_t \cdot h \cdot s^{-1} \tag{5}$$

where h is the observation model which maps the predicted quantity to the new estimate and $r_{t,m}$ is the observation noise. These outcomes are then used to update the new estimate and its variance:

$$x_t = \widehat{x}_t + k \cdot y \tag{6}$$
$$p_t = \widehat{p}_t - k \cdot s \cdot k \tag{7}$$

Missing classifier outcomes (decisions) are replaced by a measurement prior $\tilde{z}_{mt}$ equals to 0.5 and a corresponding observation noise $\tilde{r}_m$ which is set relatively high compared to the actual observation noise.

We extended the over-all classifier system by base classifiers which are additionally allowed to make use of a reject option [21]. That means, classifiers are also allowed to reject a sample and to return no class assignment. Because of the temporal structure the missing classifier outcomes can be reconstructed based on adjacent decisions. The proposed architecture is shown in Figure 2: Each of the M classifiers provide a classifier outcome and additionally a corresponding confidence value for each time step t. The confidence value is used to reject decisions with low confidence. The remaining decisions are processed by the Kalman filter which derives a combined estimate based on the current and past decisions.

3   Empirical Evaluation

The proposed combination method can be applied to any kind of multi-modal classifier fusion which is sequentially structured and where a Markov assumption

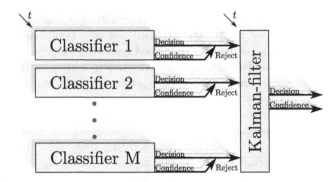

Fig. 1. Multiple classifier system using classifiers with reject option and the Kalman filter to combine the decisions. The M classifiers decisions and confidences are collected for ever time step t and temporally combined using the Kalman filter, resulting in an integrated decision and confidence for every time step.

is plausible. This section focuses on an exemplary data set for classification which has already proven to be very difficult, namely the audio/visual emotion challenge (AVEC) 2011 for human affective state recognition [6, 5].

3.1   The Data Set

The AVEC data set was introduced in 2011 in form of a challenge for the recognition of emotional user states [5, 6]. The recordings show a subject communicating with a virtual agent which endeavors to induce an affective state in the test person. Four categories have been labeled by two up to eight annotators using a continuous scale: Arousal, Expectancy, Power and Valance. Binary labels are obtained by averaging the annotations per category and recording. The final label are then found by thresholding this averaged value against the average over all recordings per category. As a result, the occurrences of each class label is balanced. Since the challenge is continued in 2012 the ground-truth labels of the test data are still not available. Therefore, the available data of the training and development set has been re-arranged to obtain a new development/test and a nested training/validation data set. The number of cross-validation folds is limited to 4×4 (number of development and trainings set, respectively) to realize subject independent tests (a person to be tested does not occur in the corresponding trainings set).

3.2   Base Classifiers

Based on the recordings, individual classifiers for audio and video are created using the partitioned cross-validation. The audio classification is performed on word-level using three bags of features:

- *Fundamental frequency*, the *energy* and *linear predictive coding* (LPC)
- *Mel frequency cepstral coefficient* (MFCC)
- *Relative spectral transform - perceptual linear prediction* (RASTA-PLP) [22].

To obtain a fixed-length feature vector from an arbitrary long sequence of features, a transformation according to Bicego et al. [23] based on HMM is implemented. The classification is conducted using five bags of random forests [24]. The final output is obtained by averaging and the standard deviation is used to calculate the confidence measure.

The classification of the video channel is based on features extracted by the computer expression recognition toolbox (CERT) [25], which is designed to recognize facial properties (such as action units or basic emotions). The output of the modules "Basic Emotions 4.4.3", "FACS 4.4", "Unilaterals" and "Smile Detector" are concatenated to form a 36-dimensional feature vector per frame. Classification and the confidence value is realized analogously to the audio classification using five naive Bayes classifiers with bagging [26]. It is worth noting that the detection of the subject's face failed in about 8% of the frames. The missing detections directly result in missing classifier outcomes. The confidence measures are obtained in the same manner by using the standard deviation of the ensemble.

Parameters of all base classifiers are optimized using the training and validation set. The results of the base classifiers are then derived from the test set. In order to optimize the parameters of the fusion algorithm, the training and validation sets partitions are used, i.e the features are replaced by the results of the base classifiers decisions.

3.3   Experimental Results

The performance of the the audio and video classifier without fusion are given in Table 1. Frames without a classifier decision, e.e. due to the subject is not speaking or the face is not recognized, are ignored in this evaluation. The Table shows the accuracies and the F_1-measures (i.e. $F_1 = 2\frac{P \cdot R}{P+R}$ where P is the precision and R the recall) for the four classes and their negations. Compared to other results of non-acted emotion recognition, e.g. studies based on the Cohn-Kanade data set [27], the achieved accuracies and the F_1-measures are remarkable low. However, compared to the challenge best performing submissions the results are already outstanding[1] (The best result of the class arousal achieved an accuracy rate of around 61%). Table 2 and Table 3 show the performance of the four categories using only uni-modal temporal fusion. Results based on the audio channel are shown in the upper part of Table 2, while the utilized parameters are listed in the lower part. The term *reject* refers to the proportion of discarded decisions, whereas q_{audio} and r correspond to the process noise and the observation noise, respectively. The measurement prior $\tilde{z}_{mt}$ is set to 0.5 and the corresponding observation noise $\tilde{r}_m$ is set a magnitude higher than the assigned

---

[1] Compare http://sspnet.eu/avec2011/ (02/21/2013).

Table 1. Audio and video classifier performances before fusion. Performance is evaluated frame-wise and only in case classifier outputs are given. Accuracies in percent with standard deviation. The F_1-measure for the class and the corresponding negation $\overline{F}_1$ is additionally provided.

AUDIO	AROUSAL	EXPECTANCY	POWER	VALANCE
↑ACC.	61.8±3.6	59.0±6.3	57.5±9.4	57.5±7.9
↑F_1	65.8±3.8	16.4±7.1	69.6±9.3	70.1±6.8
↑$\overline{F}_1$	56.7±3.4	72.6±5.2	24.7±6.6	24.9±8.4
VIDEO	AROUSAL	EXPECTANCY	POWER	VALANCE
↑ACC.	57.0±4.2	54.8±4.0	55.7±2.9	59.9±7.4
↑F_1	60.8±5.1	49.6±9.4	57.5±11.3	67.1±11.5
↑$\overline{F}_1$	51.3±9.3	56.6±10.7	48.8±12.3	43.5±7.1

Table 2. Frame-wise audio performance using Kalman filter for classifier fusion. Accuracies in percent with standard deviation and F_1-measure for the class and the corresponding negation $\overline{F}_1$. Lower part of table lists the parameter assignments.

AUDIO	AROUSAL	EXPECTANCY	POWER	VALANCE
↑ACC.	74.3±6.6	57.5±6.2	56.5±11.2	59.7±11.0
↑F_1	77.4±6.5	12.5±6.2	69.1±10.2	72.9±8.5
↑$\overline{F}_1$	69.4±8.2	71.8±5.3	22.7±8.9	16.2±14.1
REJECT	10%	90%	90%	50%
q_{AUDIO}	10^{-7}	10^{-7}	10^{-5}	10^{-4}
r	0.1	0.1	0.1	0.75

r. As a result, the estimate slowly converge to the prior in case of continuously missing measurements. Parameters have been optimized for each label independently using the aforementioned cross-validation sets. Compared to the results without temporal fusion, only Arousal and Valance are improved at first glance. However, the most important difference of the evaluation compared to Table 1 is that now a decision for every frame is available. Arousal is recognized most reliably using the audio channels which is not only confirmed by the high accuracy but, furthermore, due to the fact that the optimal portion of rejected labels is very low. In contrast, the category Power seems to be less present in the audio channel. With the exception of the category Arousal, the unbalancedness of the F_1-measures has clearly increased. The process noise determined on the training and validation set, are in general of a very low quantity, while the observation noise ranges from 0.1 to 0.75.

Table 3 shows the results for the video channel. Here, all categories are able to improve their recognition performance, compared to Table 1. However, the category Arousal is recognized better using the combined estimate based on the audio channel shown in Table 2. In general, the F_1-measures have increased as well as to the accuracies and are more balanced. Both, the process noise and the observation are chosen very similar to the ones in the audio channel, which might be related to the same range of decisions values (i.e. in the interval $[0, 1]$).

Table 3. Frame-wise video performance using Kalman filter for fusion. Accuracies in percent with standard deviation and F_1-measure for the class and the corresponding negation $\overline{F}_1$. Lower part of table lists the parameter assignments.

VIDEO	AROUSAL	EXPECTANCY	POWER	VALANCE
↑ACC.	64.5 ± 2.9	59.1 ± 4.7	58.9 ± 5.0	65.4 ± 9.8
↑F_1	67.9 ± 4.2	56.7 ± 11.1	60.6 ± 13.8	73.3 ± 11.4
↑$\overline{F}_1$	57.3 ± 12.7	56.4 ± 17.5	48.2 ± 17.1	44.2 ± 8.5
REJECT	50%	90%	0%	90%
q_{VIDEO}	10^{-7}	7.5^{-6}	5^{-5}	10^{-4}
r	0.1	0.1	0.1	0.75

Table 4. Frame-wise audio and video performance using Kalman filter for fusion. Accuracies in percent with standard deviation and F_1-measure for the class and the corresponding negation $\overline{F}_1$. Lower part of table lists the parameter assignments.

AUDIO-VISUAL	AROUSAL	EXPECTANCY	POWER	VALANCE
↑ACC.	68.5 ± 5.7	62.5 ± 4.9	61.8 ± 6.6	64.2 ± 9.3
↑F_1	72.6 ± 4.2	42.2 ± 15.7	69.1 ± 7.6	72.6 ± 10.9
↑$\overline{F}_1$	59.7 ± 15.1	71.1 ± 5.8	43.5 ± 18.0	43.7 ± 3.0
A. REJ.	0%	0%	0%	90%
V. REJ.	50%	50%	0%	10%
q_{AUDIO}	10^{-6}	5^{-6}	10^{-7}	5^{-5}
q_{VIDEO}	10^{-5}	10^{-5}	10^{-5}	10^{-4}
r	0.75	0.1	0.1	0.75

The performance of the combined audio and video channel is shown in Table 4. The multi-modal combination improves the accuracies of Expectancy and Power, whereas the category Arousal performed better using only the audio channel, and the category Valance performed better using only the video channel. The low performance of Arousal can be related to the unbalanced occurrences of outputs in video and audio channel. Decisions in the audio channel are not always available, whereas the frame-wise outputs of the video channel are almost constantly available. Therefore, the video classifier outputs are the main influence if no audio signal is present and the audio channel is practically overruled. In case of the category Valance the decrease in performance can be traced back to the weak $\overline{F}_1$-measure of the audio channel. However, although the combination of the audio and video not always achieves the best performance, it outperforms uni-modal approach. Furthermore, the uni-modal classification of emotions is far from being solved, since features from the data are in general weak and even the labels are very uncertain [18]. Multi-modal approaches are sound as using the audio channel alone leads to a high degree of uncertainty in case the subject does not make any utterances, whereas the face of the participant will not always be in the field of view. The classifier system gets more robust and the likelihood of a sensor failure is reduced, the more modalities are available.

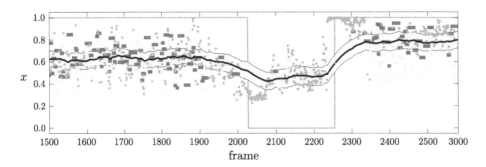

Fig. 2. Multi-modal fusion of Arousal using the Kalman filter. The orange dots and the blue squared-shaped markers correspond to the video and audio decisions. Markers in pale color have been rejected and do not contribute to the fusion. The thick black curve corresponds to the fusion result, while the area around the curve corresponds to the variance determined by the Kalman filter (scaled by 10 for illustration purposes). The light gray curve displays the ground-truth. The parameters used are $q_{\mathrm{audio}} = 10^{-6}$, $q_{\mathrm{video}} = 10^{-5}$ and $r = 0.75$.

Although it appears to be tantalizing to interpret the parameter chosen by the optimization process, a direct interpretation is difficult. For instance, the audio classifier is only able to provide outputs in case a signal is present, such that the classifier outputs of the video channel are much more frequent then the classifier outputs of the audio channel. As a result the rejection rate and the process noise for audio and video cannot be compared with each other. Furthermore, the ratio between these modalities is additionally modified by assessing the quality in terms of rejecting unreliable outputs.

4   Conclusions

The presented work proposes the application of Kalman filters for classifier fusion. The Kalman filter is a widely used and well-understood model for tracking moving objects and navigation [4, 20]. The model combines multiple measurements and can handle the absence of measurements by increasing the uncertainty of the predicted state. Within this work, we replaced the uncertain measurements that are usually processed by the Kalman filter by multi-modal classifier outputs. As a result, missing classifier outputs (e.g. caused by a missing signal, for instance of the audio channel, or by a rejected sample due to low confidence) can be handled in a unified framework.

In order to evaluate the performance of the combiner, the audio/visual emotional challenge (AVEC) 2011 data set [5, 6] has been utilized. The fusion of classifier outputs is studied uni- and multi-modally, and has clearly improved the recognition of all four categories. In addition, the filtering resulted in a reconstruction of missing classification outputs such that a class assignment is possible for all frames. Although the Kalman Filter applied can be regarded as the simplest time-series model possible (e.g. no control matrix, and assuming an identity matrix for the dynamics), the results for this data set are outstanding.

Future work will integrate the confidence, that was used here only for the rejection of samples, directly into the Kalman Filter and perform additional experiments on other data sets. Furthermore, a detailed comparison with multiple state-of-the-art approaches (e.g. MFN [10], block-averaging) is currently in preparation.

Acknowledgments. This paper is based on work done within the Transregional Collaborative Research Centre SFB/TRR 62 Companion-Technology for Cognitive Technical Systems funded by the German Research Foundation (DFG).

References

1. Beal, M.J., Attias, H., Jojic, N.: Audio-video sensor fusion with probabilistic graphical models. In: Heyden, A., Sparr, G., Nielsen, M., Johansen, P. (eds.) ECCV 2002, Part I. LNCS, vol. 2350, pp. 736–750. Springer, Heidelberg (2002)
2. Kuncheva, L.I.: Combining Pattern Classifiers: Methods and Algorithms. Wiley (2004)
3. Ruta, D., Gabrys, B.: An overview of classifier fusion methods. Computing and Information Systems 7(1), 1–10 (2000)
4. Kalman, R.E.: A new approach to linear filtering and prediction problems. Transactions of the ASME — Journal of Basic Engineering 82(Series D), 35–45 (1960)
5. Schuller, B., Valstar, M., Eyben, F., McKeown, G., Cowie, R., Pantic, M.: AVEC 2011–the first international audio/visual emotion challenge. In: D'Mello, S., Graesser, A., Schuller, B., Martin, J.-C. (eds.) ACII 2011, Part II. LNCS, vol. 6975, pp. 415–424. Springer, Heidelberg (2011)
6. McKeown, G., Valstar, M., Cowie, R., Pantic, M.: The SEMAINE corpus of emotionally coloured character interactions. In: Proceedings of the International Conference on Multimedia and Expo (ICME), pp. 1079–1084. IEEE (2010)
7. Glodek, M., Scherer, S., Schwenker, F.: Conditioned hidden Markov model fusion for multimodal classification. In: Proceedings of the Annual Conference of the International Speech Communication Association (Interspeech), ISCA, pp. 2269–2272. ISCA (2011)
8. Schwenker, F., Dietrich, C.R., Thiel, C., Palm, G.: Learning of decision fusion mappings for pattern recognition. Journal on Artificial Intelligence and Machine Learning (AIML) 6, 17–22 (2006)
9. Jeon, B., Landgrebe, D.A.: Decision fusion approach for multitemporal classification. IEEE Transaction on Geoscience and Remote Sensing 37(3), 1227–1233 (1999)
10. Glodek, M., Schels, M., Palm, G., Schwenker, F.: Multi-modal fusion based on classification using rejection option and Markov fusion network. In: Proceedings of the International Conference on Pattern Recognition (ICPR), pp. 1084–1087. IEEE (2012)

11. Glodek, M., Tschechne, S., Layher, G., Schels, M., Brosch, T., Scherer, S., Kächele, M., Schmidt, M., Neumann, H., Palm, G., Schwenker, F.: Multiple classifier systems for the classification of audio-visual emotional states. In: D'Mello, S., Graesser, A., Schuller, B., Martin, J.-C. (eds.) ACII 2011, Part II. LNCS, vol. 6975, pp. 359–368. Springer, Heidelberg (2011)
12. Picard, R.: Affective computing: Challenges. International Journal of Human-Computer Studies 59(1), 55–64 (2003)
13. Tao, J., Tan, T.: Affective computing: A review. In: Tao, J., Tan, T., Picard, R.W. (eds.) ACII 2005. LNCS, vol. 3784, pp. 981–995. Springer, Heidelberg (2005)
14. Scherer, S., Glodek, M., Layher, G., Schels, M., Schmidt, M., Brosch, T., Tschechne, S., Schwenker, F., Neumann, H., Palm, G.: A generic framework for the inference of user states in human computer interaction: How patterns of low level communicational cues support complex affective states. Journal on Multimodal User Interfaces 6(3-4), 117–141 (2012)
15. Douglas-Cowie, E., Campbell, N., Cowie, R., Roach, P.: Emotional speech: Towards a new generation of databases. Speech Communication 40(1), 33–60 (2003)
16. Frank, C., Adelhardt, J., Batliner, A., Nöth, E., Shi, R.P., Zeißler, V., Niemann, H.: The facial expression module. SmartKom: Foundations of Multimodal Dialogue Systems 1, 167–180 (2006)
17. Kim, J., André, E.: Emotion recognition based on physiological changes in music listening. IEEE Transactions on Pattern Analysis and Machine Intelligence, 2067–2083 (2008)
18. Palm, G., Glodek, M.: Towards emotion recognition in human computer interaction. In: Apolloni, B., Bassis, S., Esposito, A., Morabito, F.C. (eds.) Neural Nets and Surroundings. SIST, vol. 19, pp. 323–336. Springer, Heidelberg (2013)
19. Blackman, S., Popoli, R.: Design and Analysis of Modern Tracking Systems. Artech House Publishers (1999)
20. Bar-Shalom, Y., Li, X.R.: Estimation and Tracking: Principles, Techniques, and Software. Artech House Incorporated (1993)
21. Bishop, C.M.: Pattern Recognition and Machine Learning. Springer (2006)
22. Huang, X., Acero, A., Hon, H., et al.: Spoken language processing: A Guide to Theory, Algorithm and System Development. Prentice Hall (2001)
23. Bicego, M., Murino, V., Figueiredo, M.A.T.: Similarity-based clustering of sequences using hidden Markov models. In: Perner, P., Rosenfeld, A. (eds.) MLDM 2003. LNCS (LNAI), vol. 2734, pp. 86–95. Springer, Heidelberg (2003)
24. Breiman, L.: Random forests. Machine Learning 45(1), 5–32 (2001)
25. Littlewort, G., Whitehill, J., Wu, T., Fasel, I., Frank, M., Movellan, J., Bartlett, M.: The computer expression recognition toolbox (CERT). In: Proceedings of the International Conference on Automatic Face & Gesture Recognition and Workshops, pp. 298–305. IEEE (2011)
26. Breiman, L.: Bagging predictors. Machine Learning 24(2), 123–140 (1996)
27. Schwenker, F., Scherer, S., Schmidt, M., Schels, M., Glodek, M.: Multiple classifier systems for the recogonition of human emotions. In: El Gayar, N., Kittler, J., Roli, F. (eds.) MCS 2010. LNCS, vol. 5997, pp. 315–324. Springer, Heidelberg (2010)

Adaptive Ensemble Selection for Face Re-identification under Class Imbalance[*]

Paulo Radtke[1], Eric Granger[1], Robert Sabourin[1] and Dmitry Gorodnichy[2]

[1] Laboratoire d'imagerie, de vision et d'intelligence artificielle
École de technologie supérieure, Université du Québec, Montreal, Canada
radtke@livia.etsmtl.ca, {eric.granger,robert.sabourin}@etsmtl.ca
[2] Science and Engineering Directorate, Canada Border Services Agency
Ottawa, Canada
dmitry.gorodnichy@cbsa-asfc.gc.ca

Abstract. Systems for face re-identification over a network of video surveillance cameras are designed with a limited amount of reference data, and may operate under complex environments. Furthermore, target individuals provide a small proportion of the facial captures for design and during operations, and these proportions may change over time according to operational conditions. Given a diversified pool of base classifiers and a desired false positive rate (fpr), the Skew-Sensitive Boolean Combination (SSBC) technique allows to adapt the selection of ensembles based on changes to levels of class imbalance, as estimated from the input video stream. Initially, a set of BCs for the base classifiers is produced in the ROC space, where each BC curve corresponds to reference data with a different level of imbalance. Then, during operations, class imbalance is periodically estimated using the Hellinger distance between the data distribution of inputs and that of imbalance levels, and used to approximate the most accurate BC of classifiers among operational points of these curves viewed in the precision-recall space. Simulation results on real-world video surveillance data indicate that, compared to traditional approaches, FR systems based on SSBC allow to select BCs that provide a higher level of precision for target individuals, and a significantly smaller difference between desired and actual fpr. Performance of this adaptive approach is also comparable to full recalculation of BCs (for a specific level of imbalance), but for a considerably lower complexity. Using face tracking, a high level of discrimination between target and non-target individuals may be achieved by accumulating SSBC predictions for faces captured corresponding to a same track in video footage.

1 Introduction

Video surveillance networks found at many airport security checkpoints are comprised of a growing number of IP-based surveillance cameras. Face re-identification

---

[*] This work was supported by the Natural Sciences and Engineering Research Council of Canada, and the Centre for Security Science (Defense R&D Canada).

Z.-H. Zhou, F. Roli, and J. Kittler (Eds.): MCS 2013, LNCS 7872, pp. 95–108, 2013.
© Springer-Verlag Berlin Heidelberg 2013

consists in automatically matching facial regions captured in multiple live or pre-recorded video streams against facial models of individuals enrolled to a system [15]. Face re-identification in semi- or unconstrained video surveillance environments raises several challenges. First, FR systems must operate under complex environments with changing illumination, pose, expression, blur, occlusion, etc. Small proportion of the faces collected for design or during operation correspond to individuals of interest, although non-target faces are abundant. In addition, the covert and unobtrusive capture of video sequences provides only a limited amount of high quality reference samples to design facial models.

To avoid biasing performance towards the majority (non-target) class, classifiers applied to face matching are typically designed with balanced data, using sampling techniques or cost sensitivity analysis. Moreover, an estimate of class priors is often used to scale classifier outputs, although actual class proportions are often unknown a priori and may change over time. Specialized architectures for FR in video surveillance [14] do not exploit information on class imbalance to enhance performance. The impact of imbalance on classification performance can be observed in the *Precision-Recall Operating Characteristic* (PROC) [11] space – the precision measure allow to observe the proportion of correct target predictions over all target samples. This typically declines when the proportion of negative samples grow over the positive ones. Given its relationship to the *Receiver Operator Characteristics* (ROC) space [3], the PROC space can be exploited to adapt classification systems according to changing class imbalance.

Since the proportion of design samples per class rarely correspond to the actual distribution of operational data, the performance of systems for FR in video surveillance will differ from that achieved during the design stage. What's more, the underlying distributions change over time in video surveillance applications. For instance, a security checkpoint (inspection lane or portal) may witness peaks in the flow of target and non-target individuals. It is desirable to estimate class imbalance periodically over time, and adaptively select an operation point with design data that follows the class imbalance of the operational data.

In this paper, a BC technique is proposed to adapt the selection of classifiers ensembles given the current class imbalance, as estimated from operational data. This technique, called the Skew-Sensitive BC (SSBC) technique, exploits the PROC space. During design phases, a pool of diversified classifiers is generated, and imbalanced validation data is used to produce several BCs by successively growing the number of negative samples from the majority class. Negative samples are assumed to be available in large quantities, while the limited number of positive samples is assumed to be fixed. Each BC is optimized for one specific class imbalance level. During operations, the system relies on the Hellinger distance [8] to periodically estimate the closest class imbalance from operational data streams, from a set of known imbalanced data sets. This estimation is used to approximate the most accurate BC of classifiers among operational points of these curves viewed in the PROC space.

Proof-of-concept experiments are performed with real-world video FR data from the Carnegie Mellon University Face in Action database [7], where class proportions captured in operational video streams change over time. The performance of SSBC is assessed within a modular FR system comprised of an EoC of 2-class classifiers per person [2,14], and compared to that of BC optimized with data obtained with random under-sampling [6].

2   Face Recognition in Video Surveillance

The problem addressed in this paper is the design of accurate and robust systems for video-to-video FR from footage recorded across a network of surveillance cameras. These systems are considered for person re-identification applications, where individuals of interest must be detected within semi- or unconstrained scenes, as found at security checkpoints. Each camera captures streams of 2D images or frames, and provides a particular view of individuals populating the scene. The system first performs segmentation to isolate regions of interest (ROIs) corresponding to the faces in a frame, from which invariant and discriminant features are extracted and selected for classification and tracking functions. For classification, some features are assembled into an input pattern, $\mathbf{a}$, that corresponds to a spatial vector or an ordered sequence of measurements.

During enrolment, one or more reference patterns $\mathbf{a}$ are captured for an individual, and employed to design a user-specific facial model. Recognition is typically implemented using a template matcher or a neural or statistical classifier, mapping the input pattern space to one of N predefined classes, each corresponding to an individual enrolled to the system. During operations, input patterns $\mathbf{a}$ are matched against the models of individuals, and the system outputs a list of all possible identities.

Systems for FR in video encounter several challenges. Biometric models are poor representatives of faces to be recognized during operations. The performance of FR systems may decline because neural and statistical classifiers depend on the availability of representative reference data of users and the operational environment. In addition, underlying class distributions may change due to ageing and variation in operational environments. These factors contribute to a growing divergence between the facial model of an individual and its underlying class distribution. In addition, with FR in video surveillance, faces captured in video frames are typically lower quality and generally smaller than still images. Furthermore, faces acquired from semi- or unconstrained scenes may vary considerably due to limited control over operational conditions.

Several specialized architectures have been proposed for FR in video surveillance. The open-set Transduction Confidence Machine-kNN (TCM-kNN) algorithm [12] modified the traditional kNN, using transduction to measure strangeness between samples, and to reject samples of unknown individuals. Ekenel et al. [4] combined kNN and Gaussian Mixture Modeling with three different metrics to estimate individual frame contributions to the overall decision. Kamgar-Parsi et al. [10] proposed a morphing approach to generate new synthetic reference data and improve the separability of target and non-target classes.

These systems have also been modeled in terms of user-specific detectors, each one implemented using one or more binary (1- or 2-class) classifiers [14]. This modular approach was employed with user-specific SVMs [5] and ensembles of 2-class ARTMAP neural classifiers[14]. Binary ensembles are justified by the limited amount of positive samples for design, and by the complexity of real-world video scenes [14]. However, these architectures do not consider or exploit class imbalance information to enhance performance. This information is relevant in the context of video surveillance, owing to the potentially small number of positive samples w.r.t. the negative ones, and to changing operational conditions.

3   Binary Classification under Class Imbalance

Binary (1- or 2-class) classifiers output a crisp decision or a score that is compared to a decision threshold to provide a final crisp decision. A common assumption in pattern recognition (PR) literature is that class priors are known and that data distributions are balanced, i.e., instances of all classes are assumed to be equally present in both training and operational data. Real world problems rarely follow this ideal case – class priors are unknown and may change over time, and training samples are imbalanced and are not necessarily representative of operational data. Classifiers applied to FR in video surveillance should be designed to operate under class imbalance – limited target (positive) class samples w.r.t. non-targets (negative) class samples.

Four main approaches have been proposed in literature to train classifiers from skewed[1] reference data sets [6] – algorithm level, cost sensitive, data level and ensemble techniques. *Algorithm level approaches* modify the classifier behavior to bias toward the minority (positive) class. With *cost sensitive approaches*, the classifier training procedure minimizes the total cost of misclassified instances, instead of minimizing the number of misclassified instances. Errors have a much higher cost for the minority class, and the sum of miss-classifications costs drives the learning process. *Data level approaches* are categorized either as under-sampling or as over-sampling techniques. Data under-sampling techniques will reduce the sample number of the majority (negative) class to match that of the minority class. Finally, *ensemble learning approaches* [6] are usually performed in conjunction with one of the three other approaches to optimize the combination of classifiers. While the above techniques have been used to design monolithic classifiers and ensembles for imbalance classes, it is assumed that class imbalance observed in the design data is representative of the imbalance encountered during operations. This paper focuses on adapting the selection classification system according to the class imbalance observed during operations.

Performance of binary classifiers is commonly evaluated using the ROC analysis, which is based on two intra-class measures, the true positive rate $tpr = TP/(TP + FN)$ (proportion of correct positive class predictions) and the false positive rate $fpr = FP/(FP + TN)$ (proportion of incorrect negative class predictions). ROC graphs display the entire range of tpr and fpr values to obtain

---

[1] Data skew is the ratio of positive samples π_p to negative ones π_n, $\lambda = \pi_p/\pi_n$.

different operational points. Given a data set, each (tpr, fpr) pair in a ROC graph represents a different decision threshold for one soft classifier (an operational point or vertice), and the empirical ROC curve is obtained by connecting the observed pairs in the graph.

Given an imbalance in class distributions, the PROC space [3] (also known as the P-R space) focuses on an inter-class measure – the classifier $precision = TP/(TP+FP)$ (proportion of correct positive predictions against the total positive predictions) – which is related to classification accuracy, as well as $recall$ (the same as tpr). PROC graphs represent the impact on performance of imbalance through precision measure. A classifier has different PROC curves when evaluated on data with different class imbalance, while the ROC curves would be equivalent since both tpr and fpr are insensitive to imbalance. Davis and Goadrich discussed [3] the equivalence between dominating operational points in the ROC and PROC spaces, from which they derived a methodology to find the PROC achievable curve (analogous to the ROC convex hull).

A decision threshold γ applied to classifier scores is often selected with independent validation data (**val**) once the classifier has been designed. The optimal decision boundary for a classifier is selected to minimize the probability of error according to the Bayes theory, which is equivalent to the equal error rate (EER) when the positive and negative classes are balanced. If class imbalance changes, so does the optimal threshold. Assume that class imbalance is known, one can select a decision threshold that provides the EER for every different level of class imbalance. In video surveillance applications, an acceptable fpr is set by the human operator, projecting it to an operational point. Thus, the decision threshold is a variable defined by the desired fpr and class imbalance.

4   Boolean Combination of Classifiers

Boolean combination (BC) are versatile techniques for threshold-optimized fusion of crisp and soft 1- or 2-class classifiers at the decision level [9] (typically in the ROC space). A soft classifier c_i produces a binary decision when its normalized output score is compared to a threshold $0 \leq \gamma \leq 1$. This decision c_{i,γ_i} affects a trade off between positive and negative classes (e.g., an operational point in the ROC space). Given a set of decision thresholds Γ, the BC of two soft classifiers c_i and c_j is the fusion of all c_{i,γ_i} and c_{j,γ_j} using Boolean operations. Each resulting EoC (e.g., ROC vertices) consists of decision thresholds applied to the classifier scores and a Boolean function. Selecting the superior operational points in the decision space (for instance, the ROC convex hull or the PROC achievable curve) defines the best performance trade off. The ROCCH is the ROC curve composed of the vertices that maximize the area under the ROC curve (AUC).

A BC technique produces a set EoCs, each one corresponding to a vertice of the ROCCH. After performing BC, the next step is to define an operation point for the specific application. A general approach to select an operation point is to choose the EoC in the BC that provides the best trade off between tpr and fpr values, but for a specific application, the operation point is typically selected for a target fpr value using validation data. However, it is unlikely that an EoC produced by BC will correspond to the the target fpr value. BC of classifiers in the decision space should therefore be performed using imbalanced data corresponding to operational data, allowing to generate better operations points and performing selection and fusion of the most suitable ensembles.

Scott *et al.* proposed a method to interpolate between two consecutive vertices (EoCs) on the ROCCH, E_i and E_j, to realize an operation point E_k between the two original ones [16]. To classify input samples, the interpolation method alternates between the decisions provided by E_i and E_j for each sample. The probability of selecting one of the two vertices is determined by the distance of E_k to the vertexes E_i and E_j.

5   Adaptive Skew-Sensitive BC

Figure 1 presents the block diagram of an adaptive classification system based on the new skew-sensitive BC (SSBC) technique. It allows for adaptive selection and fusion of the best ensembles of binary classifiers, based on its estimation of class imbalance. Assume a stream of facial patterns (opd) input to some FR system, where the level of class imbalance on the input stream is estimated as λ^*, the closest level in a set Λ of known class imbalance levels. Since this estimate may change over time, and BC is a computationally intensive task, the SSBC technique is proposed to cost-effectively adapt ensembles of classifiers. In this situation, BCs are approximated from adjacent levels of class imbalance λ^i and λ^j, $\lambda^i < \lambda^* < \lambda^j$ and using validation data following the class imbalance level λ^*, the level of class imbalance estimated from the operational data opd. The approach can approximate the BC up to a maximum λ^{max} class imbalance level.

The approach uses a set of known levels of class imbalance, $\Lambda = \{1/1, \ldots, \lambda^{max}\}$, to which the system will adapt, and a subset $\Lambda_{BC} \subset \Lambda$ that is selected to optimize an initial set of BCs E used to adapt the system to class imbalance changes. The set Λ_{BC} contains evenly distributed intermediate class imbalance levels from Λ, including the initial and maximum levels of class imbalance ($\lambda^{init} = 1/1$ and λ^{max}). The SSBC approach uses OPT and VAL, data sets following the levels of class imbalance in Λ, each following a class imbalance level in Λ. The target minority class is held fixed, while those from the non-target class are grown through random sub-sampling. This allows the SSBC approach to generate OPT and VAL with data sets following any class imbalance level between $\lambda^{init} = 1/1$ and λ^{maxt}.

Once a pool of binary classifiers $C = \{c_1, \ldots, c_n\}$ is generated using balanced data, ensemble selection and fusion is performed using the Iterative Boolean Combination (IBC) technique [9]. BC of C is performed during the design phase using the levels of imbalance in Λ_{BC} using Algorithm 1. Each BC in E is optimized for one class imbalance level in Λ_{BC} with the corresponding data in OPT. During BC, the number of decision thresholds t is used to create the operations points (which provide binary decisions) for classifier fusion. After BC for is performed for all class imbalance levels in Λ_{BC}, the approach assumes that data is

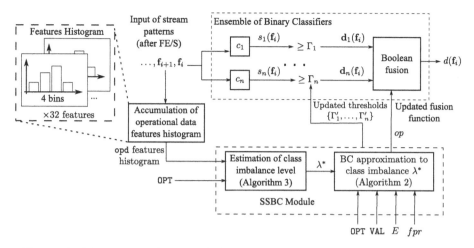

Fig. 1. Architecture to adapt a BC of classifiers to imbalanced class distributions

---

Algorithm 1. Initial BC design for the SSBC technique.

Data: Pool of classifiers C, number of decisions thresholds t, data sets OPT and
VAL and the target fpr
Result: Set of BCs E and the operation point op for $\lambda^{init} = 1/1$.

1 $E = \emptyset$;
2 **forall the** $opt \in$ OPT *matching a level of class imbalance in* Λ_{BC} **do**
3     $\lfloor$ $E = E \cup \{IBC(C, t, opt)\}$;
4 Select $E_{\lambda^{init}} \in E$ for $\lambda^{init} = 1/1$;
5 Select $op \in E_{\lambda^{init}}$ for the target fpr with $val_{init} \in$ VAL;

---

balanced and operates at $\lambda^{init} = 1/1$. An operation point op for a target fpr
from $E_{\lambda^{init}}$ is selected using the data set $val_{\lambda^{init}}$.

During system operation, the feature histogram of operational data is accumulated over time to periodically estimate the closest level of class imbalance $\lambda^* \in \Lambda$ using Algorithm 3. Once the closest class imbalance level λ^* is estimated from the levels available in Λ, the BC is approximated using Algorithm 2. Finally, an updated operation point op is selected with $val_{\lambda^{init}} \in$ VAL and used to update the decision thresholds Γ and the Boolean fusion function.

A. Approximating BCs to New Class Imbalances. Given the known set of class imbalance levels Λ, the set of BCs E created with Algorithm 1, and assuming λ^*, the level of class imbalance estimated from operational data of class imbalance levels in Λ. The procedure to approximate BCs to the class imbalance level λ^* is indicated in Algorithm 2 and graphically represented in Fig. 2, where the appoximated BC is indicated by the dashed line in Fig. 2.b.

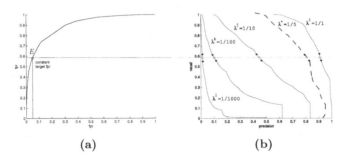

(a)   (b)

Fig. 2. ROC and inverted PROC graphs obtained with the **val** for a target $fpr = 5\%$. Given the BCs optimized for Λ (solid lines), the SSBC approximates the BC for $\lambda^* = 1/5$ (dashed line) from the adjacent BCs λ^i and λ^j.

Algorithm 2. SSBC technique for adapting BC for a new λ^* class imbalance level.

Data: set of BCs E, set of class imbalance levels Λ_{BC}, data sets OPT and VAL, the estimated class imbalance $\lambda^* \in \Lambda$ and the target fpr.

Result: Operation point op for the target fpr.

1 $E^* = \emptyset$;
2 **if** $\lambda^* \in \Lambda_{BC}$ **then**
3 $\quad\lfloor\ E^* = E_{\lambda^*}$;

4 **else**
5 $\quad$ Select $\lambda^i, \lambda^j \in \Lambda_{BC}$, such as that $\lambda^i < \lambda^* < \lambda^j$;
6 $\quad$ Select $\mathbf{opt}^* \in \mathtt{OPT}$, following λ^* ;
7 $\quad\lfloor\ E^* = ROCCH(E_{\lambda^1} \cup E_{\lambda^2}, \mathbf{opt}^*)$;

8 Select $\mathbf{val}^* \in \mathtt{VAL}$, following λ^* ;
9 Select $op \in E^*$ for the target fpr with $\mathbf{val}^*$;

When $\lambda^* \in \Lambda_{BC}$, the BC E^* is selected directly from E. Otherwise, the BC is estimated as follows. First, the adjacent class imbalance levels $\lambda^i, \lambda^j \in \Lambda_{BC}$ are determined. Next, the $\mathbf{opt}^*$ data set is selected in OPT, which was generated by random under sampling to follow the same class imbalance level as λ^*. Then EoCs (vertices in the ROCCH) in both E_{λ^i} and E_{λ^j} are combined, and only the points projected in the ROCCH using the $\mathbf{opt}^*$ data set to calculate tpr and fpr are kept in E^*. Finally, an operation point is selected for the target fpr using $\mathbf{val}^*$, validation data that was also obtained with random under sampling to follow the level of class imbalance of λ^*.

Algorithm 2 is computationally more efficient than performing full BC every time a new class imbalance levels λ^* is detected. For 2 classifiers and t decision thresholds, the worst case time complexity for IBC is $O(t^2)$. For the simulations in this paper ($t = 100$), about 200000 EoC evaluations were required with IBC. The approximation strategy in Algorithm 2 requires $O(|E_{\lambda^1}| + |E_{\lambda^2}|)$ in the worst case. In simulations, there was a significant reduction to about 1% of the original computational effort. Memory requirement is also considerably smaller with Algorithm 2, requiring $O(|E_{\lambda^1}| + |E_{\lambda^2}|)$ vertices stored in memory in the worst case, against $O(t^2)$ for IBC.

B. Estimation of Closest Class Imbalance Level λ^*. In literature, some approaches have been proposed to estimate class imbalance. Using classifier outputs to estimate class imbalance is less reliable since it is influenced by imbalance. It is however possible to use the Hellinger distance in the feature space to select, from several labeled data sets of known class imbalances, which has the closest imbalance (smallest distance) to unlabeled operational data [8]. For a given number of features and bins in the histogram, the Hellinger distance is:

$$H(\mathtt{ld}, \mathtt{opd}) = \frac{1}{f} \sum_{j=1}^{features} \sqrt{\sum_{i=1}^{bins} \left(\sqrt{\frac{|\mathtt{ld}_{j,i}|}{|\mathtt{ld}|}} - \sqrt{\frac{|\mathtt{uod}_{j,i}|}{|\mathtt{uod}|}} \right)^2} \tag{1}$$

Assume the set Λ of class imbalance levels and the set of data sets $\mathtt{OPT}$, where each data set in $\mathtt{OPT}$ follows one different class imbalance level in Λ. Algorithm 3 details the process to estimate λ^*, the class imbalance level in Λ which has the closest class proportions to unlabeled operational data $\mathtt{opd}$. Given L^+, the positive class samples in the reference data $\mathtt{OPT}$ (fixed, regardless of class imbalance), the number bins b used to calculate (1) is $b = \lfloor \sqrt{L^+} \rfloor$.

Algorithm 3. Class imbalance level λ^* estimation from an unlabeled operational data $\mathtt{opd}$ and a set of data sets $\mathtt{OPT}$.

Data: Data set $\mathtt{OPT}$, operational data features histogram $\mathtt{opd}$ and b bins
Result: Estimated class imbalance level λ^*
1 $min = \infty$;
2 $\lambda^* = 0$;
3 **forall the** $opt \in OPT$ **do**
4 $\quad$ $hd = H(\mathtt{opt}, \mathtt{opd}, b)$;
5 $\quad$ **if** $hd < min$ **then**
6 $\quad\quad$ $min = hd$;
7 $\quad\quad$ Set λ^* to class imbalance level of $\mathtt{opt}$;
8

---

6 Validation on Face Re-identification Data

A. Experimental Methodology. To validate SSBC, experiments are performed with real-world video surveillance data. They seek to detect the presence of a restrained list of individuals of interest appearing in video streams. An IP surveillance camera continuously feeds video frames to a FR system. Faces captured in the video frame are extract into ROIs. Data is then processed according to two parallel streams – a recognition stream detects the presence of individual or interest based on appearance, while the tracking stream follows the location of different individuals over successive frames. For the recognition stream, assume a modular classification architecture adapted for surveillance applications, where each target individual is modeled as a user-specific detection module. Each module is implemented with a binary classifier that is assigned to discriminate between the target (positive) and non-target (negative) classes, and with responses combined through BC [14].

Video data for this experiment is extracted from the Carnegie Mellon University – Face in Action (FIA) database [7]. This database contains 20 seconds video sequences for 244 individuals, over three different capture sessions. Each individual is captured from six views: frontal, left and right, and with two focal lengths, 2.8mm (normal) and 4.8mm (zoomed). This experiment uses the frontal camera with both focal lengths on all three sessions as the video stream for the single IP camera. The initial enrollment process considers a watch list with 10 individuals of interest selected in the database (labeled as person 2, 58, 72, 92, 147, 151, 176, 188, 190 and 209). Each individual is the positive or target class for one detector module (EoC) as described in [14].

For recognition, multi-block local binary pattern features are extracted from grey-scaled ROIs, and the 32 most discriminant features are selected through principal component analysis. Feature vectors are compared against facial models of target individuals enrolled to the system. Using a track-and-classify strategy, the classification responses corresponding to different individuals (face tracks) are accumulated over time to improve performance and reliability [13].

For design of 2-class classifiers, a Universal Background Model [1] built from unknown individuals, and a cohort model of the other target individuals. Individuals in the data base are split in two for training and test, and for each individual of interest, 100 negative class individuals are selected for training (from universal and cohort models), and other 100 negative class individuals are selected for testing (maximum class imbalance $\lambda^{max} = 1/100$).

For each experiment, a pool of diversified classifiers C is initially generated using a DPSO training strategy to co-jointly optimize all parameters of a probabilistic fuzzy ARTMAP network (PFAM) [2]. At the end of the optimization process, the local best classifiers from 7 DPSO sub-swarms is selected for the initial pool of PFAM nets. The pool is then used to optimize a set of BCs E using IBC. An initial operation point p is selected for a class imbalance level $\lambda_{init} = 1/1$ and a target $fpr = 1\%$. During operation, the detector module with the ensemble p evaluates the stream of operational data to identify the target individual in the current frame. In parallel, a CAMSHIFT algorithm is used to track the movement and location of different faces over time. The system accumulates operational data for the last 30 minutes to estimate the level of class imbalance λ^*. After the current elapsed time is higher than the update time $t_u = 15$, the operational data level of closest class imbalance λ^* in Λ is estimated with **opt** as the reference data. Then, the BC is approximated to λ^* and an operation point is selected for the target $fpr = 1\%$ to update p for operation.

Experiments use $\Lambda = \{1/1, 1/10, 1/20, 1/30, \ldots, 1/100\}$, $\Lambda_{BC} = \{1/1, 1/10, 1/50, 1/100\}$, target $fpr = 1\%$ and $\lambda^{init} = 1/1$. Class imbalance level is estimated every $t_u = 15$ minutes, over the last 30 minutes interval. The experiments are replicated 10 times using 2×5-fold cross-validation to generate training data. After the 5th replication, the 5 folds are randomly regenerated for the next five replications. FIA faces from video sequences in session one, captured with both the 2.8mm and 4.8mm frontal cameras, are used to generate the pool of diversifier classifiers C and define the BCs for the initial class imbalances in Λ_{BC}. A total of 120 facial samples per individual are randomly selected from both focal distances to build a system design data set D with $\lambda^{max} = 1/100$.

Training data folds in D are split in six folds as follows. The D_t^t uses 2 folds, with a total of 40 positive samples. Each of the remainder data sets uses one fold with 20 positive samples. The D_t^e validation data is used to stop the number of training epochs, whereas D_t^f is the validation data to evaluate the fitness function of the DPSO learning strategy. Negative data in D_t^t, D_t^e and D_t^f is balanced through random under-sampling for classifier training. The data set D_t^o is used to generate the set of data sets OPT following Λ_{BC}, while D_t^v is used to create VAL. Each fold has 2000 negative samples from the cohort and universal background models, providing class imbalance levels up to $\lambda = 1/100$.

Operational test data is extracted from FIA sessions 2 and 3 using both focal lengths. These video sequences are split in two parts of 10 seconds to produce 8 blocks of video with one target individual and 100 non-target individuals (cohort model and unseen individuals). Each block is used to simulate 30 minutes of time. During test, class imbalance in the test data changes over time in these 8 blocks of 30 minutes, with the following levels of class imbalance sequence: $1/20$, $1/35$, $1/100$, $1/65$, $1/100$, $1/80$, $1/60$ and $1/15$. Imbalance changes in test are achieved by randomly removing individuals from each block. The experiment assumes a maximum class imbalance level to adapt $\lambda^{max} = 1/100$, but actual class imbalance is known only after extracting facial regions. A stream of operational data is accumulated by SSBC to estimate the closest level of imbalance in Λ_{BC} at every 15 minutes, over the last 30 minutes. To define the closest class proportions, Hellinger distance is used with $b = \lfloor \sqrt{L^+} \rfloor = 4$ bins per feature, where $L^+ = 20$ is the positive class cardinality of the labeled reference data.

Transactional performance is measured in the ROC and PROC spaces from individual predictions on faces captured in videos: tpr, fpr, $precision$ and F_1. A FR system that integrates SSBC is compared to the one with BC that uses random under sampling (RUS) to balance the data used to optimize the BC (opt) and select the operation point (val). The pool of classifiers C is generated as described above, however, only one BC is optimized to select a single operation point for the entire simulation. Time analysis allows to evaluate the performance of FR systems over a video stream. A face tracker is used to accumulate the positive predictions for facial regions corresponding to a same individual (i.e., with a high confidence track), over a 1 second (30 frame) window.

B. Results and Discussion. Table 1 details the mean transactional performance for the compared approaches, as well as standard deviation values (between parenthesis). At time $t = 1$, both the random under sampling static approach and the proposed adaptive approach uses the same EoC, optimized for

Table 1. Average performance measures for a target $fpr = 1\%$ on test segments at different $t = 1\ldots8$ times. The standard deviation is shown in parenthesis. RUS is a static approach that uses random under sampling to balance data sets.

Approach	Measure	Update Period							
		$t=1$	$t=2$	$t=3$	$t=4$	$t=5$	$t=6$	$t=7$	$t=8$
SSBC	fpr	**4.89%** (0.024)	**1.20%** (0.008)	**1.65%** (0.008)	**1.85%** (0.012)	**1.16%** (0.006)	**1.09%** (0.008)	**0.66%** (0.005)	**0.70%** (0.006)
	tpr $recall$	65.58% (0.299)	49.66% (0.329)	54.53% (0.247)	55.67% (0.308)	53.42% (0.261)	51.00% (0.306)	47.52% (0.394)	49.85% (0.399)
	$precision$	43.68% (0.225)	55.09% (0.315)	41.15% (0.209)	45.33% (0.187)	41.99% (0.177)	47.17% (0.198)	53.59% (0.335)	67.93% (0.314)
	F_1	**0.492** (0.217)	**0.518** (0.255)	**0.446** (0.187)	**0.479** (0.212)	**0.450** (0.221)	**0.470** (0.221)	**0.498** (0.332)	**0.550** (0.344)
BC w/ RUS	fpr	4.89% (0.024)	4.32% (0.021)	5.82% (0.025)	5.93% (0.027)	4.65% (0.025)	4.57% (0.025)	3.45% (0.020)	3.63% (0.024)
	tpr $recall$	65.58% (0.299)	67.40% (0.292)	69.71% (0.186)	69.87% (0.231)	69.01% (0.153)	66.06% (0.241)	61.68% (0.320)	64.02% (0.319)
	$precision$	43.68% (0.225)	38.94% (0.211)	23.37% (0.127)	29.23% (0.109)	23.93% (0.107)	27.04% (0.108)	34.25% (0.184)	54.43% (0.237)
	F_1	0.492 (0.217)	0.470 (0.195)	0.319 (0.136)	0.382 (0.113)	0.332 (0.129)	0.349 (0.134)	0.414 (0.212)	0.550 (0.237)

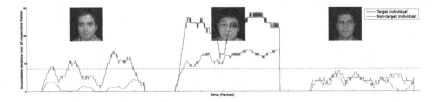

Fig. 3. Time analysis for module of individual 151, with decisions accumulated over a sliding window of 30 frames. The blue line is for a target individual of interest, while the red line is the typical of a non-target individual. Solid lines are a FR system using BC with RUS, while dotted lines are the proposed SSBC.

balanced data. After time $t = 2$ the SSBC approach uses data in the last 30 minutes to estimate the class imbalance and approximate a new BC of classifiers. The SSBC technique selects an operation point closer to the target $fpr = 1$, with smaller standard deviation values. On the other hand, the RUS approach selects an operation point with higher fpr values. The proposed SSBC technique reduces the number of false positive detections and keeps high positive performance, thereby providing better support for a human operator.

For time analysis, a face tracker allowed to accumulate positive predictions of each user-specific module (EoC) over time for improved reliability. Once a face is captured in a frame, the CAMSHIFT algorithm is initiated to track its location over time. As shown in Figure 3, when accumulated positive predictions reach a threshold (e.g., $t_{det} = 8$), target individual 151 is recognized. The FR system with SSBC provides a higher level of discrimination between target and non-target faces appearing in video but with lower fpr values. Accumulated predictions for target persons rise faster and higher than for BC with RUS.

7   Conclusions

BC are promising techniques for ensemble-based FR in video surveillance, although the impact of imbalanced class proportions is difficult to observe in the ROC space. In this paper, an adaptive SSBC technique is proposed to select the most accurate BCs according to class imbalance. Imbalanced data is used to generate several BCs in the decision space, by successively growing number of samples from the majority class. During operations, the system periodically estimates class proportions from operational data distributions using the Hellinger distance. The closest operational points on PROC curves are employed to estimate the most accurate BC of classifiers. Instead of full re-calculation of BCs, the knowledge obtained when combining classifiers for other skew levels is used to approximate the BC to new class priors, providing a significant reduction in computational complexity, and maintaining a comparable level of performance.

Experiments using real-world video data for face re-identification have allowed to compare a modular FR system that integrates the proposed SSBC technique with one that integrates static BC obtained through RUS. Results indicate the advantages of adapting the BC over time to the operation class proportions. Transaction-based analysis shows $fprs$ closer to desired values, as well as consistently higher F_1 scores when using the SSBC technique. Time-based analysis shows a high level of discrimination between target and non-target individuals. However, SSBC depends heavily on the granularity of the pre-trained λ levels, affecting a trade-off between accuracy and resources to store BC curves and reference data. Future research will focus on improving approximations of BC to estimated imbalance levels. Currently, SSBC selects ensembles from the original BCs (optimized for the adjacent imbalance levels) with a mixture of data that follows estimated imbalance level. This approximation may benefit from a strategy to combine vertices in the ROCCH by normalizing to proportions according to difference between estimated and adjacent imbalance levels.

References

1. Brew, A., Cunningham, P.: Combining cohort and ubm models in open set speaker detection. Multimedia Tools Appl. 48, 141–159 (2010)
2. Connolly, J.-F., Granger, E., Sabourin, R.: Evolution of Heterogeneous Ensembles Through Dynamic Particle Swarm Optimization for Video-Based Face Recognition. Pattern Recognition 45(7), 2460–2477 (2012)
3. Davis, J., Goadrich, M.: The Relationship Between Precision-Recall and ROC Curves. In: Proceedings of the 23rd International Conference on Machine Learning, New York, NY, USA, pp. 233–240 (2006)
4. Ekenel, H.K., Stallkamp, J., Stiefelhagen, R.: A video-based door monitoring system using local appearance-based face models. Comput. Vis. Image Underst. 114, 596–608 (2010)
5. Ekenel, H.K., Szasz-Toth, L., Stiefelhagen, R.: Open-set face recognition-based visitor interface system. In: Fritz, M., Schiele, B., Piater, J.H. (eds.) ICVS 2009. LNCS, vol. 5815, pp. 43–52. Springer, Heidelberg (2009)

6. Galar, M., Fernandez, A., Barrenechea, E., Bustince, H., Herrera, F.: A Review on Ensembles for the Class Imbalance Problem: Bagging-, Boosting-, and Hybrid-Based Approaches. IEEE Trans. on Systems, Man, and Cybernetics, Part C: Applications and Reviews 42(4), 463–484 (2012)

7. Goh, R., Liu, L., Liu, X., Chen, T.: The CMU face in action (FIA) database. In: Zhao, W., Gong, S., Tang, X. (eds.) AMFG 2005. LNCS, vol. 3723, pp. 255–263. Springer, Heidelberg (2005)

8. González-Castro, V., Alaiz-Rodríguez, R., Fernández-Robles, L., Guzmán-Martínez, R., Alegre, E.: Estimating Class Proportions in Boar Semen Analysis Using the Hellinger Distance. In: García-Pedrajas, N., Herrera, F., Fyfe, C., Benítez, J.M., Ali, M. (eds.) IEA/AIE 2010, Part I. LNCS, vol. 6096, pp. 284–293. Springer, Heidelberg (2010)

9. Granger, E., Khreich, W., Sabourin, R., Gorodnichy, D.O.: Fusion of Biometric Systems Using Boolean Combination: An Application to Iris-Based Authentication. International Journal on Biometrics 4(3), 291–315 (2012)

10. Kamgar-Parsi, B., Lawson, W., Kamgar-Parsi, B.: Toward development of a face recognition system for watchlist surveillance. TPAMI 33, 1925–1937 (2011)

11. Landgrebe, T.C.W., Paclik, P., Duin, R.P.W., Bradley, A.P.: Precision-Recall Operating Characteristic (P-ROC) Curves in Imprecise Environments. In: 18th International Conference on Pattern Recognition, pp. 123–127 (2006)

12. Li, F., Wechsler, H.: Open set face recognition using transduction. IEEE Trans. on Pattern Analysis and Machine Intelligence 27(11), 1686–1697 (2005)

13. Matta, F., Dugelay, J.-L.: Person recognition using facial video information: A state of the art. J. Vis. Lang. Comput. 20, 180–187 (2009)

14. Pagano, C., Granger, E., Sabourin, R., Gorodnichy, D.O.: Detector Ensembles for Face Recognition in Video Surveillance. In: Proc. 2012 Int'l Joint Conf. on Neural Networks, Brisbane, Australia, pp. 1–8 (2012)

15. Satta, R., Fumera, G., Roli, F.: Fast Person Re-Identification Based on Dissimilarity Representations. Pattern Recognition Letters 33(14), 1838–1848 (2012)

16. Scott, M., Niranjan, M., Prager, R.W.: Realisable classifiers: Improving operating performance on variable cost problems. In: Proc. British MV Conf. (1998)

Self-Organizing Neural Grove and Its Application to Incremental Learning

Hirotaka Inoue

Department of Electrical Engineering and Information Science,
Kure National College of Technology,
2-2-11 Agaminami, Kure, Hiroshima, 737-8506 Japan
hiro@kure-nct.ac.jp

Abstract. Recently, multiple classifier systems have been used for practical applications to improve classification accuracy. Self-generating neural networks (SGNN) are one of the most suitable base-classifiers for multiple classifier systems because of their simple settings and fast learning ability. However, the computation cost of the multiple classifier system based on SGNN increases in proportion to the numbers of SGNN. In this paper, we propose a novel pruning method for efficient classification and we call this model a self-organizing neural grove (SONG). Experiments have been conducted to compare the SONG with bagging and the SONG with boosting, the multiple classifier system based on C4.5, and support vector machine (SVM). The results show that the SONG can improve its classification accuracy as well as reducing the computation cost. Additionally, we investigate SONG's incremental learning performance.

1 Introduction

Classifiers need to find hidden information within a large amount of given data effectively and classify unknown data as accurately as possible [1]. Recently, to improve the classification accuracy, multiple classifier systems such as neural network ensembles, bagging, and boosting have been used for practical data mining applications [2]. In general, base classifiers of multiple classifier systems use traditional models such as neural networks (backpropagation network and radial basis function network) [3] and decision trees (CART and C4.5) [4].

Neural networks have great advantages such as adaptability, flexibility, and universal nonlinear input-output mapping capability. However, to apply these neural networks, it is necessary for the network structure and some parameters to be determined by human experts, and it is quite difficult to choose the right network structure suitable for a particular application at hand. Moreover, they require a long training time to learn the input-output relation of the given data. These drawbacks prevent neural networks from being the base classifier of multiple classifier systems for practical applications.

Self-generating neural networks (SGNN) [5] have a simple network design and high speed learning. SGNN are an extension of the self-organizing maps (SOM) of Kohonen [6] and utilize the competitive learning which is implemented as a self-generating neural tree (SGNT). The abilities of SGNN make it suitable for the base classifier of multiple classifier systems. In order to improve in the accuracy of SGNN, we proposed

Z.-H. Zhou, F. Roli, and J. Kittler (Eds.): MCS 2013, LNCS 7872, pp. 109–120, 2013.

ensemble self-generating neural networks (ESGNN) for classification [7] as one of multiple classifier systems. Although the accuracy of ESGNN improves by using various SGNN, the computation cost, that is the computation time and the memory capacity increases in proportion to the increase in numbers of SGNN in multiple classifier systems.

In an earlier paper [8], we proposed a pruning method for the structure of the SGNN in multiple classifier systems to reduce the computation cost. In this paper, we propose a novel pruning method for more effective processing and we call this model a self-organizing neural grove (SONG). This pruning method is constructed in two stages. In the first stage, we introduce an on-line pruning algorithm to reduce the computation cost by using class labels in learning. In the second stage, we optimize the structure of the SGNT in multiple classifier systems to improve the generalization capability by pruning the redundant leaves after learning. In the optimization stage, we introduce a threshold value as a pruning parameter to decide which subtree's leaves to prune and estimate with 10-fold cross-validation [9]. After the optimization, the SONG improve its classification accuracy as well as reducing the computation cost. We use bagging [10] and boosting [11] as a resampling technique for the SONG.

We investigate the improvement performance of the SONG by comparing it with a multiple classifier system based on C4.5 [12] using ten problems in the UCI machine learning repository [13]. Moreover, we compare the SONG with support vector machine (SVM) [14] to investigate the computational cost and the classification accuracy.

The rest of the paper is organized as follows. The next section shows how to construct the SONG. Section 3 shows the experimental results. Then section 4 is devoted to some experiments to investigate the incremental learning performance of SONG. Finally we present some conclusions, and outline plans for future work.

2   Constructing Self-Organizing Neural Grove

In this section, we describe how to prune redundant leaves in the SONG. First, we mention the on-line pruning method in the learning of SGNT. Second, we show the optimization method in constructing the SONG. Finally, we show a simple example of the pruning method for a two dimensional classification problem.

2.1   On-Line Pruning of Self-Generating Neural Tree

SGNN are based on SOM and are implemented as an SGNT architecture. The SGNT can be constructed directly from the given training data without any intervening human effort. The SGNT algorithm is defined as a tree construction problem of how to construct a tree structure from the given data which consists of multiple attributes under the condition that the final leaves correspond to the given data.

Before we describe the SGNT algorithm, we denote some notations.

- input data vector: $e_i \in \mathbb{R}^m$.
- root, leaf, and node in the SGNT: n_j.
- weight vector of n_j: $w_j \in \mathbb{R}^m$.

Table 1. Sub procedures of the SGNT algorithm

Sub procedure	Specification
$copy(n_j, e_i/w_{win})$	Create n_j, copy e_i/w_{win} as w_j in n_j.
$choose(e_i, n_1)$	Decide n_{win} for e_i.
$leaf(n_{win})$	Check n_{win} whether n_{win} is a leaf or not.
$connect(n_j, n_{win})$	Connect n_j as a child leaf of n_{win}.
$prune(n_{win})$	Prune leaves if the leaves have the same class.

- the number of the leaves in n_j: c_j.
- distance measure: $d(e_i, w_j)$.
- winner leaf for e_i in the SGNT: n_{win}.

The SGNT algorithm is a hierarchical clustering algorithm. The pseudo C code of the SGNT algorithm is given as follows:

Algorithm (SGNT Generation)

```
Input:
  A set of training examples E = {e_i},
  i = 1, ... , N.
  A distance measure d(e_i,w_j).
Program Code:
  copy(n_1,e_1);
  for (i = 2, j = 2; i <= N; i++) {
    n_win = choose(e_i, n_1);
    if (leaf(n_win)) {
      copy(n_j, w_win);
      connect(n_j, n_win);
      j++;
    }
    copy(n_j, e_i);
    connect(n_j, n_win);
    j++;
    prune(n_win);
  }
Output:
  Constructed SGNT by E.
```

In the above algorithm, several sub procedures are used. Table 1 shows the sub procedures of the SGNT algorithm and their specifications.

In order to decide the winner leaf n_{win} in the sub procedure `choose(e_i,n_1)`, competitive learning is used. This sub procedure is recursively used from the root to the leaves of the SGNT. If an n_j includes the n_{win} as its descendant in the SGNT, the weight w_{jk} ($k = 1, 2, \ldots, m$) of the n_j is updated as follows:

$$w_{jk} \leftarrow w_{jk} + \frac{1}{c_j} \cdot (e_{ik} - w_{jk}), \quad 1 \le k \le m. \tag{1}$$

In the SGNT, the input vector x_i corresponds to e_i, and the desired output y_i corresponds to the network output o_i which is stored in one of the leaf neurons, for $(x_i, y_i) \in D$. Here, D is the training data set which consists of data $\{x_i, y_i | i = 1, \ldots, N\}$, $x_i \in \mathbb{R}^m$ is the input and y_i is the desired output. After all training data are inserted into the SGNT as the leaves, the leaves each have a class label as the outputs and the weights of each node are the averages of the corresponding weights of all its leaves. The whole network of the SGNT reflects the given feature space by its topology.

We explain the SGNT generation algorithm using an simple example. In this example, m is one and the four training data (x_i, y_i) is (1,1), (2,2), (3,3), and (4,4). Hence, $e_{11} = 1, e_{21} = 2, e_{31} = 3$, and $e_{41} = 4$. Fig. 1 shows an example of the SGNT generation. First, e_{11} is just copied to a neuron n_1 as the root, and e_{11} is substituted to w_{11} (Fig. 1 (a)). In Fig. 1, the circle is the neuron, the integer in the circle is the number of neuron j, the integer of left-upper of the circle is c_j, and the integer of under the circle is w_{j1}. Next, n_2 and n_3 are generated as the children of n_1 with $w_{21} = 1, w_{31} = 2$. w_{11} is updated by e_{21} to $1 + 1/2(2 - 1) = 1.5$ (Fig. 1 (b)). Next, the winner in $\{n_1, n_2, n_3\}$ is n_3 since $d(e_3, w_1) = 1.5, d(e_3, w_2) = 2$, and $d(e_3, w_3) = 1$; and thus, n_4 and n_5 are generated as the children of n_3 with $w_{41} = 2, w_{51} = 3$. w_{31} is updated by e_{31} to $2 + 1/2(3 - 2) = 2.5$ and w_{11} is updated by e_{31} to $1.5 + 1/3(3 - 1.5) = 2$ (Fig. 1 (c)). Finally, n_6 and n_7 are generated as the children of n_5 with $w_{61} = 3, w_{71} = 4$. w_{51} is updated by e_{41} to $3 + 1/2(4 - 3) = 3.5, w_{31}$ is updated by e_{41} to $2.5 + 1/3(4 - 2.5) = 3$, and w_{11} is updated by e_{41} to $2 + 1/4(4 - 2) = 2.5$ (Fig. 1 (d)).

Note, to optimize the structure of the SGNT effectively, we remove the threshold value of the original SGNT algorithm in [5] to control the number of leaves based on the distance because of the trade-off between the memory capacity and the classification accuracy. In order to avoid the above problem, we introduce a new pruning method in the sub procedure prune(n_win). We use the class label to prune leaves. For leaves that have the n_{win}'s parent node, if all leaves belong to the same class, then these leaves are pruned and the parent node is given to the class.

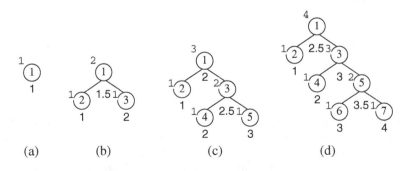

(a)   (b)   (c)   (d)

Fig. 1. An example of the SGNT generation

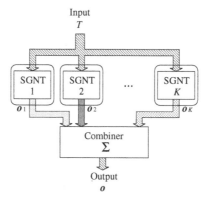

Fig. 2. The SONG which is constructed from K SGNTs. The test dataset T is entered at each SGNT, the output o_i is computed as the output of the winner leaf for the input data, and the SONG's output is decided by voting outputs of K SGNTs.

---

1 **begin   initialize** $j = $ the height of the SGNT
2   **do** for each subtree's leaves in the height j
3   **if** the ratio of the most class $\geq \alpha$,
4   **then** merge all leaves to parent node
5   **if** all subtrees are traversed in the height j,
6   **then** $j \leftarrow j - 1$
7   **until** $j = 0$
8 **end.**

Fig. 3. The merge phase

2.2   Optimization of the SONG

The SGNT has the capability of high speed processing. However, the accuracy of the SGNT is inferior to the conventional approaches, such as nearest neighbor, because the SGNT has no guarantee to reach the nearest leaf for unknown data. Hence, we construct the SONG by taking the majority of multiple SGNT's outputs to improve the accuracy (Fig. 2).

Although the accuracy of the SONG is superior or comparable to the accuracy of conventional approaches, the computational cost increases in proportion to the increase in the number of SGNTs in the SONG. In particular, the huge memory requirement prevents the use of SONG for large datasets even with the latest computers.

In order to improve the classification accuracy, we propose an optimization method of the SONG for classification. This method has two parts, the merge phase and the evaluation phase. The merge phase is performed as a pruning algorithm to reduce dense leaves (Fig. 3).

This phase uses the class information and a threshold value α to decide which subtree's leaves to prune or not. For leaves that have the same parent node, if the proportion

```
1 begin initialize α = 0.5
2    do for each α
3        evaluate the merge phase with 10-fold CV
4        if the best classification accuracy is obtained,
5        then record the α as the optimal value
6            α ← α + 0.05
7    until α = 1
8 end.
```

Fig. 4. The evaluation phase

of the most common class is greater than or equal to the threshold value α, then these leaves are pruned and the parent node is given the most common class.

The optimum threshold values α of the given problems are different from each other. The evaluation phase is performed to choose the best threshold value by introducing 10-fold cross validation (Fig. 4).

2.3   An Example of the Pruning Method for the SONG

We show an example of the pruning method for the SONG in Fig. 5. This is a two-dimensional classification problem with two equal circular Gaussian distributions that have an overlap. The shaded plane is the decision region of class 0 and the other plane is the decision region of class 1 by the SGNT. The dotted line is the ideal decision boundary. The number of training samples is 200 (class0: 100, class1: 100) (Fig. 5(a)).

The unpruned SGNT is given in Fig. 5(b). In this case, 200 leaves and 120 nodes are automatically generated by the SGNT algorithm. In this unpruned SGNT, the height is 7 and the number of units is 320. In this, we define the unit to count the sum of the root, nodes, and leaves of the SGNT. The root is the node which is of height 0. The unit is used as a measure of the memory requirement in the next section. Fig. 5(c) shows the pruned SGNT after the optimization stage in $\alpha = 1$. In this case, 159 leaves and 107 nodes are pruned away and 48 units remain. The decision boundary is the same as the unpruned SGNT. Fig. 5(d) shows the pruned SGNT after the optimization stage in $\alpha = 0.6$. In this case, 182 leaves and 115 nodes are pruned away and only 21 units remain. Moreover, the decision boundary is improved more than the unpruned SGNT because this case can reduce the effect of the overlapping class by pruning the SGNT.

In the above example, we use all training data to construct the SGNT. The structure of the SGNT is changed by the order of the training data. Hence, we can construct the SONG from the same training data by changing the input order. We investigate the pruning method for more complex problems in the next section.

3   Experimental Results

We investigate the computational cost (the memory capacity and the computation time) and the classification accuracy of the SONG with bagging for ten benchmark problems in the UCI machine learning repository [13]. We evaluate how the SONG is pruned

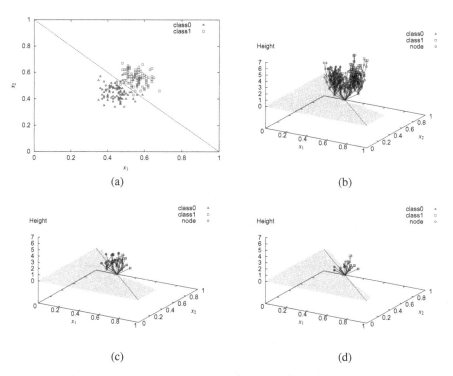

(a)   (b)

(c)   (d)

Fig. 5. An example of the SONG's pruning algorithm, (a) a two dimensional classification problem with two equal circular Gaussian distribution, (b) the structure of the unpruned SGNT, (c) the structure of the pruned SGNT ($\alpha = 1$), and (d) the structure of the pruned SGNT ($\alpha = 0.6$). The shaded plane is the decision region of class 0 by the SGNT and the dotted line shows the ideal decision boundary.

using 10-fold cross-validation for the ten benchmark problems. In this experiment, we use a modified Euclidean distance measure for the SONG. Since the performance of the SONG is not sensitive to the threshold value α, we set the different threshold values α to vary from 0.5 to 1; $\alpha = [0.5, 0.55, 0.6, \ldots, 1]$. We set the number of SGNT K in the SONG as 25 and execute 100 trials by changing the sampling order of each training set. All experiments in this section were performed on an UltraSPARC workstation with a 900MHz CPU, 1GB RAM, and Solaris 8.

Table 2 shows the average memory requirement and classification accuracy of 100 trials for the SONG. As the memory requirement, we count the number of units which is the sum of the root, nodes, and leaves of the SGNT. The average memory requirement is reduced from 65% to 96.6% and the classification accuracy is improved 0.1% to 2.9% by optimizing the SONG. This supports that the SONG can be effectively used for all datasets with regard to both the computation cost and the classification accuracy.

Table 3 shows the average classification accuracy of 10 trials for the SONG with bagging and boosting. On boosting, we implement AdaBoost [11] to the SONG. Since

Table 2. The average memory requirement and classification accuracy of 100 trials for the bagged SGNT in the SONG. The standard deviation is given inside the bracket on classification accuracy ($\times 10^{-3}$).

Dataset	memory requirement			classification accuracy		
	pruned	unpruned	ratio	pruned	unpruned	ratio
balance-scale	107.68	861.18	12.5	0.866(6.36)	0.837(7.83)	+2.9
breast-cancer-w	30.88	897.37	3.4	0.97(2.41)	0.966(2.71)	+0.4
glass	104.33	297.75	35	0.714(13.01)	0.709(14.86)	+0.5
ionosphere	50.75	472.39	10.7	0.891(6.75)	0.862(7.33)	+2.9
iris	15.64	208.56	7.4	0.962(6.04)	0.955(5.45)	+0.7
letter	6197.5	27028.56	22.9	0.956(0.77)	0.955(0.72)	+0.1
liver-disorders	163.12	471.6	34.5	0.648(12.89)	0.636(13.36)	+1.2
new-thyroid	49.45	298.21	16.5	0.958(7.5)	0.957(7.49)	+0.1
pima-diabetes	204.4	1045.03	19.5	0.749(7.05)	0.728(7.83)	+2.1
wine	15	238.95	6.2	0.976(4.41)	0.972(5.57)	+0.4
Average	693.88	3181.96	16.9	0.869	0.858	+1.1

Table 3. The average classification accuracy of 10 trials for the SONG with bagging and boosting. The standard deviation is given inside the bracket ($\times 10^{-3}$).

Dataset	SONG with bagging			SONG with boosting		
	SGNT	SONG	ratio	SGNT	SONG	ratio
breast-cancer-w	0.96(4.74)	0.975(2.86)	+1.5	0.96(6.47)	0.957(4.13)	-0.3
ionosphere	0.847(19.3)	0.89(8.23)	+4.3	0.854(18.26)	0.773(17.4)	-8.1
liver-disorders	0.571(21.4)	0.636(11.0)	+6.5	0.588(17.0)	0.572(24.3)	-1.6
pima-diabetes	0.705(9.8)	0.754(4.96)	+4.9	0.696(12.2)	0.722(6.82)	+2.6
Average	0.771	0.814	+4.3	0.775	0.756	-1.9

original AdaBoost algorithm have been proposed for binary classification problems, we use four binary classification problems in Table 3. In comparison with boosting, bagging is superior to boosting on all of the 4 datasets. In short, bagging is better than boosting in terms of the classification accuracy.

To evaluate the SONG's performance, we compare the SONG with a multiple classifier system based on C4.5. We set the number of classifiers K in the multiple classifier system as 25 and we construct both multiple classifier systems by bagging. Table 4 shows the improved performance of the SONG and the multiple classifier system based on C4.5. The results of the SGNT and the SONG are the average of 100 trials. The SONG has a better performance than the multiple classifier system based on C4.5 for 6 of the 10 datasets. Although the multiple classifier system based on C4.5 degrades the classification accuracy for iris, the SONG can improve the classification accuracy for all problems. Therefore, the SONG is an efficient multiple classifier system on the basis of both the scalability for large scale datasets and the robustly improved generalization capability for the noisy datasets comparable to the multiple classifier system with C4.5.

Table 4. The improved performance of the SONG based on pruned SGNT and the multiple classifier system (MCS) based on C4.5 with bagging

Dataset	SONG based on SGNT			MCS based on C4.5		
	SGNT	SONG	ratio	C4.5	MCS	ratio
balance-scale	0.779	**0.866**	+8.7	0.795	0.827	+3.2
breast-cancer-w	0.956	**0.97**	+1.4	0.946	0.963	+1.7
glass	0.642	0.714	+7.2	0.664	**0.757**	+9.3
ionosphere	0.852	0.891	+3.9	0.897	**0.92**	+2.3
iris	0.943	**0.962**	+1.9	0.953	0.947	−0.6
letter	0.879	**0.956**	+7.7	0.880	0.938	+5.8
liver-disorders	0.59	0.648	+5.8	0.635	**0.736**	+10.1
new-thyroid	0.939	**0.958**	+1.9	0.93	0.94	+1
pima-diabetes	0.695	0.749	+5.4	0.749	**0.767**	+1.8
wine	0.955	**0.976**	+2.1	0.927	0.949	+2.2
Average	0.823	0.869	+4.6	0.837	**0.874**	+3

To show the advantages of the SONG, we compare it with SVM on the same problems. In the SONG, we choose the best classification accuracy of 100 trials with bagging. In SVM, we use C-SVM in libsvm [14] with radial basis function kernel. We select the parameters of SVM, the cost parameters C and the kernel parameters γ, from $15 \times 15 = 225$ combinations by 10-fold cross validation; $C = [2^{12}, 2^{11}, 2^{10}, \ldots, 2^{-2}]$ and $\gamma = [2^4, 2^3, 2^2, \ldots, 2^{-10}]$. We normalize the input data from 0 to 1 for all problems in k-nearest neighbor and SVM. All methods are compiled by using gcc with the optimization level -O2 on the same workstation.

Table 5 shows the classification accuracy, the memory requirement, and the computation time achieved by the SONG and SVM. Next, we show the results for each category.

First, in view point of the classification accuracy, the SONG superior to SVM 3 of the 10 datasets and degrade 1.7% in the average. Second, in terms of the memory requirement, even though the SONG includes the root and the nodes which are generated by the SGNT generation algorithm, this is less than SVM for 8 of the 10 datasets. Although the memory requirement of the SONG is totally used K times in Table 5, we release the memory of SGNT for each trial and reuse the memory for effective computation. Therefore, the memory requirement is suppressed by the size of the single SGNT. Finally, in view of the computation time, although the SONG consumes the cost of K times of the SGNT to construct the model and test for the unknown dataset, the average computation time is faster than SVM. The SONG is slower than SVM for small datasets such as glass, ionosphere, and iris. However, the SONG is faster than SVM for large datasets such as balance-scale, letter, and pima-diabetes. Especially, in letter, the computation time of the SONG is faster than SVM about 11 times. We need to repeat 10-fold cross validation many times to select the optimum parameter for α, k, C, and γ. This evaluation consumes much computation time for large datasets such as letter.

Table 5. The classification accuracy, the memory requirement, and the computation time of ten trials for the best pruned SONG and SVM

Dataset	classification acc.		memory requirement		computation time (s)	
	SONG	SVM	SONG	SVM	SONG	SVM
balance-scale	0.885	**0.992**	109.93	**60.6**	**0.82**	4.77
breast-cancer-w	**0.976**	0.973	**26.8**	79.6	**1.18**	0.64
glass	**0.758**	0.738	**91.33**	132.4	**0.36**	0.61
ionosphere	0.912	**0.954**	**51.38**	147.9	1.93	**1.25**
iris	**0.973**	0.96	**11.34**	51.3	0.13	**0.06**
letter	0.958	**0.977**	**6208.03**	7739.7	**208.52**	2359.39
liver-disorders	0.685	**0.73**	**134.17**	214.5	**0.54**	2.07
new-thyroid	0.972	**0.977**	45.74	**44.1**	0.23	**0.22**
pima-diabetes	0.764	**0.766**	**183.57**	363.5	**1.72**	5.63
wine	0.983	**0.989**	**11.8**	62.2	0.31	**0.15**
Average	0.887	**0.904**	**687.41**	889.58	**21.57**	236.88

Therefore, the SONG based on the fast and compact SGNT is useful and practical for large datasets. Moreover, the SONG has the ability of parallel computation because each classifier behaves independently. In conclusion, the SONG is a practical method for large-scale data mining compared with SVM.

4   Considerations

In this section, we investigate the performance of the incremental learning of the SONG. We use letter in this experiment since it contains large scale data (the number of input dimension: 16, the number of classes: 26, and the number of entries: 20000).

This experiment is performed as follows. First, we divide letter dataset into ten parts. Second, we select one of the ten parts as the testing data. Third, we enter one of the remaining nine parts into the SONG for training. Forth, we test the SONG using the testing data. Finally, we continue the training and the testing until all nine parts of the dataset are entered into the SONG.

Fig. 6 shows the relation between the number of training data and the classification accuracy. The more the number of training data increases, the more the classification accuracy improves for all the number of ensembles K. The width of the improvement is wide for small K and all values of N.

As the memory requirement, we count the number of units which is the sum of the root, nodes, and leaves of the SGNT. Fig. 7 shows the relation between the number of training data N and the number of units in $\alpha = 1$. Here, the total units are the number of all units without pruning and the remaining units are the number of all units with pruning. Both of them are the average of 25 SGNTs. The number of nodes increases linearly in proportion to the increase in the number of training data. The slope of the remaining units is smaller than the slope of the total nodes. This means that the SONG has the capability for good compression for large scale data. This supports that the SONG can be effectively used for large scale datasets.

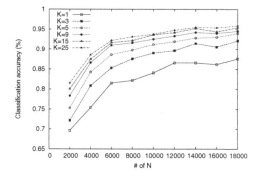

Fig. 6. The relation between the number of training data and the classification accuracy

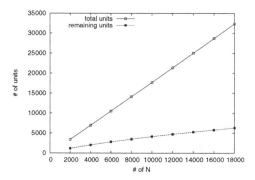

Fig. 7. The relation between the number of training data and the number of units

5   Conclusions

In this paper, we proposed a new pruning method for the multiple classifier system based on SGNT, which is called SONG, and evaluated the computation cost and the accuracy. We introduced an on-line and off-line pruning method and evaluated the SONG by 10-fold cross-validation. Experimental results showed that the memory requirement reduced remarkably, and the accuracy increased by using the pruned SGNT as the base classifier of the SONG. Additionaly, we investigated an incremental learning performance of the SONG. Experimental results showed that the SONG could be applicable to incremental learning. The SONG is a useful and practical multiple classifier system to classify large datasets. In future work, we will study a parallel and distributed processing of the SONG for large scale data mining.

Acknowledgment. The author would like to thank Kyoshiro Sugiyama for implementing AdaBoost algorithm on the SONG, Anthony Nepia and the anonymous referees for their helpful comments.

References

1. Han, J., Kamber, M.: Data Mining: Concepts and Techniques. Morgan Kaufmann Publishers, San Francisco (2000)
2. Quinlan, J.R.: Bagging, Boosting, and C4.5. In: Proceedings of the Thirteenth National Conference on Artificial Intelligence, Portland, OR, August 4-8, pp. 725–730. AAAI Press, The MIT Press (1996)
3. Bishop, C.M.: Neural Networks for Pattern Recognition. Oxford University Press, New York (1995)
4. Duda, R.O., Hart, P.E., Stork, D.G.: Pattern Classification, 2nd edn. John Wiley & Sons Inc., New York (2000)
5. Wen, W.X., Jennings, A., Liu, H.: Learning a neural tree. In: The International Joint Conference on Neural Networks, Beijing, China, November 3-6, vol. 2, pp. 751–756 (1992)
6. Kohonen, T.: Self-Organizing Maps. Springer, Berlin (1995)
7. Inoue, H., Narihisa, H.: Improving generalization ability of self-generating neural networks through ensemble averaging. In: Terano, T., Liu, H., Chen, A.L.P. (eds.) PAKDD 2000. LNCS (LNAI), vol. 1805, pp. 177–180. Springer, Heidelberg (2000)
8. Inoue, H., Narihisa, H.: Optimizing a multiple classifier system. In: Ishizuka, M., Sattar, A. (eds.) PRICAI 2002. LNCS (LNAI), vol. 2417, pp. 285–294. Springer, Heidelberg (2002)
9. Stone, M.: Cross-validation: A review. Math. Operationsforsch. Statist., Ser. Statistics 9(1), 127–139 (1978)
10. Breiman, L.: Bagging predictors. Machine Learning 24, 123–140 (1996)
11. Freund, Y., Schapire, R.E.: Boosting: Foundations and Algorithms. MIT Press, Cambridge (2012)
12. Quinlan, J.R.: C4.5: Programs for Machine Learning. Morgan Kaufmann, San Mateo (1993)
13. Frank, A., Asuncion, A.: UCI machine learning repository (2010),
    `http://archive.ics.uci.edu/ml`
14. Chang, C.-C., Lin, C.-J.: LIBSVM: A library for support vector machines. ACM Transactions on Intelligent Systems and Technology 2, 27:1–27:27 (2011), Software available at
    `http://www.csie.ntu.edu.tw/~cjlin/libsvm`

Cascaded Reduction and Growing of Result Sets for Combining Object Detectors

Uwe Knauer[1,2] and Udo Seiffert[2]

[1] Humboldt University, Unter den Linden 6, 10099 Berlin, Germany
[2] Fraunhofer IFF, Sandtorstr. 22, 39106 Magdeburg, Germany
`knauer@informatik.hu-berlin.de`

Abstract. In this paper *cascaded reduction and growing of result sets* is introduced as a principle for combining the results of different object detectors. First, different candidate operating points are selected for each object detection algorithm. This procedure is based on the analysis of precision and recall of the individual methods. Selecting an appropriate operating point prior to fusion is important because it regulates the cardinal number of the result set. As diversity and correlation between object detectors also depend on the elements of the result sets, this and the application of set operations allow to create a final set of detected objects by including missing and excluding false detections. The approach allows both diverse and correlated detectors to contribute to the performance of the combined detector. The performance of the proposed algorithm is compared to other combining algorithms. It outperforms or competes with existing state of the art combiners for several datasets. Additionally, the results provide a significant improvement in the interpretability of the combining rules. As a unique feature of the proposed algorithm, the found operating points can be used to reconfigure the object detection algorithms to adapt their individual results to the needs of the combination procedure allowing a reduction in runtime.

Keywords: object detection, combining, fusion, image processing.

1 Introduction

Object detection in digital images is an important task in many application fields such as microscopy, remote sensing, robot vision, tracking in surveillance applications, and autonomous navigation. While good solutions exist for some applications such as face recognition, it is still difficult to achieve sufficient detection rates for many other problems to fully automate the analysis of images. The main reason is the variety of the appearance due to changes of camera perspective, illumination, observation of deformable objects, and occlusion of objects. The combination of different algorithms has the potential to overcome some of these problems. First, related work on the combination of object detectors and classifiers is introduced. Second, a novel approach to late fusion is described. After introducing the dataset, the results of combination are presented. The paper concludes with a summary of the potentials of the new algorithm.

Z.-H. Zhou, F. Roli, and J. Kittler (Eds.): MCS 2013, LNCS 7872, pp. 121–133, 2013.
© Springer-Verlag Berlin Heidelberg 2013

2 Related Work

The combination of object detection algorithms is closely related to the field of multiple classifier systems, since most methods can be applied to combine object detection results. Successful approaches to classifier fusion are based on Bayes theorem [11], the Dempster-Shafer theory of evidence [2], fuzzy logic [4,5,13], voting [15], analysis of correspondence [8,18], and on machine learning approaches [9]. However, modifications are required because object detection results differ from classifier results. First, object detectors output positions and dimensions of objects in addition to the measurement, rank, or abstract level outputs [26]. Consequently, the matching of the result sets of different detectors is needed. Second, object detectors do not output a label or a measurement for each image patch, therefore, the combination algorithms must deal with missing information from some of the detectors. Third, since image patches can overlap or have different scales, matching of objects between detectors may not be unique. Fourth, the cardinal number of the result sets of different detectors for the same image can be different. Additionally, if a measurement level output is provided then a threshold on this value can be used to further reduce the number of elements in the result set of detected objects.

Current research on combining object detection approaches focuses on early or intermediate level fusion [23,21,14]. Fusion of feature vectors at an early stage has the advantage that the spatial relation of the different features are preserved at the pixel or image patch level. Its disadvantage is the curse of dimensionality, using more dimensions requires much more training data for most problems to achieve a good generalization performance for a detector. Other approaches, which claim to be late level fusion methods, still produce extended feature sets [10]. Some classifier fusion approaches have been published for the fusion of different sensor data [1], the boosting of object detection by inclusion of contextual information from classification [20], and the fusion of the decisions of multiple experts for different object categories [16]; however, it is difficult to find published work on combining the final detections of different algorithms for the same object class. In object detection, late fusion is much more related to the abstraction level of the features than in classifier fusion. The reason is obvious, because information processing in human perception works analogously and has always inspired the computer vision community.

Moreover, combining object detectors at the level of detected objects allows not only to fuse different successful approaches to achieve a better performance, but also a deeper understanding of the strengths and weaknesses of individual approaches. It also adds more levels of freedom to the selection of the combined detection methods because they can be handled as a black box.

3 Proposed Method

The basic concepts of the approach are the utilization of operating points and the construction of improved result sets. The consequences of using these concepts are summarized in section 3.1. In the following *operating points* correspond to the thresholds which are used to control the elements contained in the result set of each detector. An operating point is used to convert the measurement level output into an abstract level output. This concept is well-known from analysis of classifier performance in receiver operating characteristic (ROC) and precision recall (PR) space where it is used for the construction of characteristic curves.

3.1   Preliminaries

The operations cut ($\cap$) and union ($\cup$) are used for the fusion of result sets of different detectors. Therefore, correspondences between the elements of the result sets have to be calculated. This must be based on the positions and extends of the objects. The objective is to minimize the error related costs EC:

$$EC = (1 - TPR) \cdot p(+) \cdot C(-|+) + FPR \cdot p(-) \cdot C(+|-) \qquad (1)$$

$p(\cdot)$ denotes the apriori probability of the classes and C the costs of false positive and false negative detections. The true positive rate (TPR) and the false positive rate (FPR) depend on the operating points. These are set by thresholds t_A and t_B for the measurement level output of object detectors A and B respectively. Minimization of $EC(t_A, t_B)$ as a function of the two thresholds t_A and t_B is computationally expensive. The number of different threshold values is only bounded by the number of objects in the training dataset. For each pair (t_A, t_B) the costs EC as well as the gradient of the error cost function can be calculated by application of the set operations only.

To solve that problem, the threshold values can be quantized, such that only a limited number of k thresholds is used for each object detector. By setting $k = 3$ we achieve:

1. reduction of complexity for minimization
2. good interpretability

For the last reason the three thresholds are set such that they correspond to operating points with the attributes precise, optimal, and sensitive.

Combining object detectors is repeated until EC stops to decrease. Since the associative law does not hold for arbitrary sets:

$$(A \cap B) \cup C \neq A \cap (B \cup C) \qquad (2)$$

the order of object detectors matters. The fusion rule for the object detectors $OD^1, \ldots, OD^M$ has the form:

$$((OD^1_{o_1} \otimes OD^2_{o_2}) \ldots \otimes OD^M_{o_M}) \qquad (3)$$

where $o_1, \ldots o_M \in \{1, \ldots, k\}$ denote the operating points. The brackets denote the cascade and $\otimes$ denotes the selected fusion operation.

3.2   Matching Positions and Shapes

The matching of detection results is a necessary condition for fusion. In the learning phase the object boundaries of a reference dataset are used. In the application phase pairwise matching of the different result sets is needed.

A commonly used measure for performance evaluation of object detectors is the matching of upright rectangles [7]. The overlap of two rectangles can be checked quickly [24]. Two rectangles defined by their centers $\mathbf{M}_1$, $\mathbf{M}_2$ and the distances d^x and d^y to the corner points do not overlap, if any of the following inequations are true:

$$\left| (\mathbf{M}_1 - \mathbf{M}_2) \begin{pmatrix} 1 \\ 0 \end{pmatrix} \right| > d_1^x + d_2^x \tag{4}$$

$$\left| (\mathbf{M}_1 - \mathbf{M}_2) \begin{pmatrix} 0 \\ 1 \end{pmatrix} \right| > d_1^y + d_2^y \tag{5}$$

otherwise, the size of the overlap can be calculated easily. An efficient overlapping test for arbitrary convex polygons is known as separating axis test [19]. If a more detailed representation of the object boundaries is needed then this algorithm can be used.

If object detection is compared to other classification problems then an obvious difference is the meaning of false positives. The set of image patches which do not contain an object is typically large and not part of the reference dataset. However, for comparison and combination of different object detectors such a reference is needed. Therefore, the false positives of all methods are combined into a single reference set.

3.3   Selecting Operating Points

Selecting candidate operating points is a preprocessing step. Later, the learning algorithm selects those candidate operating points which provide the maximum error cost reduction. Operating points correspond to points on the precision recall curves of the individual object detectors. It is assumed that each object detector provides a measurement level output. The PR curves are obtained by thresholding the output values. However, if one or more object detectors provide only abstract level output then only a single operating point exist for those methods.

Fig. 1 shows the precision recall curve of an object detector. The positions i of the smallest and the largest threshold are highlighted. If the operating point corresponds to the threshold at position $i = 1$ then the cardinal number of the result set is maximized. This means that the objects of the reference dataset are reproduced best. However, at this operating point the precision has its smallest value because all false detections of the object detector also satisfy the condition that the output value is larger than the selected threshold. Therefore, in this operating range the detector is called sensitive for the presence of objects.

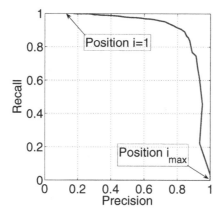

Fig. 1. Precision recall curve for which $k = 3$ operating points have to be selected

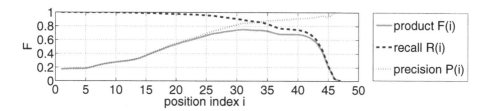

Fig. 2. Values of P, R, and F as a function of position index i

The precise or specific operating points correspond to large threshold values. Only few objects are detected, but the detectors output high confidence values for these detections. Therefore, the recall of the reference dataset is small and the precision is high.

Eq.(6) provides the measure F which is used for the selection of operating points on the precision recall curves:

$$F_i = (P_i - P_{min})(P_{max} - P_{min})^{-1} \cdot (R_i - R_{min})(R_{max} - R_{min})^{-1} \qquad (6)$$

where F_i is the product of the precision P_i and the recall R_i for the operating point with index value i. Fig. 2 shows P and R as a function of the position index i. It is one of the difference to classification problems that the recall must not reach the value 1. Hence, eq.(6) includes offset corrections for both dimensions. Fig. 3 illustrates how the sensitive operating point is found by a threshold on the value of F. Since the recall converges to one in this operating range:

$$F_1 \approx (P_1 - P_{min})(P_{max} - P_{min})^{-1} \qquad (7)$$

With decreasing recall the difference $\Delta = P - F$ grows. In Fig. 3 the threshold $\Delta \leq 0.03$ is plotted. If Δ exceeds the threshold then the operating point is determined by the corresponding position index i.

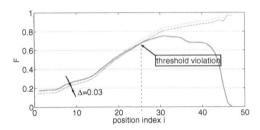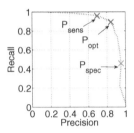

Fig. 3. Position index i of a sensitive operating point (left). Operating points P_{sens} and P_{spec} for $\Delta \leq 0.03$ (right).

Using the same approach an operating point within the high precision operating range is determined. Here, F converges to the value of the normalized recall $(R_i - R_{min})(R_{max} - R_{min})^{-1}$. Fig. 3 shows both operating points P_{spec} und P_{sens} for $\Delta \leq 0.03$. The third operating point P_{opt} was found by minimizing the Euclidian distance between the precision recall curve and the perfect operating point $P_{ideal} = (1,1)^T$. Search space is limited to the range between P_{sens} and P_{spec}.

3.4   Cascaded Reduction and Growing of Result Sets

Fig. 4 shows the search for combining rules. First, a random order of the object detectors is set. This step is repeated to avoid local minima of the error cost function. A complete search can be done for small numbers n of object detectors. Since the number of possible arrangements is $n!$, with growing n a fixed number of randomly chosen permutations is validated.

The operating point with minimum costs EC is selected for the first object detector OD^1 of the current permutation. The choice depends on the ratio of the error costs which has to be provided as an input value. Selection of the set operation and the operating point is repeated for the next object detector $OD^2, \ldots, OD^n$. By default a ratio of $1 : 1$ is assumed.

4   Datasets

The proposed algorithm has been tested with several datasets. In this paper, results for a dataset from a surveillance application and for object detection in microscopy images of plant samples are presented. The tasks are to detect individuals in surveillance camera images and the analysis of spatiotemporal fungal patterns [3], Fig. 5 shows the camera field of view as well as a leaf with two colonies.

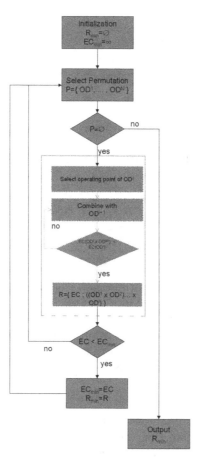

Fig. 4. Flow chart of the search for a combining rule

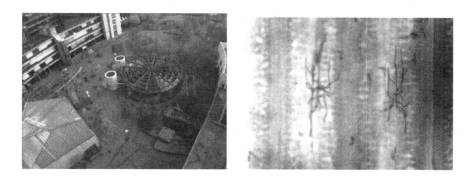

Fig. 5. Field of view of the surveillance camera with ground reference points and image of two phytopathogenic fungi

While the proposed method has been developed for combination of object detection results, it can deal with standard classification problems. The WDBC, Statlog Heart, and SPECTF datasets from the UCI machine learning repository have been selected to test the performance of the algorithm. For all comparisons 10-fold cross-validation is used.

5   Results and Discussion

For detection of people in the surveillance camera images the following algorithms were used:

1. Fixed background subtraction (BS)
2. Difference image (DIFF)
3. Lucas-Kanade tracking (LKT [17])
4. Mixture of Gaussians (MoG [22])
5. Running Average subtraction (RA [6])

Most of the algorithms belong to the class of background subtraction methods while Lucas-Kanade tracking provides a motion based object detection. To estimate the performance of the object detectors a validation set is used. Next, the performance of the combining algorithm is calculated based on 10-fold cross-validation of the outputs of the detectors for this validation set.

The following combination rule was found by cascaded reduction and growing of the result set:

$$(((\text{LKT}_1 \cup \text{BS}_2) \cup \text{DIFF}_2) \cup \text{MoG}_3) \tag{8}$$

For the given dataset the obtained rule uses only a cascade of $\cup$ operations. This indicates that the false detections of the individual object detectors are highly correlated while diversity is given for the true detections. The background subtraction with a Running Average based method does not contribute to the result set, hence, it can be excluded from the multiple object detector system. The indices of the method names are the preselected operating points (1=sensitive, 2=average, 3=specific).

Fig. 6 shows the performance of the individual and the combined detectors in more detail. The diagram shows the precision recall curves of the methods. The PR-curve of the LKT algorithm shows its great contribution to the result. The high overall precision of this detection algorithm allows the multiple detector system to operate the method with a low false positive rate at its sensitive operating point.

Fig. 7 shows a comparison between the approach of cascaded reduction and growing of result sets (CRAGORS) and a number of combiners (e.g. AdaBoost, Random Forest). The proposed method ranks second best and outperforms most of the other methods. The ranking is based on the Euclidian distance between the ideal operating point and the best operating point of each combination method. The red line marks the distance of the operating point of the best performing individual detector. For the detection of fungal patterns a good segmentation into image foreground and background is required. For each pixel 36 features from the

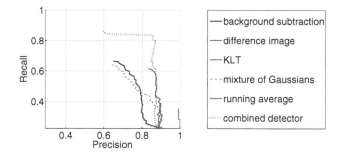

Fig. 6. Precision Recall curves for object detection in surveillance scene

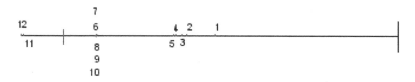

Fig. 7. Ranking of different combining methods for detection of individuals: (1) Random Forest, (2) CRAGORS, (3) CRAGORS (ROC), (4) SCANN, (5) AdaBoost, (6) Sum rule, (7) Max rule, (8) Median rule, (9) Min rule, (10) Product rule, (11) Fuzzy Templates, and (12) Voting

input RGB image are considered, which can be calculated quickly with integral images [12]. The following algorithms were used to find good segmentations of the microscopy images:

1. AdaBoost with J4.8 as weak learner (ADA, 10 stages)
2. Random Forest (RF, 10 trees)
3. Bagging classifier with kNN (BAG, 10 bags)

Finally, the different segmentation results were combined by the proposed algorithm. Only the abstract level outputs of the three segmentation approaches were used. Hence only a single operating point for each method was considered, yielding the following fusion result:

$$((ADA \cup BAG) \cup RF)$$

This combining rule can be easily implemented with binary AND and OR operations per pixel. Fig. 8 shows the improvements of combination and a ranking of different combination algorithms as well. The improved segmentation allows a better detection of objects in subsequent processing steps as well as an improved estimation of important object features such as the area of fungal patterns. The red line marks the performance of the best individual segmentation method. A number of combining algorithms failed to improve the performance (ranks 5-12). Only four algorithms including two variants of the proposed one (ranks 3,4) were capable of improving the results.

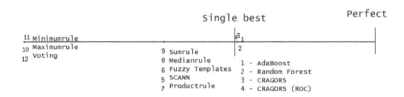

Fig. 8. Ranking of different algorithms for combining segmentation results of microscopy images

The good performance of the proposed method for the combination of detection results raises the question, whether other typical classification problems from the UCI machine learning repository can benefit or not. The following classifiers have been selected and the results of their combination have been evaluated:

1. Linear kernel SVM
2. Radial basis function kernel SVM
3. AdaBoost with decision trees (J4.8)
4. AdaBoost with kNN (k=1)
5. Random forest classifier
6. kNN classifier (k=10)

The obtained combining rules as well as corresponding precision and recall values are listed in Tab. 1.

Table 1. Combining rules, precision P, and recall R for selected datasets from UCI Machine Learning Repository

Dataset	Combining Rule	P	R
Statlog 1	$(((\text{linSVM}_2 \cap \text{RF40}_1) \cap \text{boostedJ4.8}_2) \cup \text{rbfSVM}_3)$	0.87	0.79
Statlog 2	$((((\text{linSVM}_1 \cup \text{RF40}_2) \cap \text{boostedJ48}_1) \cup \text{boostedkNN}_3) \cup \text{rbfSVM}_1)$	0.96	0.6
SPECTF	$(((\text{rbfSVM}_2 \cap \text{RF40}_2) \cup \text{linSVM}_2) \cap \text{boostedJ48}_1)$	0.86	0.74
WDBC	$(((\text{rbfSVM}_2 \cap \text{singlekNN}_2) \cup \text{RF40}_2) \cap \text{linSVM}_1)$	0.96	0.95

Fig. 9 shows the precision recall curves of the tested classifiers for the Statlog, WDBC, and SPECTF datasets. The operating points of the combined methods are indicated by circles. Each circle corresponds to a different weighting of the false positive and false negative errors. Therefore, for each circle a different combination rule for the classifiers has been calculated.

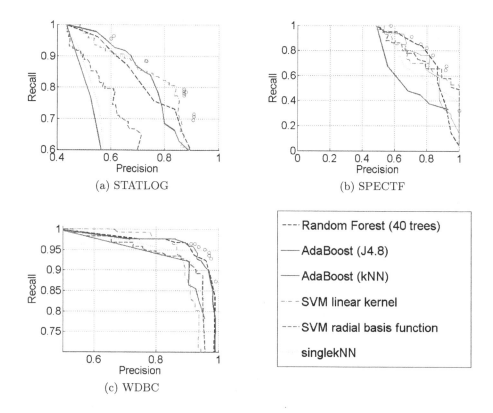

(a) STATLOG

(b) SPECTF

--- Random Forest (40 trees)

—— AdaBoost (J4.8)

—— AdaBoost (kNN)

--- SVM linear kernel

--- SVM radial basis function

singlekNN

(c) WDBC

Fig. 9. Precision/recall curves for classification performance on UCI dataset and operating points of combined classifiers (circles)

For all tested datasets the fusion algorithm improves the classification performance. This is worth mentioning because the individual classifiers are classifier ensembles such as random forests as well as adaptive boosted classifiers. It shows that the method is not limited to fusion of object detection algorithms. Its basic principle is the careful selection of operating points prior to fusion to ensure a sufficient level of diversity or compliance of the individual detectors or classifiers.

6   Conclusion

In this paper, a novel approach to the combination of object detection algorithms has been presented. The selection of operating points allows the reconfiguration of individual object detectors. Additionally, redundant detectors can be excluded automatically, as a result the time for the matching step between the result sets is greatly reduced. This is an important feature because matching is required for all combining methods. Depending on the detection algorithm, additional reductions of runtime are possible. Methods such as Viola and Jones cascade of boosted features [25] benefit from sensitive operating points because only few levels of the cascade must be evaluated. The processing time of other algorithms such as LKT [17] decreases, if only the precisely trackable objects have to be detected. Segmentation is a common preprocessing step for object detection. Showing that the proposed algorithm improves the segmentation of microscopy images of phytopathogenic fungi illustrates that combining different algorithms is beneficial at the different processing levels of an object detection chain.

References

1. Aeberhard, M.: Object existence probability fusion using dempster-shafer theory in a high-level sensor data fusion architecture. In: IEEE Intelligent Vehicles Symposium, pp. 770–775 (2011)
2. Al-Ani, A., Deriche, M.: A new technique for combining multiple classifiers using the dempster-shafer theory of evidence. Journal of Artificial Intelligence Research 17, 333–361 (2002)
3. Baum, T., Navarro-Quezada, A., Knogge, W., Douchkov, D., Schweizer, P., Seiffert, U.: Hypharea - automated analysis of spatiotemporal fungal patterns. Journal of Plant Physiology 168(1), 72–78 (2011)
4. Chen, X., Harrison, R., Zhang, Y.-Q.: Genetic fuzzy fusion of svm classifiers for biomedical data. In: IEEE Congress on Evolutionary Computing (2005)
5. Chen, X., Li, Y., Harrison, R., Zhang, Y.-Q.: Type-2 fuzzy logic-based classifier fusion for support vector machines. Applied Soft Computing 8(3), 1222–1231 (2008); Forging the Frontiers - - Soft Computing
6. Cheng, F., Huang, S., Ruan, S.: Advanced motion detection for intelligent video surveillance systems. In: ACM Symposium on Applied Computing (2010)
7. Everingham, M., Van Gool, L., Williams, C.K.I., Winn, J., Zisserman, A.: The PASCAL Visual Object Classes (VOC) Challenge. International Journal of Computer Vision 88, 303–338 (2010)
8. Greenacre, M.J.: Theory and application of correspondence analysis. Academic Press, London (1984)
9. Kang, S., Park, S.: A fusion neural network classifier for image classification. Pattern Recogn. Lett. 30(9), 789–793 (2009)
10. Khan, F.S., Anwer, R.M., van de Weijer, H., Bagdanov, A.D., Vanrell, M., Lopez, A.M.: Color Attributes for Object Detection. In: CVPR (2012)
11. Kittler, J., Hatef, M., Duin, R.P.W., Matas, J.: On combining classifiers. IEEE Transactions on Pattern Analysis and Machine Intelligence 20(3), 226–239 (1998)
12. Knauer, U., Meffert, B.: Fast computation of region homogeneity with application in a surveillance task. In: ISPRS Technical Commission V Symposium. ISPRS (2010)
13. Kuncheva, L.I.: 'Fuzzy' vs 'non-fuzzy' in combining classifiers designed by boosting. IEEE Transactions on Fuzzy Systems 11(6), 729–741 (2003)
14. Lee, D.: Multisensor fusion-based object detection and tracking using active shape model. In: Digital Information Management, pp. 108–114 (2011)
15. Leung, K.T., Parker, D.S.: Empirical comparisons of various voting methods in bagging. In: 9th ACM SIGKDD International Conference on Knowledge Discovery and Data Mining (2003)

16. Liu, N., Dellandrea, E., Zhu, C., Bichot, C.-E., Chen, L.: A Selective Weighted Late Fusion for Visual Concept Recognition. In: Fusiello, A., Murino, V., Cucchiara, R. (eds.) ECCV 2012 Ws/Demos, Part III. LNCS, vol. 7585, pp. 426–435. Springer, Heidelberg (2012)
17. Lucas, B.D., Kanade, T.: An iterative image registration technique with an application to stereo vision, pp. 674–679 (1981)
18. Merz, C.J.: Using correspondence analysis to combine classifiers. Machine Learning 36(1-2), 33–58 (1999)
19. Sevenson, A.: Separating Axis Theorem (SAT) Explanation. Internet (2009)
20. Song, Z.: Contextualizing object detection and classification. In: IEEE Conference on Computer Vision and Pattern Recognition (2011)
21. Spinello, L., Arras, K.O.: Leveraging rgb-d data: Adaptive fusion and domain adaption for object detection. In: IEEE Conference on Robotics and Automation (2012)
22. Stauffer, C., Grimson, W.E.L.: Adaptive background mixture models for real-time tracking. In: IEEE Computer Vision and Pattern Recognition, vol. 2, p. 200. IEEE (1999)
23. Suja, T.B.: Fusion based object detection. In: National Conference on Communications, pp. 1–3 (2010)
24. Teschner, M.: Algorithmen und Datenstrukturen Bereichsbume. Foliensatz (2011)
25. Viola, P., Jones, M.: Rapid object detection using a boosted cascade of simple features. In: Proceedings of the IEEE Conference on Computer Vision and Pattern Recognition, pp. 511–518 (2001)
26. Xu, L., Kryzak, A., Suen, C.V.: Methods of Combining Multiple Classifiers and Their Application to Handwriting Recognition. IEEE Transactions on Systems, Man, and Cybernetics 22(3), 418–435 (1992)

Single Classifier Based Multiple Classifications

Albert Hung-Ren Ko and Robert Sabourin

ETS, University of Quebec
1100 Notre-Dame West Street, Montreal, Quebec, H3C 1K3 Canada

Abstract. In this paper, a Single Classifier-based Multiple Classification Scheme (SMCS) is proposed as an alternative multiple classification scheme. The SMCS uses only a single classifier to generate multiple classifications for a given test data point. Because of the presence of multiple classifications, classification combination schemes, such as majority voting, can be applied, and so the mechanism may improve the recognition rate in a manner similar to that of Multiple Classifier Systems (MCS). The experimental results confirm the validity of the proposed SMCS as applicable to many classification systems.

1 Introduction

Most EoCs are created so that an abundance of diverse classifiers is generated, and subsequently an optimal subset of classifiers is selected. By partially omitting selected samples from a sample pool for each classifier training operation, we create different data subsets [4,5,8]. Then, by using these data subsets to train classifiers, every classifier will be different from the others. Multiple classifiers yield multiple class labels for a given test sample, and we can combine these multiple class labels into a single class label. Given that each classifier actually draws a boundary between classes, the MCS obtains a new boundary by applying a fusion function , that is, de facto, a combination of different boundaries drawn by different classifiers.

In this paper, we propose an unconventional Single-Classifier-based Multiple Classification Scheme (SMCS) approach that is similar to the MCS, but without the need to train multiple classifiers. We propose a mechanism that achieves multiple classifications with a single classifier, and so benefits from the logic of an MCS without repetitive classifier training and without classifier selection. Given a test sample to classify and some training samples, our method divides the training samples into two groups: one containing what we call reference samples, and the other containing what we call evaluation samples. We use different reference samples to generate different pseudo test data points, each of which constitutes a different combination of an original test sample and some reference samples (Fig. 1), and we use evaluation samples to select adequate reference samples for pseudo test data point generation. Because we use different reference samples to generate pseudo test data points, data diversity is extracted from the original training data in a way similar to that in MCS. Furthermore, because of the generation of multiple pseudo test data points for an original test sample, we can obtain multiple classifications for that test sample. Consequently, traditional classification combination schemes in the MCS, such as the majority voting fusion function, can be implemented

Z.-H. Zhou, F. Roli, and J. Kittler (Eds.): MCS 2013, LNCS 7872, pp. 134–145, 2013.
© Springer-Verlag Berlin Heidelberg 2013

to generate a final class label. The proposed method can somehow be related to local learning [1], in which local information is exploited to facilitate classification task.

Note that the generation of pseudo data points to improve classification accuracy is not new [6]. The generation of artificial training examples, known as virtual examples, have been proposed for Support Vector Machine (SVM) [2, 10]. However, this is different from the proposed methods in three perspectives: a) The virtual examples are to generate virtual training data, whereas the proposed method is to generate pseudo test data; the scopes are different. b) The virtual examples are generated so that the learning machine will extract the invariances from the artificially enlarged training data [10], whereas our proposed method is to generate pseudo test data points so that the learning machine can combine them and enhance accuracy; the purposes are different. c) Virtual examples are designed specifically for SVM, whereas the proposed method is suitable for all kinds of classifiers; the scales are different.

Also Note that there are some fundamental differences between the MCS and the SMCS. In the MCS, we benefit from the fact that each classifier has a different perception of how a test sample should be classified. Because the decision boundary made by each classifier is different, there is diversity among decision boundaries drawn by different classifiers. Given that classifiers make different errors on different test samples, diversity can actually help improve classification accuracy. So, in the MCS, one of the core issues is to generate, select, and combine multiple classifiers, such that the combined decision boundary is better than any existing single boundary. In the SMCS, we not only try to find a better decision boundary, but one with the potential to be close to the oracle and not constrained by an existing classifier boundary and the number of classifiers. In designing the SMCS, we acknowledged the fact that a decision boundary drawn by a classifier might never be optimal; so, instead of refining several existing decision boundaries by combining them, we are trying to explore and make use of information in the neighborhood of a single decision boundary. In this way, we are looking for diversity that is already present in the neighborhood, rather than trying to benefit from diversity embedded in different classifiers. Consequently, diversity is extracted not from diverse decision boundaries, but from diverse pseudo test data points. The core issue is then to adequately generate, select, and combine multiple pseudo test data points for a test sample, rather than generating, selecting, and combining multiple classifiers.

We focus on two main questions in this paper:

1. Can we extract diversity from a dataset without training multiple classifiers?
2. Can multiple classification without multiple classifiers enhance accuracy?

2   Proposed Method

Given a training dataset $\mathbf{X}$, we first divide the training samples into N reference samples $\mathbf{X_r} = \{x_1, x_2, \cdots, x_N\}$, and M evaluation samples $\mathbf{\breve{X}_e} = \{\breve{x}_1, \breve{x}_2, \cdots, \breve{x}_M\}$, a single classifier $C_\mathbf{X}$ trained by all the available training samples $\mathbf{X}$, and a test data point $\tilde{x}_t$. The mechanism involves the creation of K pseudo test data points, $\mathbf{\hat{X}_t} = \hat{x}_{1,t}, \hat{x}_{2,t}, \cdots, \hat{x}_{K,t}$, which would result in K corresponding classification outputs $\hat{y}_{1,t}, \hat{y}_{2,t}, \cdots, \hat{y}_{K,t}$ after being classified by $C_\mathbf{X}$.

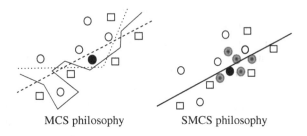

MCS philosophy   SMCS philosophy

Fig. 1. The multiple classification philosophies of the MCS and the SMCS compared. Empty circles represent reference samples of class A, and empty rectangles represent reference samples of class B. The dark circle represents a test sample to be classified, gray circles represent pseudo test data points generated by the SMCS, and lines represent the decision boundaries drawn by classifiers: a) the MCS relies on multiple classifiers to generate multiple decision boundaries, and so multiple classifications are generated to classify a test sample; b) the SMCS relies on a single classifier, and so there is only one decision boundary, but multiple pseudo test data points are generated to subsequently generate multiple classifications.

The purpose of this mechanism is to generate $\hat{\mathbf{X}}_t$, such that the combination of classification outputs on these K pseudo test data points $\hat{y}_t$ will be as close to the true class label y_t as possible. Note that:

$$\hat{y}_t = g(\hat{y}_{1,t}, \hat{y}_{2,t}, \cdots, \hat{y}_{K,t}) \tag{1}$$

where $g(\cdot)$ is the classification combination function, such as majority voting.

Here, the main problem is to design a stable mechanism that generates pseudo test data points that improve the overall classification result. We decompose this problem into two sub problems, expressed as the following two questions:

1. What is the function $f(\cdot)$ used to generate pseudo data points, given a test sample and several reference samples?
2. How do we decide which reference samples to use to generate pseudo data points, given a test sample?

We address these two sub problems in the sections below and describe them in more detail.

2.1 Define a Function to Generate Pseudo Test Data Points

There are a number of ways to solve the first component of the problem, which is to decide how to generate a pseudo test data point given a test sample and one or more reference samples.

A pseudo test data point can be generated as a combination of a test sample and a reference sample, or as a combination of a test sample and several reference samples. It can be generated in a deterministic way, or with some random factors. To gain some insight into the properties of the SMCS, we start with a simple and deterministic function to generate pseudo test data points. In our method, each pseudo test point is based on an original test sample and a single reference sample:

$$\hat{x}_{i,t} = f(x_i, \tilde{x}_t) \tag{2}$$

where $\hat{x}_{i,t}$ indicates a generated pseudo test data point, $\tilde{x}_t$ is the original test sample, and x_i is a reference sample.

For example, if each data point has L feature dimensions, then the feature l of the generated pseudo test data point $\hat{x}_{i,t}$ will simply be a weighted average of the same feature of the test sample $\tilde{x}_t$ and that of the reference sample x_i:

$$\hat{x}_{i,t,l} = \alpha x_{i,l} + (1 - \alpha)\tilde{x}_{t,l}, 1 \le l \le L, 0 \le \alpha \le 1 \tag{3}$$

where $\hat{x}_{i,t,l}$ indicates the value of the feature l of the generated pseudo test data point $\hat{x}_{i,t}$, $x_{i,l}$ indicates the value of the feature l of the reference sample x_i, and $\tilde{x}_{t,l}$ indicates the value of the feature l of the test sample $\tilde{x}_t$. Also note that α controls the noise and diversity present in pseudo test data points: the larger it is, the greater the diversity and the noise.

2.2   Select Reference Samples to Generate Pseudo Test Data Points

Not every generated pseudo data point will be adequate for classification. The fitness of a pseudo data point will largely depend on the "chemistry" between the test sample $\tilde{x}_t$ and the reference sample x_i.

In order to evaluate the fitness of each reference sample x_i for a test sample $\tilde{x}_t$ in an attempt to generate adequate pseudo data points, we propose a three-step scheme:

1. Identify valid [evaluation sample - reference sample] pairs
   Remember that we divide training samples into evaluation samples and reference samples. We will use these M evaluation samples to determine the fitness of a reference sample. Each evaluation sample will generate a pseudo data point using the reference sample, and then the pseudo data point will be classified. If the classification of this pseudo data point has the same label as the evaluation sample, then this [evaluation sample - reference sample] pair is regarded as valid; otherwise, it is regarded as invalid.
2. Assign weight to reference samples
   For a given test sample, we find the m nearest evaluation samples. Then, every reference sample is assigned a weight based on its validity with respect to these m evaluation samples, which is obtained as the sum of the m [evaluation sample - reference sample] pairs.
3. Select reference samples and generate pseudo test data points
   We set a threshold for the reference samples, and select only those with weights higher than that threshold for pseudo test data point generation.

Here we notice two things. First, each function can be manipulated independently and so is subject to optimization. This modular approach gives our proposed method a great deal of flexibility, and it can be adapted to various pattern recognition problems. Second, the step of identifying valid [evaluation sample - reference sample] pairs needs only to be performed once for all the test samples, whereas the other two steps need to be carried out for each individual test sample. Given that the first step is more time consuming, and the second and the third steps are fairly straightforward and less time consuming, the overall process can be implemented in the real world without incurring enormous cost. We provide more details below.

Identify valid [evaluation sample - reference sample] pairs. The first step is to evaluate the fitness of each reference sample by using several evaluation samples from a evaluation dataset, $\check{\mathbf{X}}_{\mathbf{e}} = \{\check{x}_1, \check{x}_2, \cdots, \check{x}_M\}$. For a reference sample x_i to be evaluated, an evaluation sample $\check{x}_k$ generates a pseudo data point $\hat{x}_{i,k}$ using this reference sample. Then, the generated pseudo data point is classified by a classifier:

$$\hat{x}_{i,k} \mapsto \hat{y}_{i,k} \tag{4}$$

Since we already know the class label y_k of each evaluation sample $\check{x}_k$, we can determine whether or not the classification of this pseudo data point is correct, meaning that it has the same class label as that of the evaluation sample. We repeat the same process between all the reference and evaluation samples, and then define a validity measure $v_{i,k}$ for each [evaluation sample $\check{x}_k$ - training data point x_i] pair, $1 \leq k \leq M$, $1 \leq i \leq N$, and set the validity to 1 for correct classification and to 0 for incorrect classification:

$$v_{i,k} = 1, \text{if } \hat{y}_{i,k} = y_k \tag{5}$$

$$v_{i,k} = 0, \text{otherwise} \tag{6}$$

Figure 2 shows the process of identifying valid [evaluation sample - reference sample] pairs. The validity measures are then used to evaluate the fitness of each training data point x_i, as described in the next section.

Assign Weight to Reference Samples. The validity measure $v_{i,k}$ for each [evaluation sample $\check{x}_k$ - reference sample x_i] pair tells us whether or not a reference sample x_i is fit to generate a pseudo test point with an evaluation sample $\check{x}_k$, but it does not tell us whether or not a reference sample x_i is fit to generate a pseudo test point with a test sample $\tilde{x}_t$.

In order to decide whether or not we should use a reference sample x_i to generate a pseudo test point with a test sample $\tilde{x}_t$, first we try to find the m nearest evaluation samples from the test sample $\tilde{x}_t$. The idea behind this action is that these evaluation samples can be seen as proxies of the test sample $\tilde{x}_t$. If they all qualify as correct classifications with the use of the reference sample x_i to generate pseudo test points, then the test sample $\tilde{x}_t$ can use this reference sample x_i to generate pseudo test points as well.

For each reference sample:

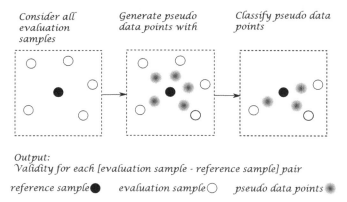

Output:
Validity for each [evaluation sample - reference sample] pair

reference sample●   evaluation sample○   pseudo data points ✳

Fig. 2. Each evaluation sample generates a pseudo data point using a reference sample. The classification result of this pseudo data point provides an indication of the fitness of the [evaluation sample - test sample] pair. Solid circles represent reference samples, white circles represent evaluation samples, and gray circles represent generated pseudo data points.

Given a test sample $\tilde{x}_t$, let us consider the nearest m evaluation samples to be trustworthy for this test sample, noting that $m \ll M$. We then use these m evaluation data points to evaluate the fitness of reference samples for the test sample $\tilde{x}_t$. The weight of a reference sample x_i is assigned as follows:

$$w_i = \sum_{k=1}^{m} \delta_{i,k} v_{i,k} \tag{7}$$

where $v_{i,k}$ is a validity measure $v_{i,k}$ for the [evaluation sample $\check{x}_k$ - reference sample x_i] pair, and $\delta_{i,k}$ is a weighting adjustment based on distance or other factors.

Figure 3 demonstrates a general scheme for assigning weight to reference samples through the aggregation of multiple validity measures between a reference sample and evaluation samples. In this paper, we define the weighting adjustment $\delta_{i,k}$ as:

$$\delta_{i,k} = \frac{d(\check{x}_k, \tilde{x}_t) + d(\check{x}_k, x_i)}{d(\tilde{x}_t, x_i)} \tag{8}$$

where $d(\cdot)$ indicates a Euclidean distance function, $\check{x}_k$ is an evaluation sample, x_i is a reference sample, and $\tilde{x}_t$ is a test sample.

So, $d(\check{x}_k, \tilde{x}_t)$ is the distance between the evaluation sample $\check{x}_k$ and the test sample $\tilde{x}_t$, $d(\check{x}_k, x_i)$ is the distance between the evaluation sample $\check{x}_k$ and the reference sample x_i, and $d(\tilde{x}_t, x_i)$ is the distance between the reference sample x_i and the test sample $\tilde{x}_t$.

In our weighting adjustment $\delta_{i,k}$, the weight w_i increases with $\frac{d(\check{x}_k, x_i)}{d(\tilde{x}_t, x_i)}$, because knowing that the reference sample produces correct pseudo data points for an evaluation sample, the distance between the evaluation sample $\check{x}_k$ and the reference sample x_i signals the robustness of the reference sample; whereas the distance between the

For each test sample:

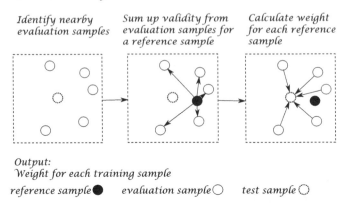

Identify nearby evaluation samples   *Sum up validity from evaluation samples for a reference sample*   *Calculate weight for each reference sample*

Output:
Weight for each training sample

reference sample●   evaluation sample○   test sample◌

Fig. 3. For each test sample, each reference sample is evaluated indirectly by aggregating the validity of [evaluation sample - reference sample] pairs from the nearby evaluation samples. Now, the weighting of reference samples can be adjusted by distances. Solid circles represent reference samples, white circles with a dotted contour represent test samples, and white circles with a solid contour represents evaluation samples.

test sample $\tilde{x}_t$ and the reference sample x_i scales down this robustness measure. The weight w_i also increases with $\frac{d(\check{x}_k, x_i)}{d(\tilde{x}_t, x_i)}$, because the ratio of the distance $d(\check{x}_k, x_i)$ to the distance $d(\tilde{x}_t, x_i)$ represents the validity to approximate the test sample $\tilde{x}_t$ using the evaluation sample $\check{x}_k$.

Note that other weighting mechanisms may be suitable as well. This is simply the one that we chose to implement.

Select Reference Samples and Generate Pseudo Test Data Points. Given a test sample $\tilde{x}_t$, once all reference samples are evaluated using the nearest m evaluation samples from that test sample, we can proceed to select adequate reference samples for the purpose of pseudo test data point generation.

Again, we can only evaluate the nearest n reference samples for the test sample. We also define a threshold θ. Therefore, the selection criterion for reference samples is:

$$\text{if } w_i \geq \theta \; s_i = 1 \tag{9}$$

$$\text{else } \; s_i = 0 \tag{10}$$

where s_i is the selection decision on reference sample x_i. The threshold θ is defined as:

$$\theta = \rho \max\{w_i\}, 0 < \rho \leq 1 \tag{11}$$

Figure 4 shows the process of reference sample selection that we use to generate pseudo data points for a test data point. Once multiple diverse pseudo test data points are generated, we can use them to feed a classifier to produce multiple classifications for a single test sample. However, these multiple classifications need to be combined in order to produce a final class label for the test sample. We describe the process below.

For each test sample:

Identify nearby   *Use weights to*   *Generate pseudo*
reference samples   *select reference samples*   *test data points*

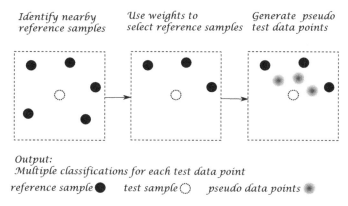

Output:
Multiple classifications for each test data point
reference sample ●   *test sample* ○   *pseudo data points* ✸

Fig. 4. Adequate reference samples are then selected by the weights that each test sample assigns to them. The selected reference samples then generate pseudo test data points for the original test data point. Solid circles represent reference samples, white circles with a dotted contour represent a test sample, and gray circles represent generated pseudo test data points.

2.3 Combine Multiple Classification Outputs

Supposing that I reference samples $x_i, 1 \leq i \leq I$ are selected for a test sample $\tilde{x}_t$, corresponding pseudo test data points can be generated:

$$\hat{x}_{i,t} = f(x_i, \tilde{x}_t) \tag{12}$$

By applying a single classifier $C_{\mathbf{X}}$ that is trained with all the available reference samples $\mathbf{X}$, multiple classification outputs can be obtained:

$$\hat{x}_{i,t} \mapsto \hat{y}_{i,t} \tag{13}$$

Once we have multiple classification outputs, a fusion function g is implemented to combine them:

$$\hat{y}_t = g(\hat{y}_{1,t}, \hat{y}_{2,t}, \cdots, \hat{y}_{I,t}) \tag{14}$$

As a result, we obtain the final class label output for the test sample concerned. Below, we provide the pseudo code for our proposed method to better illustrate the methodology.

3 Experiments

In order to verify the validity of the proposed SMCS, to understand effects of neighborhood sizes m and n and reference sample selection, and to measure the performance of the SMCS, we carried out a number of experiments on different datasets extracted from the UCI Machine Learning Repertoire. The experiments were conducted in MATLAB using PRTools [3]. The training datasets are further split into a reference dataset,

Table 1. The datasets used in our experiments

Datasets	Number of Classes	Dimension	Reference Dataset Size	Evaluation Dataset Size	Test Dataset Size	Dataset Size per Class
breast-tissue	6	9	35	36	35	6
bupa-liver	2	6	115	115	115	58
glass	6	9	72	71	71	12
iris	3	4	50	50	50	17
parkinsons	2	22	65	65	65	32
vowel	11	11	330	330	330	30
wdbc	2	30	190	189	190	95
yeast	10	8	494	495	495	49

Table 2. The error rates of a Single Classifier-based Multiple Classification Scheme with parameters $n = m = 9 \sim 21$ and $\rho = 0.8 \sim 0.95$. We show the error rates of a single classifier (denoted "single classifier") as the baseline, and those of an average SMCS without any parameter selection (denoted "average smcs").

Dataset	Method	LDC	QDC	KNN	PW	MLP	Tree
breast-tissue	single classifier	31.43%	40.00%	48.57%	51.43%	54.29%	40.00%
breast-tissue	average smcs	24.29%	29.49%	48.57%	48.57%	43.67%	32.04%
breast-tissue	error change	−22.73%	−26.28%	0.00%	−5.56%	−19.55%	−19.90%
bupa-liver	single classifier	33.04%	41.74%	32.17%	46.09%	33.91%	40.00%
bupa-liver	average smcs	33.51%	33.66%	32.02%	45.22%	30.99%	31.43%
bupa-liver	error change	1.41%	−19.35%	−0.48%	−1.89%	−8.61%	−21.43%
glass	single classifier	30.99%	36.62%	33.80%	40.85%	61.97%	25.35%
glass	average smcs	22.48%	35.97%	31.09%	40.85%	61.97%	21.83%
glass	error change	−27.44%	−1.79%	−8.04%	0.00%	0.00%	−13.89%
iris	single classifier	8.00%	4.00%	10.00%	8.00%	4.00%	24.00%
iris	average smcs	6.07%	4.43%	5.07%	6.00%	4.50%	15.43%
iris	error change	−24.11%	10.71%	−49.29%	−25.00%	12.50%	−35.71%
parkinsons	single classifier	15.38%	15.38%	20.00%	16.92%	12.31%	13.85%
parkinsons	average smcs	13.08%	13.52%	20.00%	16.92%	11.37%	10.93%
parkinsons	error change	−15.00%	−12.14%	0.00%	0.00%	−7.59%	−21.03%
vowel	single classifier	39.09%	11.21%	29.09%	3.03%	84.24%	30.61%
vowel	average smcs	33.17%	7.90%	24.10%	3.16%	83.64%	19.32%
vowel	error change	−15.14%	−29.54%	−17.15%	4.29%	−0.72%	−36.88%
wdbc	single classifier	8.42%	6.84%	6.32%	7.37%	3.68%	10.00%
wdbc	average smcs	5.62%	5.30%	6.05%	6.11%	2.18%	5.11%
wdbc	error change	−33.26%	−22.53%	−4.17%	−17.09%	−40.82%	−48.87%
yeast	single classifier	41.01%	100.00%	42.42%	42.22%	48.89%	52.53%
yeast	average smcs	37.86%	100.00%	38.97%	38.89%	47.58%	42.76%
yeast	error change	−7.67%	0.00%	−8.15%	−7.89%	−2.67%	−18.60%

to be used for pseudo data point generation, and an evaluation dataset, to be used for reference sample evaluation purposes in the SMCS. For pseudo data point generation, we implemented a weighted combination of a reference sample and a test sample, as described in Eq. (16), with equal weights for both, so $\alpha = \frac{1}{2}$. We also tested different ranges of parameters for m, n, and ρ, in an effort to gain more insight: the value of m was set equal to n, and we tested $m = n = 9$, $m = n = 11$, $\cdots$, and $m = n = 21$; for the threshold setting ρ, we tested $\rho = 0.95$, $\rho = 0.9$, $\rho = 0.85$, and $\rho = 0.80$.

Our experimental results suggest that the proposed SMCS works to some extent, with an average improvement of 16.31%. The injected diversity seems to enhance the accuracy of the recognition rates in most cases, and generally the noise that is inherent in diversity does not degrade the classification results. Table 2 provides a summary of SMCS error rates on various dataset-classifier combinations. We note that the improvement achieved with the SMCS also depends on the classification methods of the trained classifiers.

4 Conclusion

The experimental results confirm the validity of the proposed SMCS as an applicable scheme for an MCS. This is especially true when we encounter the curse of dimensionality, and can only train weak classifiers. The parameters m, n and ρ also have an impact on the number of generated pseudo points. The correlation between m, n and the number of pseudo points is 0.0432, and that between ρ and the number of pseudo points is -0.2428. Hence, we might attempt to conclude that the smaller the ρ, the larger the number of generated pseudo points. Nevertheless, more experiments may be needed to have a better understanding on the effects of m, n and ρ.

To summarize, there are several critical aspects to the potential impact of the proposed SMCS:

1. Dynamic Decision Boundary
   Unlike the MCS, which attempts to combine multiple decision boundaries from multiple classifiers in order to achieve an optimal decision boundary, the SMCS operates under the assumption that a static optimal boundary is difficult or impossible to draw by combining multiple boundaries, or may not even exist. Instead, it tends to make dynamic decisions given a static decision boundary by generating pseudo data points for multiple classifications, and therefore shifts the complexity of decision boundary optimization to pseudo data point generation optimization.
2. Compatibility with the MCS
   Selection of a suitable classification scheme does not have to be an either/or proposition, as it is feasible to apply both the MCS and the SMCS on a dataset, where each classifier trained with an MCS can further generate multiple pseudo test data points for each test sample to be classified. In this case, we actually generate a dual multiple classification system, which may further improve the performance of traditional MCS or the proposed SMCS.
3. Flexibility in Pseudo Data Points Generation
   Unlike traditional MCS, where the generation of multiple classifiers is generally quite straightforward, the proposed SMCS tends to be more flexible and it can have

almost infinite variations; such as on the choice of the pseudo data point generation function, on the evaluation of reference samples, and on the adjustment of neighborhood size, etc. Consequently, the best SMCS scheme may be different for each dataset, and this indicates more opportunity for performance enhancement.

4. Reduced Cost in Classifier Training Time
   Compared with traditional MCS, the SMCS requires the training of only one classifier. Suppose, for example, that an MCS needs to train K classifiers and requires training time T_K, the proposed SMCS would require training time T_1, and $T_1 \approx \frac{T_K}{K}$. This represents a speeding up by a factor of about K for classifier training. When K becomes large, such as $100 \sim 4000$ [11], the gain may be substantial.

5. Reduced Cost in Ensemble Selection Time
   Classifier training represents only a part of the cost of ensemble construction, because subsequent ensemble selection must be conducted to select the best subset of classifiers [7, 9]. Because the SMCS uses only one classifier, there is no need for classifier subset selection. In fact, classifier subset selection is replaced by reference sample selection in the SMCS. This operation in the proposed SMCS is quite straightforward, requiring only nearest neighbor identification and a sum operation. It is therefore less time consuming than traditional classifier selection.

Nevertheless, the classification problem is not solved without cost. Although the SMCS reduces classifier training cost considerably, it actually increases the classification cost on each test sample. In other words, the SMCS shifts the cost of classifier training to the cost of pseudo test data point classification. As a result, if the cost of classification is critical for a system performance and the training cost is negligible, the SMCS may not be suitable for the system. On the contrary, if the training cost is prohibitively high and the classification cost is less substantial, then the SMCS should be considered as a potential solution.

References

1. Bottou, L., Vapnik, V.N.: Local Learning Algorithms. Neural Computation 4(6), 888–900 (1992)
2. Decoste, D., Schoelkopf, B.: Training invariant support vector machines. Machine Learning 46(1), 161–190 (2002)
3. Duin, R.P., Juszczak, P., Paclik, P., Pekalska, E., de Ridder, D., Tax, D., Verzakov, S.: PRTools4.1, a matlab toolbox for pattern recognition (2007), http://prtools.org
4. Ko, A.H.R., Sabourin, R., de Souza Britto Jr., A.: Combining diversity and classification accuracy for ensemble selection in random subspaces. In: Proceedings of the International Joint Conference on Neural Networks, pp. 2144–2151 (2006)
5. Ko, A.H.R., Sabourin, R., de Souza Britto Jr., A.: Compound Diversity Functions for Ensemble Selection. International Journal of Pattern Recognition and Artificial Intelligence 23(4), 659–686 (2009)
6. Ko, A.H.R., Cavalin, P., Sabourin, R., de Souza Britto Jr., A.: Leave-One-Out-Training and Leave-One-Out-Testing Hidden Markov Models for a Handwritten Numeral Recognizer: the Implication of a Single Classifier and Multiple Classifications. IEEE Transactions on Pattern Analysis and Machine Intelligence 31(12), 2168–2178 (2009)
7. Ko, A.H.R., Sabourin, R., de Souza Britto Jr., A.: From dynamic classifier selection to dynamic ensemble selection. Pattern Recognition 41(5), 1718–1731 (2008)

8. Kuncheva, L.I., Skurichina, M., Duin, R.P.W.: An Experimental Study on Diversity for Bagging and Boosting with Linear Classifiers. International Journal of Information Fusion 3(2), 245–258 (2002)
9. Kuncheva, L.I., Rodriguez, J.J.: Classifier ensembles with a random linear oracle. IEEE Transactions on Knowledge and Data Engineering 19(4), 500–508 (2007)
10. Poggio, T., Vetter, T.: Recognition and structure from one 2D model view: Observations on prototypes, object classes and symmetries. A.I. Memo No. 1347, Artificial Intelligence Laboratory, Massachusetts Institute of Technology (1992)
11. Polikar, R., DePasquale, J., Mohammed, H.S., Brown, G., Kuncheva, L.I.: Learn++ MF: A random subspace approach for the missing feature problem. Pattern Recognition (43), 3817–3832 (2010)

Selective Ensemble of Classifier Chains

Nan Li[1,2] and Zhi-Hua Zhou[1]

[1] National Key Laboratory for Novel Software Technology
Nanjing University, Nanjing 210023, China
[2] School of Mathematical Sciences, Soochow University, Suzhou 215006, China
{lin,zhouzh}@lamda.nju.edu.cn

Abstract. In multi-label learning, the relationship among labels is well accepted to be important, and various methods have been proposed to exploit label relationships. Amongst them, *ensemble of classifier chains* (ECC) which builds multiple chaining classifiers by random label orders has drawn much attention. However, the ensembles generated by ECC are often unnecessarily large, leading to extra high computational and storage cost. To tackle this issue, in this paper, we propose *selective ensemble of classifier chains* (SECC) which tries to select a subset of classifier chains to composite the ensemble whilst keeping or improving the performance. More precisely, we focus on the performance measure F1-score, and formulate this problem as a convex optimization problem which can be efficiently solved by the stochastic gradient descend method. Experiments show that, compared with ECC, SECC is able to obtain much smaller ensembles while achieving better or at least comparable performance.

Keywords: multi-label, classifier chains, selective ensemble.

1 Introduction

In traditional supervised learning, one instance is associated with *one* concept; but in many real-world applications, an object is naturally associated with *multiple* concepts simultaneously. For examples, a document may belong to multiple topics [14], an image or a music can be annotated with more than one words [1,24]. Obviously, one label per instance is not capable of dealing with such tasks, and *multi-label learning* which associates each instance with multiple labels simultaneously has become an active research topic during the past few years [17,6,8,28,16,2].

A straightforward approach to multi-label learning is the binary relevance method, which decomposes the task into a number of binary classification problems, each for one label [1]. However, such a method is usually not able to achieve good performance, since it neglects the relationships between class labels, which have been widely accepted to be important [4,27]. In the literature, many methods have been proposed to exploit label relationships. Read *et al.* [16] introduced *classifier chain* (CC) to incorporate class relationships, whilst trying to keep the computational efficiency of the binary relevance method. Specifically, like the binary relevance method, each CC is also consist of multiple binary classifiers each for one label, but these binary classifiers are linked in a chain such that each incorporates the class predicted by the previous classifier as additional attributes. Obviously, the quality of CC is dependent on the label

Z.-H. Zhou, F. Roli, and J. Kittler (Eds.): MCS 2013, LNCS 7872, pp. 146–156, 2013.

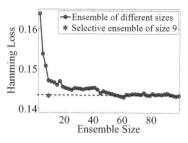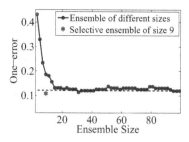

Fig. 1. The performance of ECC with different ensemble sizes on *CAL500*, where hamming loss and one-error are considered. For the selective ensemble, the nine CCs are selected based on the performance on test data from the first twenty CCs via exhaustive search.

orders in the chain. Rather than selecting one good label order, ECC [16] constructs multiple CCs by using different *random* label orders, and it combines them for prediction by a voting scheme. This method has drawn much attention [4,26,11].

Generally speaking, by combining more diverse CCs, the performance of ECC tends to improve and converge. For example, as illustrated in Fig.1, both hamming loss and one-error decrease when the ensemble size increases, and they converge after ensemble size reaches about fifty and thirty, respectively. However, one problem is that the ensembles constructed by ECC tend to be unnecessarily large, which requires large amount of memory storage and also decreases the response time for prediction. As an example, in Fig.1, we can see that similar even better performance can be achieved by using only nine out of the first twenty classifiers, this indicates that the original ECC is quite redundant and it can be largely pruned.

In this paper, we propose the *selective ensemble of classifier chains* (SECC) method, which tries to reduce the ensemble size of ECC whilst keeping or improving the performance. Specifically, by focusing on the frequently-used performance measure F1-score, we try to optimize an upper bound of the empirical risk, and formulate the problem into a convex optimization problem with ℓ_1-norm regularization, also an efficient stochastic optimization method is presented to solve the problem. Experiments on image and music annotation tasks show that SECC is able to obtain much smaller ensembles than ECC, while its performance are better or at least comparable to ECC.

The remainder of the paper is organized as follows. Section 2 briefly reviews some related work. Section 3 presents our proposed SECC method. Section 4 reports the experiment results, which is followed by the conclusion in Section 5.

2   Related Work

In the past few years, many multi-label learning methods have been proposed in the literature, which generally fall into two groups. The first group include algorithm adaptation based methods, which try to modify an algorithm to make multi-label predictions; representatives include BoostTexter [17], ML-kNN [28], HMC tree [25], etc. The other group of methods try to perform *problem transformation*, where a multi-label problem is transformed to one or more other problems, which can be solved by using existing methods. For example, it was transformed into binary classification problems [1,4,16], multi-class classification problems [23,15], and ranking problems [6,7]. In practice, problem transformation based methods are attractive due to its scalability and flexibility, that is, any off-the-shelf methods can be used to suit the requirements.

The *binary relevance* (BR) method [1] is a common problem transformation method, and it transforms a multi-label problem into multiple binary problems, each is to predict the relevance of one label. It is obviously that BR treats each label independently, and does not model relationships existing between class labels. Due to this, its performance is not satisfying, and many methods have been proposed to improve it by exploiting label relationships. Read *et al.* [16] proposed the CC model which is an important improvement over BR. By building chaining classifiers, CC overcomes the disadvantages of BR and achieves higher performance, while retaining the computational complexity. In practice, the performance of CC heavily depends on the label order in the chain; thus ECC is proposed to combine multiple CCs which are based on *random* labels orders. However, a potential problem of ECC is that it may generate unnecessarily large ensembles, reducing its efficiency.

In traditional supervised learning, selective ensemble (*a.k.a.* ensemble pruning or ensemble selection) [30, Chapter 6] is an active research topic, and many technologies has been used to build selective ensembles, such as genetic algorithm [31], semi-definite programming [29], clustering [9], sparse optimization [13]. Obviously, ECC is an ensemble of multi-label classifiers, making the current work essentially different.

3   The SECC Method

In this section, we present our proposed SECC method. Before describing the problem formulation, we begin with some notations and preliminaries. Then, we transform the problem to a convex optimization problem, and give a stochastic optimization solution.

3.1   Preliminaries and Problem Formulation

Let $\mathcal{X}$ be the instance space and $\mathcal{L}$ be a set of l labels. In multi-label learning, each instance $\mathbf{x}_i \in \mathcal{X}$ is associated with multiple labels in $\mathcal{L}$, which is represented as an l-dimensional binary vector $\mathbf{y}_i$ with element 1 indicating $\mathbf{x}_i$ is associated with the k-th label and -1 otherwise. Given a set of training examples $S = \{(\mathbf{x}_i, \mathbf{y}_i)\}_{i=1}^{m}$, the task of multi-label learning is to learn a multi-label classifier $h : \mathcal{X} \mapsto \mathcal{Y}$, where $\mathcal{Y} \subseteq \{-1, 1\}^l$ is the set of feasible label vectors, such that it can be used predict labels for unseen instances. In practice, it is often to learn a vector-valued function $f : \mathcal{X} \mapsto \mathbb{R}^l$ which determines the label of $\mathbf{x}$ as

$$\hat{\mathbf{y}} = \text{argmax}_{\mathbf{y}' \in \mathcal{Y}} \ \mathbf{y}'^{\top} f(\mathbf{x}) . \tag{1}$$

It is clear that how the argmax is computed depends on the structure of $\mathcal{Y}$; for example, if $\mathcal{Y} = \{-1, 1\}^l$, $\hat{\mathbf{y}}$ is simply $\text{sign}[f(\mathbf{x})]$.

Instead of learning one multi-label classifier, ECC constructs a group of CCs based on random label orders, and combines them via voting for prediction. Without loss of generality, we denote the group of CCs as $\{h^{(t)} : \mathcal{X} \mapsto \mathcal{Y}\}_{t=1}^k$, and ECC combines them to produce the vector-valued function $f : \mathcal{X} \mapsto \mathbb{R}^l$ as

$$f(\mathbf{x}; \mathbf{w}) = \sum_{t=1}^k w_t h^{(t)}(\mathbf{x}) , \qquad (2)$$

where $\mathbf{w} = [w_1, \ldots, w_k]^\top$ is the weighting vector; for example, they are simply set to $1/k$ for the simple voting.

As discussed above, one problem with ECC is that the CCs generated based on random label orders are redundant, which results in a large ensemble size, $i.e.$, large value of k. It can be found that in (2) the CC classifier $h^{(k)}$ will be excluded from the ensemble if w_k is zero, and the size of ensemble is exactly $\|\mathbf{w}\|_0$. Based on this, our the task of reducing the size of ECC becomes to finding a weighting vector $\mathbf{w}$, such that the performance of corresponding ensemble is better than or comparable to ECC, while the ensemble size $\|\mathbf{w}\|_0$ is small.

3.2   A Convex Formulation

Here, we find the weighting vector $\mathbf{w}$ based on the principle of empirical risk minimization. That is, given the training example $S = \{(\mathbf{x}_i, \mathbf{y}_i)\}_{i=1}^m$, we consider to solve the vector $\mathbf{w}$ by solving the following optimization problem.

$$\min_{\mathbf{w} \in \mathcal{W}} \frac{1}{m} \sum_{i=1}^m \Delta(\mathbf{y}_i, f(\mathbf{x}_i; \mathbf{w})) \quad \text{s.t. } \|\mathbf{w}\|_0 \leq b , \qquad (3)$$

where $\mathcal{W}$ is the feasible space of $\mathbf{w}$, $0 < b \leq K$ is the budget of ensemble size, and $\Delta(\mathbf{y}_i, f(\mathbf{x}_i; \mathbf{w}))$ is the empirical risk function measuring the loss of determining $\mathbf{x}_i$'s label vector by $f(\mathbf{x}_i; \mathbf{w})$ while its true label is $\mathbf{y}_i$. More specifically, in current work, we consider the frequently-used performance measure F1-score, and define the loss function Δ as

$$\Delta(\mathbf{y}, f(\mathbf{x}; \mathbf{w})) = 1 - F1(\mathbf{y}, \hat{\mathbf{y}}) , \qquad (4)$$

where $\hat{\mathbf{y}}$ is the label vector determined by $f(\mathbf{x}; \mathbf{w})$, and $F1(\cdot, \cdot)$ is

$$F1(\mathbf{y}, \hat{\mathbf{y}}) = \frac{2tp}{l + tp - tn} ,$$

where $tp = \sum_{i=1} \mathbb{I}(y_i = 1 \wedge \hat{y}_i = 1)$ the number of groundtruth labels in the predicted relevant labels, $tn = \sum_{i=1} \mathbb{I}(y_i = -1 \wedge \hat{y}_i = -1)$ counts how many predicted non-relevant labels are truly non-relevant, and l is the number of possible class labels.

Convex Relaxation. It is easy to see that the problem is not easy to solve, mainly because the risk function Δ is non-decomposable over labels, non-convex and non-smooth. Inspired by some works on directly optimizing performance measures [10,12] which is generally based on structured prediction [19,21], instead of directly minimize the empirical risk as (3), in SECC, we consider to optimize one of its convex upper bounds, which is based on the following proposition.

Proposition 1. *Given a multi-label classifier $f : \mathcal{X} \mapsto \mathbb{R}^L$ in (2) and the empirical risk function Δ in (4), define the loss function $\ell(\mathbf{y}, f(\mathbf{x}))$ as*

$$\ell(\mathbf{y}, f(\mathbf{x})) = \max_{\mathbf{y}' \in \mathcal{Y}} \left[(\mathbf{y}' - \mathbf{y})^\top f(\mathbf{x}) + \Delta(\mathbf{y}, \mathbf{y}') \right] , \tag{5}$$

where $\mathcal{Y}$ is the set of feasible label vectors, then the loss function $\ell(\mathbf{y}, f(\mathbf{x}))$ provides a convex upper bound over $\Delta(\mathbf{y}, f(\mathbf{x}))$.

Proof. It is obvious the loss function $\ell(\mathbf{y}, f(\mathbf{x}))$ is convex in f, because it is pointwise maximum of linear functions. Let $\hat{\mathbf{y}} = \text{sign}[f(\mathbf{x})]$, *i.e.*, the prediction determined by $f(\mathbf{x})$, it is easy to find that $\hat{\mathbf{y}}$ is the maximizer of $\mathbf{y}^\top f(\mathbf{x})$. Then, we can get

$$\ell(\mathbf{y}, f(\mathbf{x})) \geq \hat{\mathbf{y}}^\top f(\mathbf{x}) - \mathbf{y}^\top f(\mathbf{x}) + \Delta(\mathbf{y}, \hat{\mathbf{y}}) \geq \Delta(\mathbf{y}, \hat{\mathbf{y}}) ,$$

thus $\ell(\mathbf{y}, f(\mathbf{x}))$ is an upper bound over $\Delta(\mathbf{y}, \hat{\mathbf{y}})$, which completes the proof.   □

From the optimization problem (3), by replacing the risk function $\Delta(\cdot, \cdot)$ with its upper bound $\ell(\cdot, \cdot)$, and $\|\mathbf{w}\|_0$ with its continuous relaxation $\|\mathbf{w}\|_1$, we can obtain the optimization problem of SECC as

$$\min_{\mathbf{w}} \sum_{i=1}^{m} \ell(\mathbf{y}_i, H_i \mathbf{w}) + \lambda \|\mathbf{w}\|_1 , \tag{6}$$

where λ is the regularization parameter and $H_i \in \mathbb{R}^{l \times k}$ is the matrix collecting the predictions of $\{h^{(t)}\}_{t=1}^{k}$ on instance $\mathbf{x}_i$, i.e.,

$$H_i = [h^{(1)}(\mathbf{x}_i), \cdots, h^{(k)}(\mathbf{x}_i)] .$$

Obviously, the problem (6) is an ℓ_1-regularized convex optimization problem, and we solve it via stochastic optimization subsequently.

3.3   Stochastic Optimization

To solve the ℓ_1-regularized convex optimization problem (6), we employ the state-of-the-art stochastic optimization algorithm presented in [18], and the key is how to compute the subgradient of the loss function $\ell(\mathbf{y}_i, H_i \mathbf{w})$ with respect to $\mathbf{w}$. The following proposition provides the method to compute the subgradient of $\ell(\mathbf{y}_i, H_i \mathbf{w})$.

Proposition 2. *Given an example $(\mathbf{x}_i, \mathbf{y}_i)$, a set of multi-label classifiers $\{h^{(t)}\}_{t=1}^{k}$ and a weighing vector $\mathbf{w}_0 \in \mathbb{R}^K$, denote $\mathbf{p}_i = H_i \mathbf{w}_0$ be the ensemble's prediction on example $(\mathbf{x}_i, \mathbf{y}_i)$, then the vector*

$$\mathbf{g} = (\tilde{\mathbf{y}} - \mathbf{y}_i)^\top H_i ,$$

is a subgradient of $\ell(\mathbf{y}_i, H_i \mathbf{w})$ at $\mathbf{w}_0$, where $\tilde{\mathbf{y}}$ is the solution of the argmax problem

$$\tilde{\mathbf{y}} = \text{argmax}_{\mathbf{y}' \in \mathcal{Y}} \left[\mathbf{y}'^\top \mathbf{p}_i + \Delta(\mathbf{y}_i, \mathbf{y}') \right] . \tag{7}$$

Proof. It is obvious that the function $\ell(\mathbf{y}_i, H_i \mathbf{w})$ is a pointwise maximum of linear functions in $\mathbf{w}$, then it is straightforward to obtain its subgradient if the maximizer of (5) at $\mathbf{w}_0$ can be obtained. Actually, the argmax problem (7) solves the maximizer, which completes the proof.   □

Algorithm 1. Stochastic optimization algorithm for SECC

Input: $S = \{(\mathbf{x}_i, \mathbf{y}_i)\}_{i=1}^m$, CC classifiers $\{h^{(t)}\}_{t=1}^k$, regularization
parameter λ, step size η

Output: weighting vector $\mathbf{w}$

1: let $\mathbf{w} = 0$, $\varrho = 0$ and $p = 2 \ln k$
2: **repeat**
3:     select $(\mathbf{x}_i, \mathbf{y}_i)$ uniformly at random from S
4:     let $\mathbf{H}_i = [h^{(1)}(\mathbf{x}_i), \cdots, h^{(k)}(\mathbf{x}_i)]$ and $\mathbf{p}_i = \mathbf{H}_i \mathbf{w}$
5:     solve the argmax, $i.e.$, $\tilde{\mathbf{y}} \leftarrow \mathrm{argmax}_{\mathbf{y}' \in \mathcal{Y}} \left[\mathbf{y}'^\top \mathbf{p}_i + \Delta(\mathbf{y}_i, \mathbf{y}') \right]$
6:     compute the sub-gradient, $i.e.$, $\mathbf{g} = (\tilde{\mathbf{y}} - \mathbf{y}_i)^\top \mathbf{H}_i$
7:     let $\tilde{\varrho} = \varrho - \eta \mathbf{g}$
8:     $\forall t$, let $\varrho_t = \mathrm{sign}(\tilde{\varrho}_t) \max(0, |\tilde{\varrho}_t| - \eta \lambda)$
9:     $\forall t$, let $w_t = \mathrm{sign}(\varrho_t) |\varrho_t|^{p-1} / \|\varrho\|_p^{p-2}$
10: **until** convergence

Algorithm. Based on above proposition, we can present the stochastic optimization method for solving the optimization problem (6), whose pseudocode is summarized in Algorithm 1. Specifically, this algorithm is a stochastic subgradient descend method. At each iteration, it first samples an example $(\mathbf{x}_i, \mathbf{y}_i)$ uniformly at random from data S, and then compute the subgradient of $\ell(\mathbf{y}_i, \mathbf{H}_i \mathbf{w})$ (lines 4-6). Since the example $(\mathbf{x}_i, \mathbf{y}_i)$ is chosen at random, the vector $\mathbf{g}$ is an unbiased estimate of the gradient of the empirical risk $\sum_{i=1}^m \ell(\mathbf{y}_i, \mathbf{H}_i \mathbf{w})$. Next, the dual vector ϱ is updated with step size η (line 7) so that the empirical risk is decreased; and also it is truncated to decrease the regularizer $\lambda \|\mathbf{w}\|_1$ (line 8). Finally, the updates of ϱ are translated to the variable $\mathbf{w}$ via a link function in line 9. This procedure iterates until convergence. It is worth noting that, at each iteration of Algorithm 1, we even do not need to compute the predictions on all examples, and all the operations are very efficient as long as the argmax problem (7) can be efficiently solved.

Solving the Argmax. To make Algorithm 1 practical, it is obvious that the argmax problem (7) needs to be solved efficiently. Noting that the feasible space of $\tilde{\mathbf{y}}$ is of exponential size ($i.e.$, 2^l), and it is infeasible to solve it by exhaustive search. Fortunately, this problem has been studied in [10], and an efficient solution has been proposed. Roughly speaking, the solution is based on fact that F1-score can be computed from the contingency table ($i.e.$, four numbers: true positive, true negative, false positive, false negative) which take at most $O(l^2)$ different values. Algorithm 2 summarizes the pseudocode, and it is easy to see that its computational complexity is $O(l^2)$, where l is the number of class labels.

Convergence & Complexity. Based on Theorem 3 in [18], we can find that the number of iterations of Algorithm 1 to achieve ϵ-accuracy is bounded by $O(\log k / \epsilon^2)$, where k is the number of CC classifiers. Moreover, in each iteration, all the operations are performed on one single example, and we can see that the computational complexity is dominated by the argmax which as shown above can be solved by Algorithm 2 in $O(l^2)$ time. As a consequence, we can see that the method can converge to a ϵ-accurate

---

Algorithm 2. Solve the argmax problem (7)   [10]

Input: true label vector $\mathbf{y}$, current prediction $\mathbf{p}$
Output: label vector $\tilde{\mathbf{y}}$
1: $\{i_1^+, \ldots, i_{pos}^+\} \leftarrow$ sort $\{i \mid y_i = +1\}$ in descending order of p_i's
2: $\{i_1^-, \ldots, i_{neg}^-\} \leftarrow$ sort $\{i \mid y_i = -1\}$ in descending order of p_i's
3: **for** $tp = 0$ to pos **do**
4:     $fn = pos - tp$
5:     set $y'_{i_1^+}, \ldots, y'_{i_{tp}^+}$ to $+1$ and set $y'_{i_{tp+1}^+}, \ldots, y'_{i_{pos}^+}$ to -1
6:     **for** $tn = 0$ to neg **do**
7:         $fn = neg - tn$
8:         set $y'_{i_1^-}, \ldots, y'_{i_{fn}^-}$ to $+1$ and set $y'_{i_{fn+1}^+}, \ldots, y'_{i_{neg}^+}$ to -1
9:         let $v \leftarrow \Delta(\mathbf{y}, \mathbf{y}') + \mathbf{y}'^\top \mathbf{p}$
10:        **if** v is the largest so far **then**
11:            $\tilde{\mathbf{y}} = \mathbf{y}'$
12:        **end if**
13:    **end for**
14: **end for**

---

solution in at most $O(l^2 \cdot \log k/\epsilon^2)$ time. It worth noting that this computational complexity is independent of m, *i.e.* the number of training examples, this constitutes one of the appealing properties of SECC.

4   Experiments

In this section, we perform a set of experiments to evaluate the effectiveness of SECC. Specifically, we first show the effectiveness of optimizing the upper bounds, and then compare SECC with ECC and other state-of-the-art multi-label classifiers.

4.1   Configuration

The experiments are performed on image and music annotation tasks. Specifically,
 – **Image annotation** tasks are *Corel5k* [5] which has 5000 images and 374 possible labels, and *Scene* [1] has 2407 images and 6 possible labels;
 – **Music annotation** tasks are *CAL500* [24] which has 502 songs and 174 possible labels, and *Emotions* [20] which has 593 songs and 6 possible labels.
The information of these tasks are also summarized in Table 1.

In the experiments, we first train an ECC of size 100, then build SECC upon it. We compare SECC with BSVM [1] which trains one SVM for each label, and state-of-the-art methods including the lazy method ML-kNN [28], label ranking method CLR [7] and the full ECC combining all classifiers (denoted as ECC$_{\text{full}}$). It is also compared with the *random strategy* which selects classifiers randomly (denoted as ECC$_{\text{rand}}$) . Specifically, LIBSVM [3] is used to implement the base classifier in BSVM and ECC; the default implementation of ML-kNN and CLR in MULAN [22] are used.

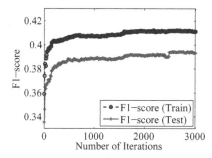 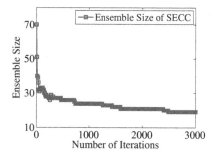

Fig. 2. The F1-score (left) and ensemble size (right) of SECC on the CAL500 task, during the optimization procedure. Both the F1-score on training and test set are ploted.

For each task, these comparative methods are evaluated by 30 times random holdout test, *i.e.*, in each time, 2/3 for training and 1/3 for testing; finally, the mean F1-score over 30 times and the standard derivation are reported; also the sizes of ensembles generated by SECC are reported. In each time of holdout test, SECCs choose the regularization parameter λ by 5 fold cross validation on training set; ECC_{rand} randomly selects N CC classifiers to form the ensemble, where N is the size of SECC in this time.

4.2   Effectiveness of Optimizing the Upper Bound

A question naturally raised about SECC is whether it is effective to optimize the upper bound of the empirical risk; in other words, whether optimizing the upper bound will improve the performance?

To answer this question, we record the training and test performance on the CAL500 task where the parameter λ is to 2^{-8}, and the results are shown in the left plot of Fig.2. It can be seen that both the F1-score on training and test data improve as the number of iterations increases. This optimizing the upper bound is effective in improving the performance. Also, we record the ensemble sizes of SECC at different iterations, and plot then in the right plot of Fig.2. It can be seen that the ensemble size of SECC decreases with the number of iterations increases. This results shows the effectiveness of SECC in obtaining a smaller ensemble.

Moreover, we can see from Fig.2 that the performance tends to converge after some iterations, *i.e.*, the F1-scores converge after about 1500 iterations. Noting there are 502 examples in CAL500, this means we need to scan the data set for only three times, also the operations in each iteration are very simple, so this method is efficient. For example, on the CAL500 task, it takes only 1.5 seconds for 3000 iterations on a PC with Intel Core 1.7GHz CPU, which is very efficient.

4.3   Performance Comparison Results

The F1-scores achieved by comparative methods are shown in Table 1, where the ensemble size of ensemble methods, including SECC, ECC_{full} and ECC_{rnd}, are also

Table 1. Experimental results (mean±std.) achieved by comparative methods, which incudes the F1-score and the ensemble sizes for ensemble methods (SECC, ECC$_{full}$ and ECC$_{rnd}$). Also, the information of each task is summarized, where *#F*, *#L* and *#N* indicate the number of features, labels and examples, respectively. The mark '•'('○') indicates that SECC is significantly better (worse) than the corresponding method based on paired *t*-tests at 95% significance level.

	SECC	ECC$_{full}$	ECC$_{rand}$	BSVM	ML-kNN	CLR
Corel5k: #F=499, #L=374, #N=5000						
F1-score	.149±.006	.134±.006•	.127±.008•	.135±.005•	.016±.003•	.034±.001•
Size	15.3±5.6	100.0	15.3±5.6	–	–	–
Scene: #F=294, #L=6, #N=2407						
F1-score	.672±.002	.668±.012	.657±.014•	.595±.014•	.675±.018	.629±.009•
Size	22.0±10.8	100.0	22.0±10.8	–	–	–
CAL500: #F=68, #L=174, #N=502						
F1-score	.384±.013	.323±.010•	.333±.023•	.314±.029•	.322±.011•	.405±.003○
Size	19.6±4.7	100.0	19.6±4.7	–	–	–
Emotions: #F=72, #L=6, #N=593						
F1-score	.619±.017	.618±.015	.612±.016	.614±.014	.602±.029•	.622±.016
Size	63.7±21.2	100.0	63.7±21.2	–	–	–

given. For better comparison, we perform paired *t*-tests at 95% significance level to compare SECC with other methods, and the results are indicated in Table 1 by '•'('○').

It can be seen from these results that the performance of SECC is quite promising. Compared with the full ensemble ECC$_{full}$, it achieves 3 significant better F1-scores, while the ensemble size is much smaller. For examples, on *Corel5k*, the F1-score is improved from 0.134 to 0.149, while the ensemble size is reduced to 15.3; on *CAL500* the F1 score is improved from 0.323 to 0.384, whiles the ensemble size of reduced from 100 to less than 20; even on *Scene* and *Emotions*, SECC still achieves comparable performance to ECC, but it still reduces the average ensemble size to 22.0 and 63.7, respectively. Compared with ECC$_{rand}$ which choose CCs randomly, SECC achieves significantly better performance on *Corel5k*, *CAL500* and *Scene*, and comparable F1-score on *Emotions*. All of these results show the effectiveness of SECC, in both reducing the ensemble sizes and improving the performance.

Comparing with other methods, we can see that the performance of SECC is significantly better than BSVM and ML-kNN on three tasks, also its performance is comparable against the state-of-the-art methods CLR (*i.e.*, 2 wins and 1 loss). It is not hard to find that the superiority of SECC mainly inherits from the good performance of ECC.

In summary, SECC inherits from the good performance of ECC while improve its efficiency by generating smaller ensembles.

5   Conclusion and Future Work

Noticing that ECC tends to generate unnecessarily large ensembles, in this paper, we propose *selective ensemble of classifier chains* (SECC) and it is expected that it can reduce the size of ECC whilst keeping or improving its performance. Specifically, after obtaining multiple CCs, instead of combining all of them by voting like ECC, SECC tries to select some of them to composite the ensemble. In this paper, by focusing on the frequently-used performance measure F1-score, we formulate this selection problem as an ℓ_1-norm regularized convex optimization problem, and present a stochastic subgradient descend method to solve it. Experiments on image and music annotation tasks shows the effectiveness of the proposed method.

In this work, we consider the CC classifier as a common multi-label classifier which has multiple outputs. In practice, CC itself is a group of single-label classifiers, hence it will be interesting to consider SECC at the level of such binary classifiers.

Acknowledgements. We want to thank anonymous reviewers for helpful comments. This research was supported by the National Science Foundation of China (61021062, 61105043) and the Jiangsu Science Foundation (BK2011566).

References

1. Boutell, M., Luo, J., Shen, X., Brown, C.: Learning multi-label scene classification. Pattern Recognition 37(9), 1757–1771 (2004)
2. Cesa-Bianchi, N., Re, M., Valentini, G.: Synergy of multi-label hierarchical ensembles, data fusion, and cost-sensitive methods for gene functional inference. Machine Learning 88(1), 209–241 (2012)
3. Chang, C.-C., Lin, C.-J.: LIBSVM: A library for support vector machines. ACM Transactions on Intelligent Systems and Technology 2(3), 1–27 (2011)
4. Dembczynski, K., Cheng, W., Hüllermeier, E.: Bayes optimal multilabel classification via probabilistic classifier chains. In: Proceedings of the 27th International Conference on Machine Learning, Haifa, Israel, pp. 279–286 (2010)
5. Duygulu, P., Barnard, K., de Freitas, J.F.G., Forsyth, D.A.: Object recognition as machine translation: Learning a lexicon for a fixed image vocabulary. In: Heyden, A., Sparr, G., Nielsen, M., Johansen, P. (eds.) ECCV 2002, Part IV. LNCS, vol. 2353, pp. 97–112. Springer, Heidelberg (2002)
6. Elisseeff, A., Weston, J.: A kernel method for multi-labelled classification. In: Advances in Neural Information Processing Systems 14, pp. 681–687. MIT Press, Cambridge (2002)
7. Fürnkranz, J., Hüllermeier, E., Loza Mencía, E., Brinker, K.: Multilabel classification via calibrated label ranking. Machine Learning 73(2), 133–153 (2008)
8. Ghamrawi, N., McCallum, A.: Collective multi-label classification. In: Proceedings of the 14th ACM International Conference on Information and Knowledge Management, Bremen, Germany, pp. 195–200 (2005)
9. Giacinto, G., Roli, F., Fumera, G.: Design of effective multiple classifier systems by clustering of classifiers. In: Proceedings of the 15th International Conference on Pattern Recognition, Barcelona, Spain, pp. 160–163 (2000)
10. Joachims, T.: A support vector method for multivariate performance measures. In: Proceedings of the 22nd International Conference on Machine Learning, Bonn, Germany, pp. 377–384 (2005)
11. Kumar, A., Vembu, S., Menon, A.K., Elkan, C.: Learning and inference in probabilistic classifier chains with beam search. In: Flach, P.A., De Bie, T., Cristianini, N. (eds.) ECML PKDD 2012, Part I. LNCS, vol. 7523, pp. 665–680. Springer, Heidelberg (2012)

12. Li, N., Tsang, I.W., Zhou, Z.-H.: Efficient optimization of performance measures by classifier adaptation. IEEE Transactions on Pattern Analysis and Machine Intelligence (2013) (preprint)
13. Li, N., Zhou, Z.-H.: Selective ensemble under regularization framework. In: Benediktsson, J.A., Kittler, J., Roli, F. (eds.) MCS 2009. LNCS, vol. 5519, pp. 293–303. Springer, Heidelberg (2009)
14. McCallum, A.: Multi-label text classification with a mixture model trained by EM. Working Notes of AAAI 1999 Workshop on Text Learning (1999)
15. Read, J., Pfahringer, B., Holmes, G.: Multi-label classification using ensembles of pruned sets. In: Proceedings of the 8th IEEE International Conference on Data Mining, Pisa, Italy, pp. 995–1000 (2008)
16. Read, J., Pfahringer, B., Holmes, G., Frank, E.: Classifier chains for multi-label classification. Machine Learning 85(3), 333–359 (2011)
17. Schapire, R., Singer, Y.: BoosTexter: A boosting-based system for text categorization. Machine Learning 39(2-3), 135–168 (2000)
18. Shalev-Shwartz, S., Tewari, A.: Stochastic methods for l1-regularized loss minimization. Journal of Machine Learning Research 12, 1865–1892 (2011)
19. Taskar, B., Guestrin, C., Koller, D.: Max-margin markov networks. In: Advances in Neural Information Processing Systems 16, pp. 25–32. MIT Press, Cambridge (2003)
20. Trohidis, K., Tsoumakas, G., Kalliris, G., Vlahavas, I.: Multilabel classification of music into emotions. In: Proceedings of 2008 International Conference on Music Information Retrieval, Philadelphia, PA, pp. 325–330 (2008)
21. Tsochantaridis, I., Joachims, T., Hofmann, T., Altun, Y.: Large margin methods for structured and interdependent output variables. Journal of Machine Learning Research 6, 1453–1484 (2005)
22. Tsoumakas, G., Spyromitros-Xioufis, E., Vilcek, J., Vlahavas, I.: MULAN: A Java library for multi-label learning. Journal of Machine Learning Research 12, 2411–2414 (2011)
23. Tsoumakas, G., Vlahavas, I.: Random k-labelsets: An ensemble method for multilabel classification. In: Kok, J.N., Koronacki, J., Lopez de Mantaras, R., Matwin, S., Mladenič, D., Skowron, A. (eds.) ECML 2007. LNCS (LNAI), vol. 4701, pp. 406–417. Springer, Heidelberg (2007)
24. Turnbull, D., Barrington, L., Torres, D., Lanckriet, G.: Semantic annotation and retrieval of music and sound effects. IEEE Transactions on Audio, Speech and Language Processing 16(2), 467–476 (2008)
25. Vens, C., Struyf, J., Schietgat, L., Džeroski, S., Blockeel, H.: Decision trees for hierarchical multi-label classification. Machine Learning 73(2), 185–214 (2008)
26. Zaragoza, J.H., Sucar, L.E., Morales, E.F., Bielza, C., Larrañaga, P.: Bayesian chain classifiers for multidimensional classification. In: Proceedings of the 22nd International Joint Conference on Artificial Intelligence, Barcelona, Spain, pp. 2192–2197 (2011)
27. Zhang, M.-L., Zhang, K.: Multi-label learning by exploiting label dependency. In: Proceedings of the 16th ACM SIGKDD Conference on Knowledge Discovery and Data Mining, Washington, DC, pp. 999–1007 (2010)
28. Zhang, M.-L., Zhou, Z.-H.: ML-KNN: A lazy learning approach to multi-label learning. Pattern Recognition 40(7), 2038–2048 (2007)
29. Zhang, Y., Burer, S., Street, W.: Ensemble pruning via semi-definite programming. Journal of Machine Learning Research 7, 1315–1338 (2006)
30. Zhou, Z.-H.: Ensemble Methods: Foundations and Algorithms. Chapman & Hall/CRC, Boca Raton, FL (2012)
31. Zhou, Z.-H., Wu, J., Tang, W.: Ensembling neural networks: Many could be better than all. Artificial Intelligence 137(1-2), 239–263 (2002)

The Link between Multiple-Instance Learning and Learning from Only Positive and Unlabelled Examples

Yan Li, David M.J. Tax, Robert P.W. Duin, and Marco Loog

Pattern Recognition Laboratory, Delft University of Technology
Mekelweg 4, 2628 CD Delft, The Netherlands
http://prlab.tudelft.nl

Abstract. This paper establishes a link between two supervised learning frameworks, namely multiple-instance learning (MIL) and learning from only positive and unlabelled examples (LOPU). MIL represents an object as a bag of instances. It is studied under the assumption that its instances are drawn from a mixture distribution of the concept and the non-concept. Based on this assumption, the classification of bags can be formulated as a classifier combining problem and the Bayes classifier for instances is shown to be closely related to the classification in LOPU. This relationship provides a possibility to adopt methods from LOPU to MIL or vice versa. In particular, we examine a parameter estimator in LOPU being applied to MIL. Experiments demonstrate the effectiveness of the instance classifier and the parameter estimator.

1   Introduction

Multiple-instance learning (MIL) [1] is a generalised supervised-learning framework that represents an object as a bag consisting of many feature vectors called instances. Only some of the instances in the bag are informative about the label of the object, while others share the same probability distribution for objects from different classes. In the training phase, only the labels of bags (not instances) are known, and a classifier is trained to separate bags into different classes. Many problems can be formulated as MIL problems, such as image annotation [2, 3], text categorization [2, 4], and visual tracking [5].

Learning with only positive and unlabelled examples (LOPU) deals with another limitation of supervised learning [6]. Here one is given a set of examples which are all positive, and another set of examples which include both positive and negative examples. It is the task of the classifier to learn from the unlabelled examples a model of the negative class. For example, to classify users' preference of web pages, an user's bookmarks can be seen as positive examples, while other web pages in the internet include both negative and potentially positive examples. LOPU has been applied to information retrieval [7], document or web page classification [8–10], and biomedical classification [11], among others.

Z.-H. Zhou, F. Roli, and J. Kittler (Eds.): MCS 2013, LNCS 7872, pp. 157–166, 2013.

In MIL, it is typically assumed that there is an underlying concept, whose instances distinguish between positive bags an negative bags [1, 12, 2]. The classical assumption of MIL is that a bag is classified to be positive if *at least one* of its instances is from the concept [1]. Many MIL algorithms (e.g. Diverse Density [12], MI-SVM [2], and the method in [13]), however, usually use the information of only one concept instance from each positive bag [14]. In order to exploit the information from concept instances more effectively, numerous new assumptions have been proposed for MIL [15].

We study MIL based on the assumption proposed by the authors in [16]. This assumption helps to effectively combine the information from all concept instances, which can be formulated as a classifier combining problem. The MIL model and method have been studied in [16]. The focus of this paper is on the link between this MIL model and LOPU. We examine how the instance classifier of this MIL model is related to the classification in LOPU, and how to adapt methods from LOPU to MIL or vice versa. In particular, a parameter estimator from LOPU is modified and applied to MIL.

The paper is organized as follows. Section 2 introduces MIL with our assumption and derives the instance classifier. The relationship between MIL and LOPU is elaborated in Section 3. Section 4 presents experiment results with the derived instance classifier and parameter estimator. Finally, Section 5 concludes the paper.

2   Instance Classifier for MIL

The MIL model with the mixture assumption is introduced, based on which a Bayes classifier is derived for instance classification.

2.1   The Mixture Assumption

Denote an object as a bag $B_i = \{\mathbf{B}_{i1}, \mathbf{B}_{i2}, \cdots, \mathbf{B}_{iJ_i}\}$, which contains J_i feature vectors $\mathbf{B}_{ij}$ of dimensionality D. Assume that instances in a bag are conditionally statistically independent [17]. That is, given the label of a bag, its instances are drawn independently: $p(\mathbf{B}_{i1}, \mathbf{B}_{i2}, \cdots, \mathbf{B}_{iJ_i} | \omega(B_i)) = \prod_{j=1}^{J_i} p(\mathbf{B}_{ij} | \omega(B_i))$, where $\omega(B_i)$ is the label of the bag B_i and can be either positive '+' or negative '−'.

Denote the probability density function (pdf) $p(\mathbf{x}|+)$, from which the instances $\mathbf{x}$ in positive bags are drawn, as $f^+(\mathbf{x})$ and the pdf for negative bags as $f^-(\mathbf{x})$. Assume further that a concept $\mathcal{C}$ exists, which defines the difference between positive and negative bags. The non-concept $\bar{\mathcal{C}}$ denotes the background region which is shared by both positive and negative bags. Formally, there are two distinct distributions to generate instances, one is $f^{\mathcal{C}}(\mathbf{x})$ for the concept $\mathcal{C}$ and the other is $f^{\bar{\mathcal{C}}}(\mathbf{x})$ for the non-concept $\bar{\mathcal{C}}$.

We follow the mixture assumption for MIL in [16]. It assumes that the instances in a negative bag are all drawn from $\bar{\mathcal{C}}$, while the instances in a positive bag are drawn from both $\mathcal{C}$ and $\bar{\mathcal{C}}$,

$$f^-(\mathbf{x}) = f^{\bar{\mathcal{C}}}(\mathbf{x}),$$
$$f^+(\mathbf{x}) = \alpha f^{\mathcal{C}}(\mathbf{x}) + (1 - \alpha)f^{\bar{\mathcal{C}}}(\mathbf{x}), \quad 0 < \alpha < 1, \tag{1}$$

where the mixing coefficient α represents the fraction of instances sampled from the concept $\mathcal{C}$ in a positive bag.

2.2 The Instance Classifier

Based on the mixture assumption, the Bayes classifier for instances has been briefly presented in [16]. For completeness, a more detailed derivation is provided, which is followed with an analysis.

Denote the prior probabilities for positive and negative *bags* as β and $1 - \beta$ respectively,

$$P(+) = \beta,$$
$$P(-) = 1 - \beta, 0 < \beta < 1. \tag{2}$$

Then the prior probabilities for the concept $\mathcal{C}$ and the non-concept $\bar{\mathcal{C}}$ are

$$P(\mathcal{C}) = \alpha\beta,$$
$$P(\bar{\mathcal{C}}) = (1 - \beta) + (1 - \alpha)\beta = 1 - \alpha\beta. \tag{3}$$

From Bayesian decision theory, we know that an instance should be classified to the class with the largest posterior. That is, for a new test instance $\mathbf{x}$,

assign $\mathbf{x}$ to the *concept* $\mathcal{C}$, if
$$P(\mathcal{C}|\mathbf{x}) \geq P(\bar{\mathcal{C}}|\mathbf{x}) \Longleftrightarrow$$
$$p(\mathbf{x}|\mathcal{C})P(\mathcal{C}) \geq p(\mathbf{x}|\bar{\mathcal{C}})P(\bar{\mathcal{C}}) \Longleftrightarrow \tag{4}$$
$$f^{\mathcal{C}}(\mathbf{x}) \cdot \alpha\beta \geq f^{\bar{\mathcal{C}}}(\mathbf{x}) \cdot (1 - \alpha\beta)$$

From Eq. (1), the density of the concept $f^{\mathcal{C}}(\mathbf{x})$ can be obtained as

$$f^{\mathcal{C}}(\mathbf{x}) = \frac{f^+(\mathbf{x}) - (1 - \alpha)f^-(\mathbf{x})}{\alpha}. \tag{5}$$

Substituting it into (4), the decision rule becomes

$$\frac{f^+(\mathbf{x}) - (1 - \alpha)f^-(\mathbf{x})}{\alpha} \cdot \alpha\beta \geq f^-(\mathbf{x}) \cdot (1 - \alpha\beta) \Longleftrightarrow$$
$$f^+(\mathbf{x}) \geq \left(\frac{1}{\beta} + 1 - 2\alpha\right) f^-(\mathbf{x}) \tag{6}$$

Since $f^+(\mathbf{x}) = p(\mathbf{x}|+) = \frac{p(\mathbf{x})P(+|\mathbf{x})}{P(+)} = \frac{P(+|\mathbf{x})}{\beta}p(\mathbf{x})$ and similarly $f^-(\mathbf{x}) = \frac{P(-|\mathbf{x})}{1-\beta}p(\mathbf{x})$, the decision rule (6) can be expressed using the posteriors $P(+|\mathbf{x})$ and $P(-|\mathbf{x})$:

$$\frac{P(+|\mathbf{x})}{\beta}p(\mathbf{x}) \geq \left(\frac{1}{\beta} + 1 - 2\alpha\right) \frac{P(-|\mathbf{x})}{1 - \beta}p(\mathbf{x}) \Longleftrightarrow$$
$$P(+|\mathbf{x}) \geq \frac{1 + \beta - 2\alpha\beta}{1 - \beta}P(-|\mathbf{x}) \tag{7}$$

Note that as $(1 + \beta - 2\alpha\beta) - (1 - \beta) = 2\beta(1 - \alpha) > 0$, $\frac{1+\beta-2\alpha\beta}{1-\beta} > 1$ always holds.

Equation (7) provides a way to adapt traditional classifiers for instance classification. Applying a traditional classifier to instances labelled according to their bag label ('+' or '−'), the posteriors $P(+|\mathbf{x})$ or $P(-|\mathbf{x})$ can be obtained. By weighting the obtained posteriors according to (7), the instances can then be classified to the concept $\mathcal{C}$ or the non-concept $\bar{\mathcal{C}}$.

Traditional supervised classifiers have been shown to work well on many MIL problems, despite the fact that they ignore the assumptions underlying MIL [14]. By taking such MIL assumptions into account, our approach can improve the performance of standard supervised classifiers when applied to MIL. In a comparable setting, [14] proposed to use higher cost for false positives in order to improve the classification performance. A higher cost for false positives is analogous to the term $\frac{1+\beta-2\alpha\beta}{1-\beta}$ in (7). The reason is that increasing the cost for false positives is equivalent to increasing the threshold of posteriors to classify an instance to the concept, which has the same meaning as (7).

The k-NN for MIL. To illustrate the decision rule (7), we present an instance classifier using k-NN (k-nearest neighbour) [18], which is also used in our experiments. If k nearest neighbours are found for an instance $\mathbf{x}$, k_+ of them are from positive bags and $k_-(= k - k_+)$ from negative bags, then the estimates of posteriors are

$$\hat{P}(+|\mathbf{x}) = \frac{k_+}{k}, \text{ and } \hat{P}(-|\mathbf{x}) = \frac{k_-}{k}. \tag{8}$$

From (7) the decision rule for the concept becomes

assign $\mathbf{x}$ to the *concept* $\mathcal{C}$, if

$$k_+ \geq \frac{1 + \beta - 2\alpha\beta}{1 - \beta} \cdot k_-. \tag{9}$$

3   Relation between MIL and LOPU

The instance classifier (7) is trained by assigning bag labels to their instances. The instances from negative bags are known to be from the non-concept, while those from positive bags are partly from the concept and partly from the non-concept. Consequently, this problem turns out to be very similar to LOPU. In LOPU, all labelled examples are positive, while unlabelled examples may be positive or negative. In terms of LOPU, the classification of instances in MIL can be considered as a learning problem with *only unlabelled and negative examples* and we can use this relation to our benefit.

One of the widely used methods in LOPU is to identify unlabelled examples that are most likely to to be negative with some heuristics, and then train a classifier based on the identified negative examples and the given positive examples [9, 11]. Similar heuristics have been used in Diverse Density to explicitly search for the concept area [12], in MI-SVM to identify the instance in each positive bag which is mostly likely to be from the concept [2], and in the disambiguation method to identify concept instances in each positive bag [13].

The LOPU method proposed in [11] shares additional similarities with our instance classifier. It is based on the so-called "selected completely at random" assumption [6], which means that any positive example has a constant probability to be chosen by the user and then labelled. Consequently, the unlabelled examples include other positive examples which are not chosen by the user and all the negative examples. The pdf of the unlabelled examples can thus be expressed as a mixture of distributions, which is similar to the pdf $f^+(\mathbf{x})$ in (1). In addition, a central result in that paper is a lemma which relates the classifier trained by *treating all unlabelled examples as negative* and the "ideal" classifier if the labels of all unlabelled examples are provided. This lemma can be derived from the decision rule (7). Besides, they proposed a method to assign weights to unlabelled examples in LOPU, which can also be used to weight instances from positive bags in MIL.

Based on the relationship between LOPU and MIL, a parameter estimator of α has been proposed in [16]. This estimator is based on another estimator proposed in [11], but modified to make it robust in the MIL setting. The basic idea is as follows. In MIL terms, the estimator in [11] is for the parameter

$$\theta = \frac{1-\beta}{(1-\beta) + \beta(1-\alpha)}, \tag{10}$$

which is the probability that a non-concept instance is from a negative bag. It is estimated as the average posteriors of instances in negative bags B^- from a validation set

$$\hat{\theta} = \underset{\mathbf{x} \in B^-}{\mathrm{mean}}\, P(-|\mathbf{x}). \tag{11}$$

This results in an estimator

$$\hat{\alpha} = \frac{\hat{\theta} + \hat{\beta} - 1}{\hat{\theta} \cdot \hat{\beta}}, \tag{12}$$

which is, however, is very sensitive to estimation errors, as θ and β appear in the denominator. A more robust estimator is proposed in [16]. It is based on the assumption that in positive bags, the posteriors of concept instances should be large (close to one) since these instances occur only in positive bags, and the posteriors of non-concept instances have an expectation of $1 - \theta$. Therefore, by definition, α can be estimated as the fraction of instances in positive bags whose posteriors are larger than a threshold τ,

$$\hat{\alpha} = \frac{\#\left\{\mathbf{x} \mid \mathbf{x} \in B^+, P(+|\mathbf{x}) > \tau\right\}}{\#\left\{\mathbf{x} \mid \mathbf{x} \in B^+\right\}}, \tag{13}$$

where $\#$ returns the number of elements in a set and τ takes value as the mean of $1 - \hat{\theta}$ and the maximum posterior of instances in positive bags $\max_{\mathbf{x} \in B^+} P(+|\mathbf{x})$.

MIL was linked to semi-supervised learning (SSL) in [19], by viewing MIL as a problem with unlabelled data but positive constraints. The relation with LOPU makes it clearer that in MIL, there is no labelled positive instances. Besides, the link with SSL is based on the classical MIL assumption, while the mixture assumption is used to relate it to LOPU.

The crucial difference between MIL and LOPU is that the instances in MIL are from bags and are not individual objects. Therefore, MIL has to consider the problem of bag classification in addition to instance classification.

4   Experiments

The derived instance classifier (7) and parameter estimator (13) are tested with a synthetic data and various benchmark MIL datasets. To obtain the label of a bag from its instances, the method derived in [16] is used, which is based on classifier combining. Its basic idea is to compute the fraction of a bag's instances which are classified to the concept and label the bag as positive if the fraction is large than a threshold. The threshold is derived to be $\frac{(1-\alpha\beta)(1-2\beta)}{2(1-\beta)J_i} + \alpha\beta$, which can be approximated by $\alpha\beta$ if the number of instances in a bag J_i is sufficiently large or is $\alpha\beta$ if $\beta = 0.5$. We use k-NN as the instance classifier, with k set to half of the average number of instances in a bag. The corresponding MIL algorithm is as follows:

> **Training data construction.** For all training bags, assign the bag label to its instances and use all instances for training.
> **Instance classification.** For a test bag $B_i = \{\mathbf{B}_{i1}, \mathbf{B}_{i2}, \cdots, \mathbf{B}_{iJ_i}\}$, classify each instance $\mathbf{B}_{ij}$ according to (9).
> **Bag classification by Combining.** Calculate the fraction of concept instances in the bag B_i, and classify it as positive if the fraction is larger than the threshold $\frac{(1-\alpha\beta)(1-2\beta)}{2(1-\beta)J_i} + \alpha\beta$, and negative otherwise.

4.1   Synthetic Data

We use the synthetic data in [20, 21]. Figure 1(a) shows the ground-truth of the problem, where the black area is the non-concept and the white area is the concept. With $\alpha = 0.3$ and $\beta = 0.5$, one realisation is shown in Figure 1(b).

Trained on the data in Fig. 1(b), the decision boundaries of the instance classifier (9) is shown in Fig. 1(c), where the cyan area is for the concept and the purple area is for the non-concept. We can see that the instance classifier separates the concept and the non-concept very well. In comparison, the decision boundaries for the classifier without the weighting factor $\frac{1+\beta-2\alpha\beta}{1-\beta}$ in (9) is shown in Fig. 1(d), where much non-conept area is misclassified to the concept.

The parameter estimator (13) and the instance classifier (9) are tested with α changing from 0.05 to 0.9. The prior β is fixed to 0.5, and there are 30 positive and 30 negative bags, with 40 instances in each bag. The average results of ten times 10-fold cross-validation are reported. Figure 2(a) shows the estimated α. Overall, the estimator works very well for all different αs, though it seems that

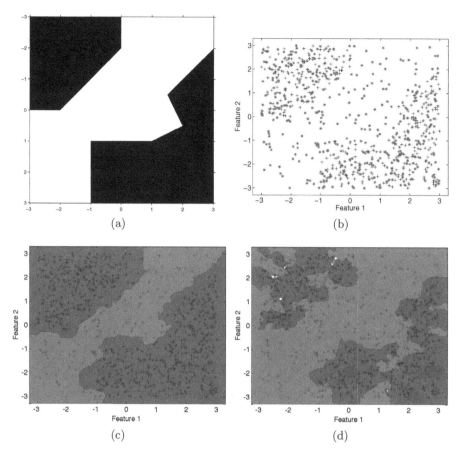

Fig. 1. A synthetic MIL dataset. (a) The ground-truth decision boundary between the concept (white) and the non-concept (black). (b) Scatter plot of all instances, where red instances (*) are from positive bags and blue ones (+) are from negative bags. (c) Decision boundaries of the instance classifier (9). (d) Decision boundary of (9), but without the weighting factor $\frac{1+\beta-2\alpha\beta}{1-\beta}$. The cyan area is for the concept and the purple area (the dark area, if viewed in black and white) is for the non-concept.

there is a small tendency of underestimation (around 0.03). Figure 2(b) shows the classification error of instances with our approach (9) and the traditional k-NN, i.e., (9) without the weighting factor $\frac{1+\beta-2\alpha\beta}{1-\beta}$. The results clearly demonstrate the necessity of this weighting factor, especially when α is small. When α is very large (e.g. 0.8 or 0.9), this weighting factor goes close to one and the difference between the two methods becomes small. Based on our instance classification, the results of bag classification are shown in Figure 2(c). It shows that when α is large than 0.2, perfect classification are obtained. When α is very small (e.g. 0.05), the error is relatively high, as there are only one or two concept instances

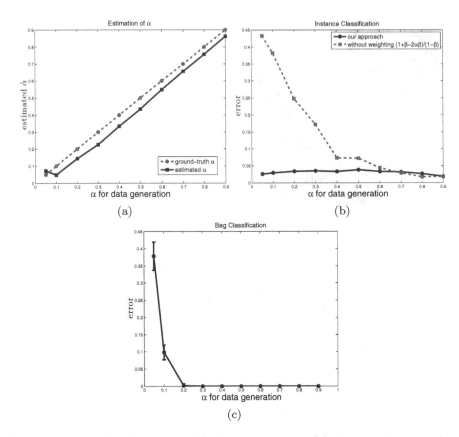

Fig. 2. Results with different αs. (a) The estimated α. (b) The classification of instances. (c) The classification of bags.

in a positive bag. As a result, there are very few concept instances in the training set. Using k-NN with a smaller k, or increasing the number of bags for training, may improve the performance.

4.2   Benchmark MIL Datasets

Our method has been applied to a few mostly tested MIL datasets in [16]. The data sets are the MUSK1 and MUSK2 collected in [1], and the Elephant, Tiger, and Fox from the COREL dataset [2]. A support vector machine with RBF kernel is used as the instance classifier. It has been shown that our method, though relatively simple, achieves results comparable to other state-of-the-art methods.

Table 1 shows the classification results of our method, and another two methods which output estimations of a parameter very similar to α. This parameter is the so-called witness rate, which is the fraction of "true positive" instances in all the positive bags. The witness rate in [21] was automatically estimated, while

Table 1. Accuracy on the five benchmark MIL datasets. The results of our method are comparable to other state-of-the-art methods [16].

	MUSK1	*MUSK2*	*Elephant*	*Tiger*	*Fox*
ALP-SVM [20]	86.3	86.2	83.5	86.0	66.0
witness rate	1.00	0.28	0.58	0.6	0.71
SVR-SVM [21]	87.9 (1.7)	85.4 (1.8)	85.3 (2.8)	79.8 (3.4)	63.0 (3.5)
witness rate	1.00	0.895	0.378	0.427	1.00
Our method	88.38 (1.08)	84.92 (2.18)	84.35 (0.88)	80.75 (1.16)	62.80 (0.86)
estimated α	0.82	0.77	0.80	0.51	0.88

that in [20] was tuned manually. We can see that the estimated witness rates and αs are quite close to each other (except for Elephant). In addition, their values are quite large, which indicates that there are usually more than one concept instance in a positive bag and thus justifies our assumption to some extent. The large values of α may also explain why using traditional supervised classifiers without taking MIL assumptions can work well for some data sets [14]. When α is very large, the weighting factor in the instance classifier (7) becomes close to one, and thus the traditional classifier without this weighting factor can already work relatively well (c.f. Figure 2(b)).

5   Conclusion

Based on the assumption that instances from positive bags follow a mixture distribution, the Bayes classifier is derived for instance classification. A relationship is then established between the classification of instances in MIL and another learning framework called LOPU. It is shown how numerous results in both fields can be linked together. This link also makes it possible to apply methods from MIL to LOPU, or the other way around. In particular, it is studied how to adopt a parameter estimator proposed in LOPU for estimating the parameter α in MIL. The derived instance classifier and the parameter estimator are shown to perform well in the experiments.

References

1. Dietterich, T., Lathrop, R., Lozano-Pérez, T.: Solving the multiple instance problem with axis-parallel rectangles. Artificial Intelligence 89(1-2), 31–71 (1997)
2. Andrews, S., Tsochantaridis, I., Hofmann, T.: Support vector machines for multiple-instance learning. In: Adv. Neu. Inf. Proc. Sys., pp. 577–584 (2003)
3. Chen, Y., Bi, J., Wang, J.: MILES: Multiple-instance learning via embedded instance selection. IEEE Trans. PAMI 28(12), 1931–1947 (2006)
4. Zhou, Z., Sun, Y., Li, Y.: Multi-instance learning by treating instances as non-IID samples. In: Proc. 26th ICML, pp. 1249–1256 (2009)
5. Babenko, B., Yang, M., Belongie, S.: Visual tracking with online multiple instance learning. In: IEEE CVPR, pp. 983–990 (2009)

6. Denis, F.: PAC learning from positive statistical queries. In: Richter, M.M., Smith, C.H., Wiehagen, R., Zeugmann, T. (eds.) ALT 1998. LNCS (LNAI), vol. 1501, pp. 112–126. Springer, Heidelberg (1998)

7. Lee, W., Liu, B.: Learning with positive and unlabeled examples using weighted logistic regression. In: Proc. 20th ICML, pp. 448–455 (2003)

8. Liu, B., Dai, Y., Li, X., Lee, W., Yu, P.: Building text classifiers using positive and unlabeled examples. In: Proc. Int'l Conf. Data Mining, pp. 179–188 (2003)

9. Yu, H., Han, J., Chang, K.: PEBL: Web page classification without negative examples. IEEE Trans. Know. and Data Eng. 16(1), 70–81 (2004)

10. Zhou, K., Xue, G., Yang, Q., Yu, Y.: Learning with Positive and Unlabeled Examples Using Topic-Sensitive PLSA. IEEE Trans. on Knowledge and Data Engineering 22(1), 46–58 (2010)

11. Elkan, C., Noto, K.: Learning classifiers from only positive and unlabeled data. In: Proc. 14th ACM Conf. Knowledge Discovery and Data Mining, pp. 213–220 (2008)

12. Maron, O., Lozano-Pérez, T.: A framework for multiple-instance learning. In: Adv. Neu. Inf. Proc. Sys., pp. 570–576 (1998)

13. Li, W.J., Yeung, D.Y.: MILD: Multiple-instance learning via disambiguation. IEEE Transactions on Knowledge and Data Engineering 22(1), 76–89 (2010)

14. Ray, S., Craven, M.: Supervised versus multiple instance learning: An empirical comparison. In: Proc. 22nd Int'l Conf. Mach. Learn., pp. 697–704 (2005)

15. Foulds, J., Frank, E.: A review of multi-instance learning assumptions. The Knowledge Engineering Review 25(01), 1–25 (2010)

16. Li, Y., Tax, D., Duin, R., Loog, M.: Multiple-instance learning as a classifier combining problem. Pattern Recognition 46(3), 865–874 (2013)

17. Blum, A., Kalai, A.: A note on learning from multiple-instance examples. Machine Learning 30(1), 23–29 (1998)

18. Bishop, C.: Pattern Recognition and Machine Learning. Springer, New York (2006)

19. Zhou, Z., Xu, J.: On the relation between multi-instance learning and semi-supervised learning. In: Proc. 24th ICML, pp. 1167–1174 (2007)

20. Gehler, P., Chapelle, O.: Deterministic annealing for multiple-instance learning. In: Proc. 11th Int'l Conf. AISTAT, pp. 123–130 (2007)

21. Li, F., Sminchisescu, C.: Convex Multiple-Instance Learning by Estimating Likelihood Ratio. In: Adv. Neu. Inf. Proc. Sys., pp. 1–8 (2010)

Stable L2-Regularized Ensemble Feature Weighting

Yun Li[1], Shasha Huang[1], Songcan Chen[2], and Jennie Si[3]

[1] College of Computer Science, Nanjing University of Posts and Telecommunications,
Nanjing 210023, China
[2] College of Computer Science and Technology, Nanjing University of Aeronautics
and Astronautics, Nanjing 210016, China
[3] School of Electronic Computer and Energy Engineering, Arizona State University,
Tempe 85281, USA
liyun@njupt.edu.cn, s.chen@nuaa.edu.cn, si@asu.edu

Abstract. When selecting features for knowledge discovery applications, stability is a highly desired property. By stability of feature selection, here it means that the feature selection outcomes vary only insignificantly if the respective data change slightly. Several stable feature selection methods have been proposed, but only with empirical evaluation of the stability. In this paper, we aim at providing a try to give an analysis for the stability of our ensemble feature weighting algorithm. As an example, a feature weighting method based on L2-regularized logistic loss and its ensembles using linear aggregation is introduced. Moreover, the detailed analysis for uniform stability and rotation invariance of the ensemble feature weighting method is presented. Additionally, some experiments were conducted using real-world microarray data sets. Results show that the proposed ensemble feature weighting methods preserved stability property while performing satisfactory classification. In most cases, at least one of them actually provided better or similar tradeoff between stability and classification when compared with other methods designed for boosting the stability.

1 Introduction

High dimensional data poses challenges into learning tasks due to the curse of dimensionality. In the presence of many irrelevant features, learning models tend to overfit and become less comprehensible. Feature selection has been an active area in machine learning for decades. It is an important and frequently used technique in data mining for dimension reduction via removing irrelevant and redundant features. Various studies show that features can be removed without performance deterioration [1]. Moreover feature selection brings the immediate effects of speeding up a data mining algorithm, enhancing generalization performances and allowing insights into the problem through the interpretation of the most relevant features [2]. Feature selection has been widely applied to many research fields such as genomic analysis [3], text mining [4], etc. A feature selection algorithm is usually associated with two important aspects: search strategy and evaluation criterion. Algorithms designed with different strategies broadly fall into three categories: filter, wrapper and embedded models [5]. Alternatively, according to the outcomes, feature selection algorithms can be divided into either feature weighting (ranking) algorithms or subset selection algorithms. A comprehensive surveys of

Z.-H. Zhou, F. Roli, and J. Kittler (Eds.): MCS 2013, LNCS 7872, pp. 167–178, 2013.

existing feature selection techniques and a general framework for their unification can be found in [2, 5–7].

Various feature selection algorithms have been developed with a focus on improving classification accuracy while reducing dimensionality [2, 5, 6]. Besides high accuracy, another important issue is stability of feature selection - the insensitivity of the result of a feature selection algorithm to variations of the training set [8–10]. This issue is particularly critical for applications where feature selection is used as a knowledge discovery tool for identifying characteristic markers to explain the observed phenomena. For example, in microarray analysis, biologists are interested in finding a small number of features (genes or proteins) that explain the mechanisms driving different behaviors of microarray samples. A feature selection algorithm often selects largely different subsets of features under variations to the training data, although most of these subsets are as good as each other in terms of classification performance [10]. Such instability dampens the confidence of domain experts when experimentally validating the selected features. More concrete example, in analyzing cancer biomarkers such as leukemia, the available data sets usually are high dimensional yet with small sample size. Among the thousands of genetic expression levels, a critical subset is to be discovered that links to two leukemia labels. It is therefore necessary that the selected predictive genes are common to variations of training samples. Otherwise the results will lead to less confident diagnosis. Stable feature selection has been demonstrated in biomarker identification via empirical process [11, 12]. Moreover, stability is desirable in algorithm designing since it is believed to lead to good generalization ability [13, 14].

For the stable feature selection methods, ensemble technique is among the most powerful to improve the stability of feature selection [8, 10, 15, 16]. Similar to the case of supervised learning, the general idea is to repeat the feature selection process on many randomly perturbed training sets (e.g., by bootstrapping the samples in the original training set), and aggregate the outputs in this procedure. Indeed, in large feature/small sample size domains it is often reported that several different feature subsets may yield equally optimal results [17], and ensemble feature selection may reduce the risk of choosing an unstable subset. Furthermore, different feature selection algorithms may yield feature subsets that can be considered local optima in the feature subsets, and ensemble feature selection might give a better approximation to the optimal subset or ranking of features. Finally, the representational power of a particular feature selector might constrain its search space such that optimal subsets cannot be reached. Ensemble feature selection could help in alleviating this problem by aggregating the outputs of several feature selectors [8].

However, the existing stable feature selection algorithms are always short of one important aspect, that is a theoretic stability analysis of the selection algorithm. It is imperative to go beyond simple and empirical evaluation. Therefore in this study, as an example, a special feature weighting algorithm is introduced, which is based on L2-regularized logistic regression loss for describing the local data structure. And its ensemble version using linear aggregation strategy is also proposed. Moreover, the proof for the stability of ensemble feature weighting about small changes (changing or removal of one sample) of data set is also presented based on uniform stability.

The introduced feature weighting algorithm is under embedded model and outputs a feature weights (measuring features' relevance) vector.

The paper is organized as follows, a feature weighting algorithm based on L2-regularized logistic loss and its ensemble version is introduced in section 2. Section 3 provides the proof for stability of ensemble feature weighting. The experimental results are shown in section 4, the paper ends with conclusion in section 5.

2   Ensemble Feature Weighting

To train a ensemble model for feature weighting, we are given a training sample set D, which contains n samples, $D = \{X, Y\} = \{x_i, y_i\}_{i=1}^n$, where x_i is the input for the i-th training sample, and y_i is the label, and each sample is represented by a d-dimensional vector $x_i = (x_{i1}, x_{i2}, \cdots, x_{id}) \in R^d$.

2.1   Feature Weighting Algorithm

In general, in order to achieve good generalization, the nearest neighbors with the same label to a sample (i.e., target samples) always should be closer to the sample, while other samples from different classes are separated by a large margin. Based on local learning, for sample x_i, it should be close to the nearest target sample (i.e., nearest hit sample $NH(x_i)$) and away from the nearest neighbor sample with different class label (i.e., near miss sample $NM(x_i)$). For the purposes of this paper, we use the Manhattan distance to define the nearest neighbors and their closeness, while other standard definitions may also be used. The logistic regression loss is adopted to model the fit of data for its simplicity and effectiveness. To prevent from overfitting and improve the robustness of feature weighting, the L2-regularization is used for its rotational invariance [1] and strong stability property [18]. Thus, the evaluation criterion for feature weighting is defined as follows,

$$L_D(w) = \frac{1}{n} \sum_{i=1}^n \log(1 + \exp(-w^T z_i)) + \gamma.||w||^2 \tag{1}$$

where γ is the cost parameter balancing the importance of the two terms, T is the transpose, w is the feature weight vector, $z_i = |x_i - NM(x_i)| - |x_i - NH(x_i)|$ and $|.|$ is an element-wise absolute operator. The z_i can be considered as the mapping point of x_i [9].

In the Eqn. (1), $w^T z_i$ is the local margin for x_i, which belongs to hypothesis margin [19] and an intuitive interpretation of this margin is a measure of the proportion of the features in x_i that can be corrupted by noise (or how much x_i can move in the feature space) before being misclassified [20]. By the large margin theory [21], a classifier that minimizes a margin-based error function usually generalizes well on unseen test data. Then one natural idea is to scale each feature, and thus obtain a weighted feature space parameterized by a vector w, so that a margin-based error function in the induced feature space is minimized. In the end, feature selection aims to find the target model w, which minimizes the loss function in Eqn.(1) through gradient descent-based techniques.

2.2 Weight-Based Ensemble Feature Weighting

Similar to the ensemble models for supervised learning, there are two essential steps in ensemble feature selection. The first step involves creating a set of different base feature selectors, each provides its output, while the second step aggregates the results of all feature selectors [8]. We adopt a subsampling based strategy and linear aggregation. Then m subsamples of size $\alpha n (0 < \alpha < 1)$ are drawn randomly from D, where the parameters m and α can be varied. Subsequently, feature weighting is performed on each of the m subsamples. Therefore, we obtain feature weighting results ensemble $En = \{w_1, w_2, \cdots, w_m\}$, where $w_t(t = 1, 2, \cdots, m)$ represents the outcome of the t-th base feature selector trained on t-th subsample. Specifically, in our case, each feature selection result $w_t(t = 1, 2, \cdots, m)$ is a feature weighting vector. And we obtain the final ensemble feature weighting result $w_e = \frac{1}{m} \sum_{t=1}^{m} w_t$, where $w_t \in En$. This ensemble method belongs to weight-based ensemble model (WEn).

The proposed ensemble feature weighting is also corresponding to the recognition that when estimating an unknown function from data, one needs to find a tradeoff between bias and variance [13]. Indeed, besides the regularization, another idea is to use statistical procedures to reduce the variance without altering the bias and lead to high stability. One such technique is the bagging approach [22], which consists in averaging several estimators built from random subsamples of the data.

3   Stability Analysis

Now, we will firstly show the rotation invariance for our proposed feature weighting algorithm. Based on the Proposition 4.2 in [1], let H be a rotational matrix $\{H \in \mathcal{R}^{d \times d}, H^T H = HH^T = I, |H| = 1\}$, then Hx is x rotated through some angle around the origin. It is evident that loss function $L_D(w)$ is rotational invariance with respect to H. In other words, $L_D(w) = \frac{1}{n} \sum_{i=1}^{n} \log(1 + \exp(-(Hw)^T (Hz_i))) + \gamma.||Hw||^2$, and $Hz_i = H(|x_i - NM(x_i)| - |x_i - NH(x_i)|) = |Hx_i - H.NM(x_i)| - |Hx_i - H.NH(x_i)|$, which means the proposed feature weighting algorithm is rotational invariance for sample x_i. The linear aggregation strategy is employed, then intuitively the ensemble feature weighting is also rotational invariance.

3.1   Uniform Stability Definition

On the other hand, a stable algorithm is one whose output does not change significantly with small changes in the input. The stability of classification, regression and sample ranking has been deeply analyzed [13, 14, 23], however, the stability of feature selection has not been explicitly introduced in theory. Similarly, we also consider changes to such a sample that consist of replacing a single example in the sequence with a new example or the exclusion of the sample. For a given training set D of size n, we will denote $D^{\backslash i}$ as the training set obtained by removing point (x_i, y_i) for all $i \in \{1, \cdots, n\}$. And we denote by D^i the training set obtained by changing one point (x_i, y_i) into (x_i', y_i'), which is assumed to be independent from D.

Definition 1. (Uniform weighting stability) For a feature selection algorithm A whose outputs on data set D and $D^{\backslash i}$ are weight vectors denoted by w_D and $w_{D^{\backslash i}}$, respectively. Algorithm A has uniform weighting stability β ($\beta \geq 0$) if for all D and any $i \in \{1, \cdots, n\}$, we have

$$||w_D - w_{D \backslash i}|| \leq \beta.$$

A smaller value of β corresponds to greater weighting stability. More formally, point (x_i, y_i) is replaced by the empty set which we assume the learning method treats as having this point simply removed, and the D^i can be regarded as one data set is firstly replaced by the empty set and then the empty set is replaced by point (x'_i, y'_i). So an feature weighting algorithm with uniform stability β has also the following property: For all D and $i \in \{1, \cdots, n\}$, $||w_D - w_{D^i}|| \leq ||w_D - w_{D \backslash i}|| + ||w_{D^i} - w_{D \backslash i}|| \leq 2\beta$. In other words, stability with respect to the exclusion of one point implies stability with respect to changes of one point. Then we only focus on stability analysis for the exclusion case in the follows.

3.2 Stability for Ensemble Feature Weighting

For ensemble, bootstrap strategy is used to train the same feature weighting algorithm on a number m of different bootstrap sets of a training set D and by averaging the obtained solutions. We denote these bootstrap sets by $D(r_t)$ for $t = 1, \cdots, m$, where the $r_t \in R = \{1, \cdots, n\}^p (p < n)$ are instances of a random variable corresponding to sampling without replacement of p elements from the training set D. And R is a space containing elements r that model the randomization of the subsampling. We will use the shorthand $w_{D(r_t)}$ to denote the outcome of the feature weighting algorithm applied on the t-th bootstrap training set $D(r_t)$. And the ensemble result is $\frac{1}{m} \sum_{t=1}^{m} w_{D(r_t)}$. The uniform weighting stability of ensemble feature weighting is defined as follows: For all D and $i \in \{1, \cdots, n\}$.

$$\beta_e = \mathbb{E}_{r_1, \cdots, r_m} [||\frac{1}{m} \sum_{t=1}^{m} w_{D(r_t)} - \frac{1}{m} \sum_{t=1}^{m} w_{D \backslash i(r_t)}||]$$

where $\mathbb{E}$ is the expectation and $r_1, \cdots, r_m$ are i.i.d. random variables modeling the random sampling and having the same distribution as r. Then

$$\beta_e \leq \frac{1}{m} \sum_{t=1}^{m} \mathbb{E}_{r_t} [||w_{D(r_t)} - w_{D \backslash i(r_t)}||]$$

$$= \mathbb{E}_r [||w_{D(r)} - w_{D \backslash i(r)}||] = \mathbb{E}_r [|| \triangle w_{D(r)}||]$$

The stability for the removal of $x_i (i \in \{1, 2, \cdots, n\})$ case is considered, then the random sampling containing sample x_i should be determined.

$$\beta_e \leq \mathbb{E}_r [|| \triangle w_{D(r)}||(\mathbb{I}(i \in r) + \mathbb{I}(i \notin r))]$$

$$= \mathbb{E}_r [|| \triangle w_{D(r)}||\mathbb{I}(i \in r)] + \mathbb{E}_r [|| \triangle w_{D(r)}||\mathbb{I}(i \notin r)]$$

$$= \mathbb{E}_r [|| \triangle w_{D(r)}||\mathbb{I}(i \in r)]$$

where $\mathbb{I}(.)$ is indicator function. Note that the second part of the last second equation is equal to zero because when i is not in r, which means point x_i does not belong to $D(r)$

and, thus, $D(r) = D^{\setminus i}(r)$. The size of subsample $D(r)$ is p, then $\mathbb{E}_r(\mathbb{I}(i \in r)) = \frac{p}{n}$ because this subsampling is done without replacement,

$$\beta_e \leq \frac{p}{n} \| \triangle w_{D(r)} \|$$

where $\| \triangle w_{D(r)} \|$ is the uniform stability of base feature weighting on bootstrap set $D(r)$ that contains sample x_i, and $\triangle w_{D(r)} = w_{D(r)} - w_{D^{\setminus i}(r)}$ where $w_{D(r)}$ and $w_{D^{\setminus i}(r)}$ is the minimizer for the convex objective function $L_{D(r)}(w)$ and $L_{D^{\setminus i}(r)}(w)$ respectively. According to Eqn.(1), these objective functions are defined as follows,

$$L_{D(r)}(w) = \frac{1}{p} \sum_{j=1}^{p} \log(1 + \exp(-w^T z_j)) + \gamma. \|w\|^2$$

$$L_{D^{\setminus i}(r)}(w) = \frac{1}{p} \sum_{j=1, j \neq i}^{p} \log(1 + \exp(-w^T z_j)) + \gamma. \|w\|^2$$

Due to the convexity of the objective functions, for any $a \in [0, 1]$, we get

$$L_{D(r)}(w_{D(r)}) - L_{D(r)}(w_{D(r)} - a \triangle w_{D(r)}) \leq 0$$
$$L_{D^{\setminus i}(r)}(w_{D^{\setminus i}(r)}) - L_{D^{\setminus i}(r)}(w_{D^{\setminus i}(r)} + a \triangle w_{D(r)}) \leq 0$$

So summing the two equations above, we get that

$$\frac{1}{p} \sum_{j=1, j \neq i}^{p} \log(1 + \exp(-w_{D(r)}^T z_j)) + \frac{1}{p} \log(1 + \exp(-w_{D(r)}^T z_i))$$

$$- \frac{1}{p} \sum_{j=1, j \neq i}^{p} \log(1 + \exp(-(w_{D(r)} - a \triangle w_{D(r)})^T z_j))$$

$$- \frac{1}{p} \log(1 + \exp(-(w_{D(r)} - a \triangle w_{D(r)})^T z_i))$$

$$+ \frac{1}{p} \sum_{j=1, j \neq i}^{p} \log(1 + \exp(-w_{D^{\setminus i}(r)}^T z_j)) \tag{2}$$

$$- \frac{1}{p} \sum_{j=1, j \neq i}^{p} \log(1 + \exp(-(w_{D^{\setminus i}(r)} + a \triangle w_{D(r)})^T z_j))$$

$$+ \gamma. \|w_{D(r)}\|^2 - \gamma. \|w_{D(r)} - a \triangle w_{D(r)}\|^2$$

$$+ \gamma. \|w_{D^{\setminus i}(r)}\|^2 - \gamma. \|w_{D^{\setminus i}(r)} + a \triangle w_{D(r)}\|^2$$

$$\leq 0$$

Since logistic loss is the convex function, then by Jensen's inequality,

$$\log(1 + \exp(-(w_{D(r)} - a \triangle w_{D(r)})^T z_j)$$
$$= \log(1 + \exp(-((1-a)w_{D(r)}^T z_j + a w_{D^{\setminus i}(r)}^T z_j))$$
$$\leq \log(1 + \exp(-w_{D(r)}^T z_j))$$
$$- a(\log(1 + \exp(-w_{D(r)}^T z_j)) - \log(1 + \exp(-w_{D^{\setminus i}(r)}^T z_j)))$$

Similarly, we also can get

$$\log(1 + \exp(-(w_{D\backslash i(r)} + a \triangle w_{D(r)})^T z_j))$$
$$\leq \log(1 + \exp(-w_{D\backslash i(r)}^T z_j)$$
$$+ a(\log(1 + \exp(-w_{D(r)}^T z_j)) - \log(1 + \exp(-w_{D\backslash i(r)}^T z_j)))$$

The two equations above are plugged into (2), then

$$||w_{D(r)}||^2 - ||w_{D(r)} - a \triangle w_{D(r)}||^2 - ||w_{D\backslash i(r)} + a \triangle w_{D(r)}||^2 + ||w_{D\backslash i(r)}||^2$$
$$\leq \frac{a}{p\gamma}(\log(1 + \exp(-w_{D\backslash i(r)}^T z_i)) - \log(1 + \exp(-w_{D(r)}^T z_i)))$$
$$\leq \frac{a}{p\gamma}| \triangle w_{D(r)}^T z_i|$$

and the last line above is gotten because it is proved in [24] that the logistic loss function is a Lipschitz function with Lipschitz constant 1. If we set $a = 1/2$, the left side of previous equation approximately amounts to

$$||w_{D(r)}||^2 + ||w_{D\backslash i(r)}||^2 - \frac{1}{2}||w_{D(r)} + w_{D\backslash i(r)}||^2$$
$$= \frac{1}{2}||w_{D(r)}||^2 + \frac{1}{2}||w_{D\backslash i(r)}||^2 - w_{D(r)}^T w_{D\backslash i(r)} = \frac{1}{2}|| \triangle w_{D(r)}||^2$$

Thus,

$$|| \triangle w_{D(r)}||^2 \leq \frac{1}{p\gamma}| \triangle w_{D(r)}^T z_i|$$

and based on Cauchy-Schwarz inequality

$$| \triangle w_{D(r)}^T z_i| \leq || \triangle w_{D(r)}||||z_i||$$

Then combine the two equations above and the samples are normalized, it can be shown $||z_i|| \leq 2$, we obtain the stability for our base feature weighting.

$$||w_{D(r)} - w_{D\backslash i(r)}|| = || \triangle w_{D(r)}|| \leq \frac{2}{p\gamma}$$

and the uniform stability for ensemble feature weighting is

$$\beta_e \leq \frac{2}{n\gamma}$$

3.3 Remarks and Discussions

The analysis results show that a larger regularization parameter γ leads to better stability. And the ensemble feature weighting owns better stability bounds than base feature weighting. The ensemble algorithm has a uniform stability bound goes to zero as $\frac{1}{n\gamma}$, and it is stable because of the wide acceptance that the algorithms have a uniform stability bound that decreases as $O(\frac{1}{n})$, and are hence stable [13, 14, 23, 18]. To our best knowledge, this work provides the first uniform stability-style analysis on the stability of feature selection. Although in [2], the analysis for robustness of spectral feature selection against noise is presented, and in [9], the experimental results also show the variance reduction leads to stable feature selection. It is obvious that our work is significantly different because formal stability notion is considered explicitly, and we mainly focus on sampling randomness instead of noise, and we thus are interested in how changes to the training data influence the result of feature weighting algorithm.

Moreover, the L2-norm is used as stability metric in the paper, this is only for ease of presentation. Other norm can be adopted, such as $L\infty$-norm, which is employed to measure the uniform stability of classification and regression algorithm [13, 14]. Certainly, the proof for the stability also makes sense because of $L\infty - norm \leq L2 - norm$ in most cases. And it should be noted that the stability bound is loose, and we only like to prove that the proposed ensemble feature weighting algorithm is stable because its stability scales like $\frac{1}{n}$.

Finally, it is evident the theoretical analysis results still hold true for other ensemble feature weighting algorithms where base feature weighting algorithm is based on L2-regularized convex Lipschitz loss functions and linear aggregation strategy is employed.

4   Experiments

In order to validate the performance of our ensemble algorithm, the experiments are conducted on several real-world data sets to show its stability and classification power. The data sets consist of small samples with high dimension, medium samples and large samples with low dimension. The chosen data sets are Sonar, Arcene, Musk, Ionosphere, which are taken from UCI ML repository [25], and Colon cancer diagnosis data set is introduced in [26] and Lung cancer is introduced in [27]. Colon, Arcene and Lung owns small samples (62,200,203) with extremely high dimensionality (2000,10000,12600). The small sample problem is one of the most challenging problem for feature selection, particularly on its output stability.

Note that feature weighting is almost never directly used to measure the stability of feature selection, and instead converted to a ranking based on the weights [8]. Because the feature weights are always changed to feature ranks, then another ensemble strategy should be considered: instead of the feature weights linear combination, the feature weight vectors outputted from the m subsamples are firstly changed to feature rank vectors (Noting that the ranking value for a feature is set as follows: The best feature with the largest weight is assigned rank 1, and the worst one rank d), then linear combination of these feature rank vectors is adopted to obtain the ensemble ranking results as in [8]. And we call this ensemble strategy as rank-based ensemble(REn). Other chosen stable feature weighting algorithms for comparison are ensemble Relief (En-Relief) [8] and newly proposed stable feature selection strategy based on variance reduction, which is to assign different weights to different samples based on margin, and then to obtain high stability for feature selection [9]. We combine the sample weighting strategy with the newly proposed feature weighting algorithm-Lmba [28] and named as VR-Lmba.

4.1   Experimental Results for Stability

To measure the stability of feature weighting algorithms, we also adopt a subsampling based strategy-bootstrap without replacement. For a data set, 10 subsamples containing 90% of the data are randomly drawn without replacement. This percentage was chosen as in [8] to assess robustness with respect to relatively small changes in the data set. Of course, the sampling rate and the number of subsamples can be varied. Subsequently, the proposed ensemble algorithms (WEn and REn) with $\alpha = 0.9$ and $\gamma = 1$, En-Relief and VR-Lmba is performed on each subsample, which is considered as the data set D described in section 2, and output a feature rank vector (if the output is a feature weight vector, it should be changed to feature rank vector). Then the similarity between feature ranking result pairs is calculated using Spearman rank correlation coefficient [8], and the stability is the average similarity over all pairwise similarity between the different feature ranking results [8]. The stability of these feature weighting algorithms for

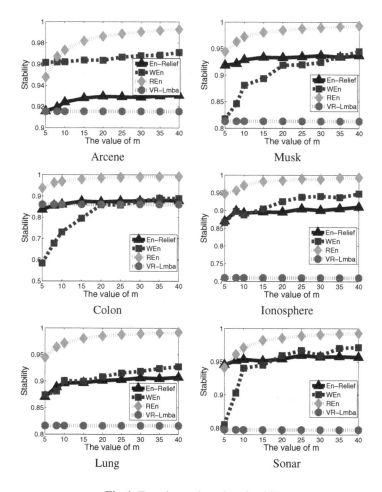

Fig. 1. Experimental results of stability

different data sets is shown in Fig.1. The X-axis is the number of base feature selectors m. Note that VR-Lmba is not a ensemble method, then its stability does not change along with m.

4.2   Balance between Stability and Classification

Besides the stability, classification performance is another important issues for feature selection. In order to validate the tradeoff between the stability and classification accuracy, a F-Measure is employed, which is defined as $\frac{2 \times stability \times accuracy}{stability + accuracy}$ [8]. In this part of experiments, the number of base selectors for ensemble feature weighting is constant and set as 20 for all ensemble algorithms, i.e., $m = 20$. 10-cross validation is used and the linear SVM with C=1 and 3-nearest neighbors(3NN) classifier is adopted. The experimental results are shown in Fig. 2 and 3 corresponding to 3NN and SVM. For space constraints, only the experimental results of two data sets for each classifier are shown in the figures.

4.3   Observations and Discussions

From the results, we can observe that the stability value of rank-based ensemble-REn is the highest among all stable feature weighting algorithms, and weight-based ensemble-WEn always gets higher or similar stability to En-Relief and VR-Lmba. In addition,

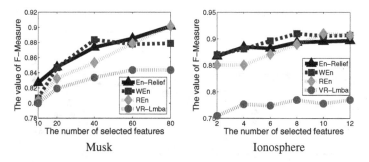

Musk   Ionosphere

Fig. 2. Experimental results of F-Measure for 3NN classifier

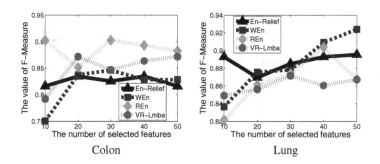

Colon   Lung

Fig. 3. Experimental results of F-Measure for SVM

at least one of our proposed ensemble methods (REn or WEn) always obtain higher or similar balance between the stability and classification accuracy to other stable ones.

For the higher stability of rank-based ensemble-REn than weight-based ensemble-WEn, this can be explained intuitively by the fact that the stability is measured based on the feature ranks. Consider the scenarios if the feature weights produced by base feature weighting algorithm change due to the data variation, however, their ranks may not change, which leads to higher stability for rank-based ensemble than weight-based ensemble. Of course, if the feature weights do not change, then the feature ranks surely stable. Thus it means that the stable weight-based ensemble leads to stable rank-based ensemble, then the theoretic analysis of stability for weight-based ensemble hold true for the rank-based ensemble. And the above analysis also give some reasons for the high efficiency of En-Relief, which also belongs to rank-based ensemble model.

5 Conclusion

The stability of feature selection is attracted much attention. Our major contribution is presenting the theoretical analysis for the uniform stability of ensemble feature weighting algorithm. In the paper, as an example, a logistic loss-based feature weighting algorithm via L2-regularization is introduced. And its weight-based ensemble version-WEn is presented and is formally analyzed on the stability. The experimental results on some real-world data sets including microarray data (small sample size problem) have also shown the proposed ensemble feature weighting algorithms (weight-based ensemble-WEn or rank-based ensemble-REn) get higher stability and better or comparable tradeoff between classification and stability to other stable algorithms in most cases. In our analysis, the linear combination is adopted in ensemble feature weighting, other combination scheme is our future work.

Acknowledgments. This work is, in part, supported by NSFC Grant (60973097, 61035003, 61073114 and 61105082) and Jiangsu Government Scholarship.

References

1. Ng, A.Y.: Feature selection, l1 vs. l2 regularization, and rotational invariance. In: Proceedings of International Conference on Machine Learning, Banff, Canada (2004)
2. Zhao, Z.: Spectral Feature Selection for Mining Ultrahigh Dimensional Data. PhD thesis, Arizona State University (2010)
3. Inza, I., Larranaga, P., Blanco, R., Cerrolaza, A.J.: Filter versus wrapper gene selection approaches in dna microarray domains. Artificial Intelligence in Medicine 31, 91–103 (2004)
4. Forman, G.: An extensive empirical study of feature selection metrics for text classification. Journal of Machine Learning Research 3, 1289–1305 (2003)
5. Liu, H., Yu, L.: Toward integrating feature selection algorithms for classification and clustering. IEEE Trans. Knowledge and Data Engineering 17, 494–502 (2005)
6. Guyon, I., Gunn, S., Nikravesh, M., Zadeh, L.: Feature Extraction, Foundations and Applications. Springer, Physica-Verlag, New York (2006)
7. Guyon, I., Elisseeff, A.: An introduction to variable and feature selection. Journal of Machine Learning Research 31, 1157–1182 (2003)

8. Saeys, Y., Abeel, T., Van de Peer, Y.: Robust feature selection using ensemble feature selection techniques. In: Daelemans, W., Goethals, B., Morik, K. (eds.) ECML PKDD 2008, Part II. LNCS (LNAI), vol. 5212, pp. 313–325. Springer, Heidelberg (2008)

9. Han, Y., Yu, L.: A variance reduction for stable feature selection. In: Proceedings of the International Conference on Data Mining, pp. 206–215 (2010)

10. Loscalzo, S., Yu, L., Ding, C.: Consensus group stable feature selection. In: Proceedings of ACM SIGKDD Conference on Knowledge Discovery and Data Mining, pp. 567–575 (2009)

11. Abeel, T., Helleputte, T., Van de Peer, Y., Dupont, P., Saeys, Y.: Robust biomarker identification for cancer diagnosis with ensemble feature selection methods. Bioinformatics 26, 392–398 (2010)

12. Yu, L., Han, Y., Berens, M.E.: Stable gene selection from microarray data via sample weighting. IEEE/ACM Trans. Computational Biology and Bioinformatics 9, 262–272 (2012)

13. Bousquet, O., Elisseeff, A.: Stability and generalization. Journal of Machine Learning Research 2, 499–526 (2002)

14. Elisseeff, A., Evgeniou, T., Pontil, M.: Stability of randomized learning algorithm. Journal of Machine Learning Research 6, 55–79 (2005)

15. Li, Y., Gao, S.Y., Chen, S.C.: Ensemble feature weighting based on local learning and diversity. In: AAAI Conference on Artificial Intelligence, pp. 1019–1025 (2012)

16. Woznica, A., Nguyen, P., Kalousis, A.: Model mining for robust feature selection. In: Proceedings of ACM SIGKDD Conference on Knowledge Discovery and Data Mining, pp. 913–921 (2012)

17. Saeys, Y., Inza, I., Larranaga, P.: A review of feature selection techniques in bioinformatics. Bioinformatics 23, 2507–2517 (2007)

18. Xu, H., Caramanis, C., Mannor, S.: Sparse algorithm are not stable: A no-free-lunch theorem. IEEE Trans. Pattern Analysis and Machine Intelligence 34, 187–193 (2012)

19. Crammer, K., Bachrach, R.G., Navot, A., Tishby, N.: Margin analysis of the lvq algorithm. In: Advances in Neural Information Processing Systems, pp. 462–469 (2002)

20. Sun, Y.J., Todorovic, S., Goodison, S.: Local learning based feature selection for high dimensional data analysis. IEEE Trans. Pattern Analysis and Machine Intelligence 32, 1–18 (2010)

21. Schapire, R.E., Freud, Y., Bartlett, P., Lee, W.S.: Boosting the margin: A new explanation for the effectiveness of voting methods. The Annals of Statistics 26, 1651–1686 (1998)

22. Breiman, L.: Bagging predictors. Machine Learning 26, 123–140 (1996)

23. Agarwal, S., Niyogi, P.: Generalization bounds for ranking algorithm via algorithmic stability. Journal of Machine Learning Research 10, 441–474 (2009)

24. Anthony, M., Bartlett, P.L.: Neural Network Learning: Theoretical Foundations. Cambridge University Press, Cambridge (1999)

25. Frank, A., Asuncion, A.: UCI machine learning repository (2010),
http://archive.ics.uci.edu/ml

26. Alon, U., Barkai, N., Notterman, D.A., Gish, K., Ybarra, S., Mack, D., Levine, A.J.: Broad patterns of gene expression revealed by clustering analysis of tumor and normal colon cancer tissues probed by oligonucleotide arrays. Proceedings of the National Academy of Sciences of the United States of America, 6745–6750 (1999)

27. Bhattacharjee, A., Richards, W.G., Staunton, J., Li, C., Monti, S.: Classification of human lung carcinomas by mrna expression profiling reveals distinct adenocarcinoma subclasses. Proceedings of the National Academy of Sciences of the United States of America 98, 13790–13795 (2001)

28. Li, Y., Lu, B.L.: Feature selection based on loss margin of nearest neighbor classification. Pattern Recognition 42, 1914–1921 (2009)

Selective Clustering Ensemble Based on Covariance

Xuyao Lu, Yan Yang⋆, and Hongjun Wang

School of Information Science & Technology
Southwest Jiaotong University
Chengdu, 610031, P.R. China
lxy@my.swjtu.edu.cn, {yyang,wanghongjun}@home.swjtu.edu.cn

Abstract. Clustering Ensemble effectively improves clustering accuracy, stability and robustness, which is most resulted from the diversity of the base clustering results. It is a key point to measure the diversity of clustering results. This paper proposes a method to measure diversity of base clustering results and a covariance-based selective clustering ensemble algorithm. Experiments on 20 UCI data sets show that this algorithm effectively improves the clustering performance.

Keywords: Clustering ensemble, Covariance, Selective ensemble.

1 Introduction

Clustering is the process of splitting the set of physical or abstract objects into similar object classes [1]. It is a high similarity between the same class and a great difference between the different classes. Ensemble learning gets base learners by different methods, and obtains a final results by combining base learners in some way [2]. Strehl et al. [3] proposed the clustering ensemble in 2002. Clustering ensemble is a method of aggregating the multiple division collection of one object into a final clustering result. Clustering ensemble effectively reduces the impact on noise and outliers, and increases the clustering stability and robustness.

Recent years, there are many research works in clustering ensemble. Topchy et al. [4] designed a mixture model for clustering ensemble, and they offered a probabilistic model of consensus with a finite mixture of multinomial distributions in a space of clustering. A new consensus function by the generalized mutual information was proposed in [5]. Luo et al. [6] used information theory to design a genetic algorithm to combine multiple clusterings. Hassan et al. [7] developed a ensemble method with majority voting and parallel fusion in conjunction with a neural classifier. Mohammadi et al. [8] stated an evolutionary approach to clustering ensemble, and they used an evolutionary combinational clustering method to find the number of clusters. Iqbal et al. [9] proposed the semi-supervised clustering ensemble by voting, and they introduced a flexible two parameters weighting mechanism in their algorithm. The semi-supervised

---

⋆ Corresponding author.

Z.-H. Zhou, F. Roli, and J. Kittler (Eds.): MCS 2013, LNCS 7872, pp. 179–189, 2013.
© Springer-Verlag Berlin Heidelberg 2013

cluster ensemble model based on bayesian network was designed. And the variational inference oriented semi-supervised cluster ensemble is illustrated in this paper [10]. Jia et al. [11] presented a bagging-based spectral clustering ensemble selection. Yang et al. [12] presented a semi-supervised clustering ensemble based on multi-ant colonies, and they incorporated pairwise constraints not only in each ant colony clustering process, but also in computing new similarity matrix. Iam-On et al. [13] advanced a link-based cluster ensemble approach for categorical data clustering, and they improved the conventional matrix by discovering unknown entries through similarity between clusters in an ensemble.

There are also some disadvantages when the number of base clusterings is large. For example, computing and storage overhead of system is greatly increased, and the difference between the base clusterings will continue to decrease. Zhou et al. [14] proposed the selective ensemble, and proved that the performance of the integration of some clustering results is better than the integration of all clustering results. Fern et al. [15] designed three different selection approaches of JC (Joint Criterion), CAS (Cluster and Select), CH (Convex Hull) that jointly consider quality and diversity. Azimi et al. [16] presented an adaptive cluster ensemble selection, and they proposed a novel framework that selects ensemble members for each data set based on its own characteristics. Jia et al. [17] developed a similarity-based spectral clustering ensemble selection, and they used the random scaling parameter, Nyström approximation and random initialization of k-means to perturb spectral clustering for producing the components of an ensemble system. Liu et al. [18] advanced a new selective clustering ensemble algorithm, they used the compactness and the separation to measure the quality of the clustering and defined the connectivity matrix to measure the quality and diversity.

We propose a new method based on covariance to measure the diversity. Firstly, base clustering results are generated by K-Means, AP, and FCM. Secondly, we calculate the covariance between each of the two base clustering results, and generate covariance matrix. Finally, part of base clustering results with small covariance are chosen to ensemble by CSPA.

The rest of the paper is organized as follows. Section 2 describes the related work. Section 3 introduces the principle of selective cluster ensemble based on covariance. Section 4 reports the experimental results. Section 5 provides conclusions and future work.

2   Related Work

It is a key point to measure the diversity of clustering results in selective clustering ensemble. Fern [19] used the normalized mutual information (NMI) to measure the diversity of clustering results.

$$NMI = \frac{I(X, Y)}{\sqrt{H(X)H(Y)}} \tag{1}$$

where $I(X, Y)$ is the mutual information of random variable X and Y, $I(X, Y) = \sum_{x,y} p(x, y) log \frac{p(x,y)}{p(x)p(y)}$, $H(X)$ is the he entropy of the X, $H(Y)$ is the entropy of the Y, and $H(X) = \sum_x p(x) log \frac{1}{p(x)}$. The NMI value is between 0 and 1, the value is smaller, the diversity is lager. Unlike other measure methods, NMI is not biased by large clusters.

Derek [20] used a method based on entropy to measure the diversity of clustering results.

$$div(c) = \frac{2}{N(N-1)} \sum_{i=1}^{N-1} \sum_{j=i+1}^{N} -(p_{ij} log_2 p_{ij} + (1 - p_{ij}) log_2(1 - p_{ij})) \quad (2)$$

where p_{ij} is the probability of x_i and x_j are cluster in the same class, $p(x, y) = \frac{1}{k} \sum_{h=1}^{k} \delta(\pi_h(x_i), \pi_h(x_j))$, $\pi_h(x_i)$ is the label of the x_i in the class π_h, and $\pi_h(x_j)$ is the label of the x_j in the class π_h. If $\pi_h(x_i) = \pi_h(x_j)$, δ is 1, otherwise δ is 0. The value is also between 0 and 1, the value is smaller, the diversity is smaller.

Hadjitodorov [21] proposed four methods based on the adjusted rand index to measure the diversity of clustering results, and discovered the performance of the ensemble by middle value of diversity is better than the ensemble by max value of diversity.

$$ar(\pi_a, \pi_b) = \frac{\sum_{h=1}^{k_a} \sum_{l=1}^{k_b} \binom{n_{h,l}}{2} - t_3}{\frac{1}{2}(t_1 + t_2) - t_3} \quad (3)$$

where $t_1 = \sum_{h=1}^{k_a} \binom{n_h}{2}$, $t_2 = \sum_{l=1}^{k_b} \binom{n_l}{2}$, $t_3 = \frac{2 t_1 t_2}{N(N-1)}$, k_a and k_b are the number of clusters of π_a and π_b, respectively, $n_{h,l}$ is the number of points that are the same time in the cluster h and the cluster l, n_l is the number of points in the cluster l, and n_h is the number of points in the cluster h. The value is smaller, the diversity is lager. When two clusters are completely independent, the value is 0.

Luo [22] proposed five methods to measure the diversity, including CEBDM based on conditional entropy, DFBDM based on double fault measure, CFDBDM based on coincident failure diversity and IRABDM based on measurement of inter-rater agreement. The values of five methods are smaller, the diversity is smaller. Li [23] proposed a new method based on support vector machine to measure the diversity. Zhou [24] described in details some other methods of pairwise measures and non-pairwise measures, including Q-Statistic, Kohavi-Wolpert variance and so on.

3   Selective Clustering Ensemble Based on Covariance

We propose a new method based on covariance to measure the diversity. Covariance is a method used to measure the correlation between random variables, and the clustering result is deemed to the random variable, so the covariance is used to measure the diversity of clustering results. In addition, unlike NMI and

CE also consider expectation and variance after obtaining values, the covariance uses the expectation in calculating the value, so the covariance has been considered the problem of the offset. Let (X, Y) be a two-dimensional random variable, $E(X)$ and $E(Y)$ were the expectation of X and Y, respectively. $COV(X, Y)$ is the covariance between X and Y, as follows,

$$COV(X, Y) = E[(X - E(X))(Y - E(Y))] = E(XY) - E(X)E(Y). \quad (4)$$

Let $\pi(x_i)$ be the label of the x_i, $\pi(x_j)$ be the label of the x_j. We define a formula as follows,

$$\pi(x_i) - \pi(x_j) = \begin{cases} 1 & \pi(x_i) \neq \pi(x_j) \\ 0 & \pi(x_i) = \pi(x_j) \end{cases}. \quad (5)$$

For an n-dimensional random variable $X = (X_1, X_2, ..., X_n)$, let $\sigma_{ij} = COV(X_i, X_j)$, $i, j = 1, 2, ..., n$, it defines matrix V is the covariance matrix of X, and V is an n-order symmetric matrix.

$$V = \begin{bmatrix} \sigma_{11} & \cdots & \sigma_{1n} \\ \vdots & \ddots & \vdots \\ \sigma_{n1} & \cdots & \sigma_{nn} \end{bmatrix} \quad (6)$$

where $\sigma_{11} = COV(X_1, X_1)$ is the variance of X_1.

N clustering results are deemed to an n-dimensional random variable $X = (X_1, X_2, ..., X_n)$. The covariance matrix V is a symmetric matrix, and the values on the diagonal are variance. We only consider the difference between the base clustering results, and don't consider the positive correlation and negative correlation, so we simplify V to V' that all values are non-negative and values on the diagonal are 0. And it is

$$V = \begin{bmatrix} \sigma_{11} & \cdots & \sigma_{1n} \\ \vdots & \ddots & \vdots \\ \sigma_{n1} & \cdots & \sigma_{nn} \end{bmatrix} \longrightarrow V' = \begin{bmatrix} 0 & \sigma_{12} & \cdots & \sigma_{1n-1} & \sigma_{1n} \\ 0 & 0 & \cdots & \sigma_{2n-1} & \sigma_{2n} \\ \vdots & \vdots & \ddots & \vdots & \vdots \\ 0 & 0 & \cdots & 0 & \sigma_{(n-1)n} \\ 0 & 0 & \cdots & 0 & 0 \end{bmatrix}. \quad (7)$$

The steps of select base clustering results to ensemble(SBCRE) are shown in Algorithm 1. The input is m base clustering results, and the output is m ensemble results. Firstly, we calculate the covariance between two base clustering results, and generate covariance matrix V. Secondly, V is simplified to V'. Thirdly, we select the maximum value of V', remove the row with maximum, and set the maximum to 0. Fourth, the ensemble results are obtain by ensemble the remaining base clustering results with CSPA. Finally, the output is obtained with m iterations.

Algorithm 1. SBCRE

Input: m base clustering results
Output: m ensemble results
begin

Calculate the covariance between two base clustering results according to the formula (4), and generate covariance matrix V;
Simplify V to V' according to the formula (7);
if $m \geqslant 1$ **then**

Select the maximum value of V', record row number r and column number c;
Remove the rth row;
The maximum is set to 0, update V';
$m = m - 1$;
The ensemble results are obtain by ensemble the m base clustering results with CSPA.

end
end

We uses three different cluster methods of K-Means [25,26], AP [27] and FCM [28,29]. The 60 base clustering results are generated with different initialization. We get 60 ensemble results by Algorithm 1, and calculate the F-measure between each ensemble result and the label of each data set. The final result with maximum F-measure on each data set is obtained. The steps of selective clustering ensemble based on covariance(SCEBC) are shown in Algorithm 2.

Algorithm 2. SCEBC

Input: The data set X has n samples
Output: The set has labels of n samples
begin

Generate 20 base clustering results according to the K-Means;
Generate 20 base clustering results according to the AP;
Generate 20 base clustering results according to the FCM;
Get m ensemble results according to the Algorithm 1;
Calculate the F-measure between each ensemble result and the label of data set;
The ensemble result with the maximum F-measure as the output.
end

---

4   Experiment

4.1   Data Set

The 20 UCI data sets are used in the experiment. The number of features, classes and instances on each data set are shown in Table 1.

Table 1. The number of features, classes and instances on each data set

Data Set	Features	Classes	Instances
Iris	4	3	150
Glass	9	6	214
Wine	13	3	178
Zoo	16	7	101
Ionosphere	34	2	351
Sonar	60	2	208
Balance scale	4	3	625
Pima	8	2	768
Spect-heart	22	2	267
Hepatitis	19	2	155
Bupa	6	2	345
Habermans survival	3	2	306
Wdbc	30	2	569
Statlog	19	7	2310
Vehicle	18	4	846
Breast-cancer-Wisconsin	9	2	683
Car	6	4	1728
Credit-g	20	2	1000
Vowel	13	11	990
Lymphography	18	4	148

4.2   Evaluation Criteria

F-measure is the evaluation criteria of experiment results [30], and it is shown in formula (8).

$$F(i) = \frac{2 \times precision(i, j) \times recall(i, j)}{precision(i, j) + recall(i, j)} \qquad (8)$$

where $precison(i, j) = \frac{N_{ij}}{N_i}$ is the precision, $recall(i, j) = \frac{N_{ij}}{N_j}$ is the recall, N_i is the total number of samples of correct clustering, N_j is the total number of samples of jth class in clustering results, and N_{ij} is the total number of correct clustering of jth class in clustering results. However, the formula (8) will get a lot of $F(i)$ values, so the F-measure is weighted and averaged by formula (9), as follows,

$$F(i)' = \frac{\sum_{i=1}^{k}(|i| \times F(i))}{\sum_{i=1}^{k}|i|}. \qquad (9)$$

4.3   Experiment Result

The experiment results are reported in this subsection. The F-measures of different algorithms on each data set are shown in Table 2, where ALL is directly ensemble, RSE is selective ensemble based on random, and CSEV is average

Table 2. The F-measures of different algorithms on the each data set

Data Set	Base Clustering Methods			Clustering Ensemble by CSPA		
	K-means	AP	FCM	ALL	RSE	CSEV
Iris	0.8519	0.8351	0.8644	0.8667	0.8783	**0.8812**
Wine	0.6598	0.6618	0.6622	0.6632	0.6636	**0.6689**
Zoo	0.6010	0.6183	0.6273	0.6319	0.6223	**0.6329**
Glass	0.5581	0.5268	0.5281	0.5390	0.5252	**0.5867**
Ionosphere	0.6708	0.6670	0.6792	0.6815	0.6650	**0.6902**
Sonar	0.5428	0.4636	0.5425	0.5476	0.5389	**0.5506**
Balance scale	0.5511	0.5410	0.5554	0.5629	0.5579	**0.5672**
Pima	0.5547	0.5469	0.5433	0.5656	0.5657	**0.5658**
Spect-heart	0.6112	0.6114	0.6096	0.6188	0.6169	**0.6234**
Hepatitis	0.5335	0.5232	0.5266	0.5409	0.5364	**0.5519**
Bupa	0.6429	0.5538	0.6080	0.6674	0.6623	**0.6687**
Habermans survival	0.5951	0.5934	0.5948	0.5952	**0.5953**	0.5953
Wdbc	0.8639	0.8758	0.8639	**0.8850**	0.8747	0.8850
Statlog	0.5478	0.5490	0.5456	0.5477	0.5473	**0.5495**
Vehicle	0.4691	0.4580	0.4632	0.4747	0.4749	**0.4757**
Breast-cancer-Wisconsin	0.7606	0.7592	0.7564	0.7963	0.7969	**0.7997**
Car	0.3253	0.3219	**0.4817**	0.3186	0.3209	0.3476
Credit-g	0.5009	0.4952	0.4976	0.5185	0.5029	**0.5385**
Vowel	0.1966	0.1973	0.1649	0.1992	0.1944	**0.2004**
Lymphography	0.5029	0.4820	0.4866	0.4905	0.4894	**0.5085**

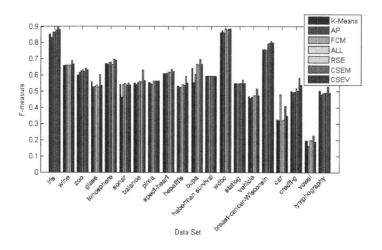

Fig. 1. The F-measure of different algorithms on the each data set

value of selective ensemble based on covariance. A F-measure value between an ensemble result and the labels of data set is obtained with one iteration, so a total of 60 F-measure values are obtained. The CSEV is the average value of the 60 F-measure values.

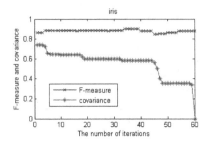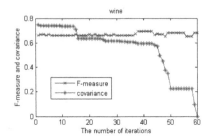

Fig. 2. The F-measure and covariance between each ensemble result and the label of Iris and Wine

Table 3. The covariances with maximum F-measure on each data set

Data Set	Covariance	maximum F-measure
Iris	0.5803	0.9000
Wine	0.5940	0.6924
Zoo	0.5822	0.6438
Glass	0.5029	0.6045
Ionosphere	0.2498	0.6977
Sonar	0.4236	0.5524
Balance scale	0.4416	0.6326
Pima	0.2360	0.5658
Spect-heart	0.2474	0.6352
Hepatitis	0.1425	0.5927
Bupa	0.1817	0.6990
Habermans survival	0.3827	0.5953
Wdbc	0.2493	0.8870
Statlog	0.1846	0.5698
Vehicle	0.4248	0.5152
Breast-cancer-Wisconsin	0.2229	0.8113
Car	0.2135	0.4109
Credit-g	0.1489	0.5834
Vowel	0.5993	0.2268
Lymphography	0.2950	0.5285

From the Table 2, we can see that the F-measures of clustering ensemble are better than base clustering on 16 data sets except Glass, Statlog, Car, and Lymphography. The F-measure of CSEV equals RSE on the Habermans survival, the F-measure of CSEV equals ALL on the Wdbc, the F-measure of FCM is better than CSEV on the Car, and the F-measures of CSEV are better than base clustering, ALL, and RSE on other 17 data sets.

We can obtain two conclusions based on above results. Firstly, the clustering ensemble result is better than base clustering. Secondly, the CSEV is better than base clustering, ALL, and RSE, which can also be seen from Fig. 1, where CSEM is max value of selective ensemble based on covariance. The x axis of Fig. 1 are 20 data sets and the y axis are the F-measures.

The covariances with maximum F-measure on each data set are shown in Table 3. From the Table 3, we can see that the covariance is between 0.1 and 0.6 on 20 data sets. Therefore, we will be directly select base clustering results that covariance in this interval to ensemble in the practical applications.

We can clearly see all F-measures and covariances of each selection on iris, wine, zoo, and glass from Fig. 2 to Fig. 3. The x axis is the number of base clustering results that does not use to ensemble. The y axis are F-measures and covariances between each ensemble result and the label of each data set.

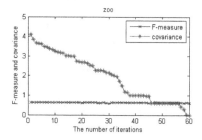 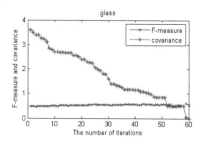

Fig. 3. The F-measure and covariance between each ensemble result and the label of Zoo and Glass

5 Conclusion

In this paper, we propose the selective clustering ensemble based on covariance. We measure diversity based on covariance. Our work may prove the better performance of our algorithm with experiments on the 20 UCI data sets, and get a covariance interval that is between 0.1 and 0.6. In future work, we will try to study more on selective clustering ensemble based on covariance and use them to the practical applications. We also will try to add semi-supervised information in this algorithm and achieve the parallelization of this algorithm.

Acknowledgements. This work is supported by the National Science Foundation of China (Nos. 61170111, 61003142, 611734002), the Fundamental Research Funds for the Central Universities(No. SWJTU11ZT08) and the Research Fund of Traction Power State Key Laboratory, Southwest Jiaotong University (No. 2012TPL_T15).

References

1. Han, J.W., Kamber, M.: Data Mining: Concepts and Techniques. Morgan Kaufmann (2007)
2. Tang, W., Zhou, Z.H.: Selective Clustering Ensemble Based on Bagging. Journal of Software 16(4), 496–502 (2005)
3. Strehl, A., Ghosh, J., Cardie, C.: Cluster Ensembles: A Knowledge Reuse Framework for Combining Multiple Partitions. Journal of Machine Learning Research (3), 583–617 (2002)
4. Topchy, A., Jain, A.K., Punch, W.: A Mixture Model for Clustering Ensembles. In: Proc. of the 4th SIAM International Conference on Data Mining, pp. 379–390 (2004)
5. Topchy, A., Jain, A.K., Punch, W.: Clustering Ensembles: Models of Consensus and Weak Partitions. IEEE Trans. on Pattern Analysis and Machine Intelligence 21(12), 1866–1881 (2005)
6. Luo, H.L., Jing, F.R., Xie, X.B.: Combining Multiple Clusterings Using Information Theory Based Genetic Slgorithm. In: IEEE International Conference on Computational Intelligence and Security, vol. 1, pp. 84–89 (2006)
7. Hassan, S.Z., Verma, B.: Decisions Fusion Strategy: Towards Hybrid Cluster Ensemble. In: Intelligent Sensors, Sensor Networks and Information, pp. 377–382 (2007)
8. Mohammadi, M., Nikanjam, A., Rahmani, A.: An Evolutionary Approach to Clustering Ensemble. In: The Fourth International Conference on Natural Computation, vol. 3, pp. 77–82 (2008)
9. Iqbal, A.M., Moh'd, A., Khan, Z.A.: Semi-supervised Clustering Ensemble by Voting. In: The International Conference on Information and Communication System, pp. 1–5 (2009)
10. Wang, H.J., Li, Z.S., Qi, J.H.: Semi-Supervised Cluster Ensemble Model Based on Bayesian Network. Journal of Software 21(11), 2814–2825 (2010)
11. Jia, J.H., Xiao, X., Liu, B.X.: Bagging-Based Spectral Clustering Ensemble Selection. Pattern Recognition Letters 32(10), 1456–1467 (2011)
12. Yang, Y., Wang, H., Lin, C., Zhang, J.: Semi-supervised Clustering Ensemble Based on Multi-ant Colonies Algorithm. In: Li, T., Nguyen, H.S., Wang, G., Grzymala-Busse, J., Janicki, R., Hassanien, A.E., Yu, H. (eds.) RSKT 2012. LNCS (LNAI), vol. 7414, pp. 302–309. Springer, Heidelberg (2012)
13. Iam-On, N., Boongoen, T., Garrett, S., Price, C.: A Link-Based Cluster Ensemble Approach for Categorical Data Clustering. IEEE Transactions on Knowledge and Data Engineering 24(3), 413–425 (2012)
14. Zhou, Z.H., Wu, J., Tang, W.: Ensembling Neural Networks: Many Could be Better Than All. Artificial Intelligence 137(1-2), 239–263 (2002)
15. Fern, X.Z., Lin, W.: Cluster ensemble selection. Statistical Analysis and Data Mining 1(3), 128–141 (2008)
16. Azimi, J., Fern, X.Z.: Adaptive cluster ensemble selection. In: Proceedings of the Twenty-First International Joint Conference on Artificial Intelligence (IJCAI 2009), pp. 992–997 (2009)
17. Jia, J.H., Xiao, X., Liu, B.X.: Similarity-based Spectral Clustering Ensemble Selection. In: Proceedings of the 9th International Conference on Fuzzy Systems and Knowledge Discovery (FSKD 2012), pp. 1071–1074 (2012)
18. Liu, L.M., Fan, X.P.: A New Selective Clustering Ensemble Algorithm. In: Proceedings of the 9th International Conference on e-Business Engineering (ICEBE), pp. 45–49 (2012)

19. Fern, X.Z., Brodley, C.E.: Random Projection for High Dimensional Data Clustering: A Cluster Ensemble Approach. In: Proceedings of the 20th International Conference on Machine Learning (ICML 2003), pp. 186–193 (2003)
20. Derek, G., Alexey, T., Nadia, B.: Ensemble Clustering in Medical Diagnostics. In: Proceedings of the 17th IEEE Symposium on Computer-Based Medical Systems, CBMS, pp. 576–581 (2004)
21. Hadjitodorov, S.T., Kuncheve, L.I., Todorova, L.P.: Moderate Diversity for Better Cluster Ensembles. Information Fusion 7(3), 264–275 (2006)
22. Luo, H.L., Kong, F.S., Li, Y.X.: Diversity Measure of Cluster Ensemble. Journal of Computers 30(8), 1315–1324 (2007)
23. Li, K., Gao, H.T.: A Novel Measure of Diversity for Support Vector Machine Ensemble. In: Proceedings of the Third International Symposium on Intelligent Information Technology and Security Informatics (IITSI), pp. 367–370 (2010)
24. Zhou, Z.H.: Ensemble Methods: Foundations and Algorithms, Boca Raton, FL (2012)
25. MacQueen, J.: Some Methods for Classification and Analysis of Multivariate Observations. In: Proceedings of the Fifth Berkeley Symposium on Mathematical Statistics and Probability, pp. 281–297 (1967)
26. Hartigan, J.A., Wong, M.A.: Algorithm AS 136: A K-Means Clustering Algorithm. Applied Statistics 28(1), 100–108 (1979)
27. Frey, J.B., Dueck, D.: Clustering by Passing Messages Between Data Points. Science 315(5814), 972–976 (2007)
28. Dunn, J.C.: A Graph Theoretic Analysis of Pattern Classification via Tamura's Fuzzy Relation. Journal of Cybernetics 3(3), 32–57 (1973)
29. Bezdek, J.C.: A Convergence Theorem for the Fuzzy ISODATA Clustering Algorithms. IEEE Transactions on Pattern Analysis and Machine Intelligence 2(1), 1–8 (1980)
30. Yang, Y., Kamel, M.: An Aggregated Clustering Approach Using Multi-Ant Colonies Algorithms. Pattern Recognition 39(7), 1278–1289 (2006)

Semi-supervised Clustering Ensemble for Web Video Categorization

Amjad Mahmood, Tianrui Li, Yan Yang, Hongjun Wang, and Mehtab Afzal

School of Information Science and Technology, Southwest Jiaotong University,
Chengdu 610031, China
{amjad.pu,mehtabafzal}@gmail.com, {trli,yyang,wanghongjun}@swjtu.edu.cn

Abstract. Recently, web video categorization has been an ever interesting research with the popularity of web videos. Clustering ensemble has become a good alternative for categorization. Semi-supervised clustering ensemble has shown a better performance since it may incorporate the known prior knowledge, e.g., pairwise constraints. In this paper, we propose a Semi-supervised Cluster-based Similarity Partitioning Algorithm (SS-CSPA) to categorize the videos containing textual data provided by their up-loaders. The feature of this algorithm is the introduction of an unsupervised learning, consensus between clustering and additional support of pairwise constraints to formulate semi-supervised clustering ensemble paradigm. Experimental results on the real-world web videos show that the proposed algorithm outperforms existing methods for categorization of web videos.

Keywords: Clustering, Cluster Ensemble, Pairwise Constraints, Video Categorization.

1 Introduction

Multimedia advancement in digital world has provided an easy path to produce abundant videos by its users. This abundance of videos has made the selection criteria quite complicated for a user to search and get the desired video. Web video categorization is principally a procedure of assigning web videos to pre-defined categories (such as Sports, Autos & Vehicle, Animals, Education, etc). It performs a critical role in many information retrieval tasks. On social web sites (such as YouTube [1]), extreme load of web video data obstructs the users to comprehend them effectively. Allocation of certain categories to these videos is a primary step. However, the diversity of web videos ranges from professional high quality videos to non-professional low quality videos, it makes web video categorization task more difficult. Conventionally, web videos are classified by using audio, textual, visual low-level features or their combinations [2,3]. These methods depend mostly on building models through machine learning techniques (e.g., SVM, HMM, GMM) to map visual low-level features to the high-level semantics. Due to unsatisfactory results of present high-level concept detection

Z.-H. Zhou, F. Roli, and J. Kittler (Eds.): MCS 2013, LNCS 7872, pp. 190–200, 2013.
© Springer-Verlag Berlin Heidelberg 2013

methods [4,5] and the expense of feature extraction, the content based categorization could not achieve the expected results. In addition, one of features of web videos is that the up-loaders generally use three ways, i.e., title, tag and description, to label their web videos. Therefore, the additional textual information associated with the web videos may become feasible features for the categorization.

Clustering ensemble is a framework for combining multiple base clustering results without accessing the original features of the objects. Base clusters can be generated either by using different clustering algorithms or the same algorithm with different parameters. Consensus function formulation is the most critical part in this scheme. Several efficient consensus functions have been derived from statistical, graph-based and information theoretic principles, e.g., co-association matrix [6], hyper-graph cuts [7], mutual information [8], mixture models [9] and voting process [10]. Ensemble models have been validated to improve the accuracy and robustness of single clustering methods [11] and successfully applied in many domains [12]. Recently, the semi-supervised clustering ensemble has emerged as an important variant of clustering ensemble since it incorporates the known prior knowledge and achieves a better performance [13].

In this paper, we aim to deal with the categorization problem of web videos by using their textual data based on the semi-supervised clustering ensemble. The rest of the paper is organized as follows. In Section 2, a brief survey of related work is described. Section 3 demonstrates the proposed framework together with the algorithm for web video categorization. Section 4 shows the experimental details along with evaluation of results. Concluding remarks and future work are stated in Section 5.

2   Related Work

2.1   Web Video Categorization

Automatic categorization of web videos is a crucial task in the field of multimedia indexing. Numerous studies have been conducted so far on this critical subject [3]. Ramchandran et al. [16] proposed a consensus learning approach using YouTube categories for multi-label video categorization. However, the specific categories and the amount of data are not described in their work. Schindler et al. [17] categorized the videos using bag-of-words representation but the classification results are unsatisfactory. Zanetti et al. [18] used 3000 YouTube videos to explore existing video classification techniques. Wu et al. used textual and social information for web video categorization that consist of user upload habits (They determined that users upload videos usually related to the same category) and YouTube related videos (specified by YouTube) [15]. Liu et al. suggested a technique for video topic retrieval using 'related video' links that YouTube relates to each web video to improve its textual information [19]. Ballen et al. proposed the use of social knowledge to suggest video tag and temporal localization [20]. Chen et al. used Wikipedia categories (WikiCs) and content duplicated open resources (CDORs) for web video categorization [21]. They also proposed a voting scheme that categorizes videos into the space spanned by enriched WikiC instead of raw title and tag.

2.2 Clustering Ensemble

Clustering ensemble has become a good alternative in dealing with cluster analysis problems. Dimitriadou et al. proposed a voting based ensemble method using cluster alignment technique [22]. Fred et al. presented a clustering ensemble method by considering co-association matrix as the similarity matrix [11]. Strehl et al. developed a hypergraph partitioning based the ensemble method [7]. Topchy et al. designed an adaptive scheme for integration of multiple non-independent clustering and extended the ensemble framework for generation of partitions [23]. Zhang et al. solved the ensemble problem by reducing it to a graph partitioning problem [24].

Recently, semi-supervised clustering ensemble has been proposed and shown a better performance by incorporating the known prior knowledge, e.g., pairwise constraints. Most commonly used constraints are must-link (ML) and cannot-link (CL). A must-link constraint enforces that two objects must belong to the same cluster while a cannot-link constraint enforces that two objects must belong to the different clusters [13]. Zhou et al. proposed disagreement-based semi-supervised learning paradigm, where multiple learners are trained for the task and the disagreements among the learners are exploited during the semi-supervised learning process [25]. Zhou et al. pointed out that most semi-supervised ensemble methods work by training learners using the initial labeled data first, and then using the learners to assign pseudo-labels to unlabeled data [26]. Iqbal et al. solved semi-supervised clustering ensemble by voting [27]. Wang et al. explored a semi-supervised cluster ensemble model based on semi-supervised learning and ensemble learning utilizing Bayesian network and EM algorithm [28]. Yang et al. presented a novel semi-supervised consensus clustering ensemble algorithm based on multi-ant colonies [29].

3   Proposed Framework

3.1 System Overview

Our system builds on top of the bag-of-words paradigm which is one of text information retrieval approaches. This technique assumes that the set of words in a document is a representative of its content and meaning. Firstly, we identify the meaningful words (terms) from textual data (e.g., the title, tag and description of a web video) of web videos, and apply the $TF - IDF$ scheme to find the weights of terms in a video. This step is crucial due to the noisy, mixed and inaccurate nature of textual data of web videos. We use the *Vector Space Model* with *Cosine Similarity* for the comparison of vectors, which ultimately leads to find the desired similarity between videos. Related videos information provided by the YouTube algorithm is used as a strong support for the clustering ensemble process. The important fact is that two related videos do not necessarily belong to same category. We use this information as must-link constraints. Using the above calculated similarity matrix and must-link information, we execute different clustering algorithms to identify the group of similar videos, leading to assign the category label to a video. Here we select three algorithms, graph partitioning, spectral clustering and affinity propagation [30] for the clustering purpose.

Next, we apply different clustering ensemble algorithms, e.g., CSPA, MCLA and HGPA [7], to integrate the base clustering results and obtain the consensus solution.

Finally, we incorporate pairwise constraints into the clustering ensemble algorithm and get the ultimate results. We denote this algorithm as a Semi-supervised Cluster-based Similarity Partitioning Algorithm (SS-CSPA). The framework of the proposed SS-CSPA algorithm is shown in Fig. 1.

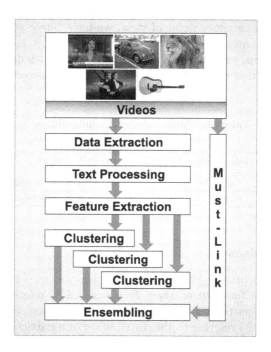

Fig. 1. Proposed framework

3.2   Pairwise Constraint Rules

Related video information can be viewed from two different angles. First, from the configuration point of view, we have *Star* and *Mesh* schemes. In *Star* scheme, we consider only the first video to be related with all remaining videos in a given

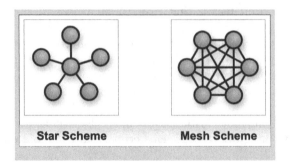

Fig. 2. Different configurations of pairwise constraints

set, whereas in *Mesh* scheme, each video is considered to be related with every other video within the group, as shown in Fig. 2.

Second, from the category point of view, we have *Same* and *Cross* category membership. If the category of both videos is the same, we refer it as the *Same* category membership, whereas if there exist two videos with different categories, we refer it as the *Cross* category membership, as summarized in Table 1.

Table 1. Must-link matrix configuration

Relation Type	Membership Type	Reference	Membership Type	Reference
Star	Same Category	ML-1	Cross Category	ML-3
Mesh	Same Category	ML-2	Cross Category	ML-4

3.3   Similarity Measure

A simple two-fold heuristics [32] based on frequency is used to score each component directly as a function of *Term Frequency* (TF) referring to the number of occurrences of a particular term in a specific document, and *Inverse Document Frequency* (IDF) refering to the distribution of a particular term across all documents.

$$IDF(t) = 1 + \log[\frac{n}{k}] \tag{1}$$

where n = Total number of documents, and k = Number of documents with term t appearing at least once.

The basic theme of $TF - IDF$ scheme is that, if a word appears frequently in a document, it must be an important keyword, unless it also appears frequently in other documents. The final weight of a term in a document can be calculated as

$$Weight(t, d) = TF(t, d) * IDF(t). \tag{2}$$

In order to accommodate different lengths of documents, we use relative counts. To find the similarity between documents, we represent each document in term space as a vector of term weights. For example, for two documents D_i and D_j, we have

$$D_i = (w_{i1}, w_{i2}, ...w_{iN}) \text{ and } D_j = (w_{j1}, w_{j2}, ...w_{jN}). \tag{3}$$

The similarity between these two documents can be calculated by using normalized Cosine function between them,

$$Sim(D_i, D_j) = \frac{\sum_{i=1}^{N}(w_{it} * w_{jt})}{\sqrt{\sum_{i=1}^{N}(w_{it})^2 * \sum_{i=1}^{N}(w_{jt})^2}}. \tag{4}$$

Using the above calculated similarity matrix and a set of must-link matrices we are ready to use any similarity based clustering algorithm.

3.4   The Algorithm

The proposed SS-CSPA Algorithm is outlined as follows.

Input: (i) Dataset, containing textual part of videos (UtVd).
            (ii) Related Video Information (RVi).
Calculate all possible configurations of pairwise constraints M-(1,2,3,4) from RVi.
for $i \in \{DataSets\ UtVd\ (Dec - 08(3), Jan - 09(6), Feb - 09(9))\}$
   for $j \in \{DataSet\ Copies\ C_j\}$
      for $k \in \{Title, Tag, Description\}$
         Text pre-processing for extraction of unique and meaningful terms.
         Apply $TF - IDF$ scheme to find term weights.
         Calculate the initial similarity matrix S_i.
            for $m \in \{M - 0, M - (1, 2, 3, 4)\}$ M-0 is without must-link
            Calculate net similarity $S_n = S_i + m$.
            Execute different clustering algorithms and get labels.
            Ensemble the labels with and without must-link.
         End m
         Ensemble the labels with and without must-link.
      End k
      Ensemble the labels with and without must-link.
   End j
End i
Output: Micro precision accuracy (SS-CSPA vs Ground truth labels).

---

4   Experiments

4.1   Datasets and Evaluation Criteria

Among the different available datasets, we select MCG-WEBV [33] as a benchmark containing most viewed videos for four months (Dec 2008 to Mar 2009).

Datasets. We perform a number of experiments on textual part by considering the basic textual features like title, tag and description. Related videos data is also included as must-link constraint. Some basic facts about three considerable datasets are stated in Table 2.

Table 2. Dataset description

DataSet UtVd	Number of Categories	Copies C_i	Instances	Features		
				Title	Tag	Des
Dec-08(3)	3	3	1000	1846	3917	5601
Jan-09(6)	6	3	1821	3089	6672	7983
Feb-09(9)	9	1	2504	3992	8502	9998

Evaluation Criteria. For evaluation, we use micro-precision [34] to measure accuracy of the consensus cluster with respect to the true labels: the micro-precision is defined as

$$MP = \sum_{h=1}^{K} [\frac{a_h}{n}], \tag{5}$$

where K is the number of clusters and n is the number of objects, a_h denotes the number of objects in consensus cluster h that are correctly assigned to the corresponding class. We identify the corresponding class for consensus cluster h as the true class with the largest overlap with the cluster, and assign all objects in cluster h to that class. Note that $0 \leq MP \leq 1$ with 1 indicating the best possible consensus clustering which has to be in full agreement with the class labels.

4.2   Results

Using the above stated scheme, we first perform the three clustering algorithms with must-link constrains and find the clustering labels. For each dataset, we select at least three subsets for experiments and take their average. The results are compared with true labels to find the accuracy. The average accuracy for three datasets with spectral clustering is shown in Table 3.

Table 3. Average clustering accuracies for three datasets with spectral clustering

Data	M-0	M-1	M-2	M-3	M-4	M-0	M-1	M-2	M-3	M-4	M-0	M-1	M-2	M-3	M-4
SubSection	Dataset Dec-08(3)					Dataset Jan-09(6)					Dataset Feb-09(9)				
Title	**0.75**	0.65	0.50	0.73	0.60	0.42	0.42	**0.45**	0.38	0.43	0.30	0.34	0.33	0.34	**0.37**
Tag	0.90	0.90	**0.92**	0.88	0.83	0.58	**0.61**	0.56	0.56	0.43	0.46	0.44	**0.47**	0.44	0.38
Des	0.64	0.75	**0.83**	0.66	0.63	0.48	**0.54**	0.49	0.46	0.46	0.31	0.36	**0.40**	0.34	0.31
All	0.92	0.82	**0.94**	0.88	0.93	0.57	0.57	**0.65**	0.56	0.50	0.44	0.46	**0.49**	0.47	0.42

Table 4. Clustering ensemble for dataset Dec-08(3)

Data	M-0	M-1	M-2	M-3	M-4	M-0	M-1	M-2	M-3	M-4	M-0	M-1	M-2	M-3	M-4
	CSPA					MCLA					HGPA				
SubSection	Dataset Dec-08(3)-C_1														
Title	0.41	0.43	0.42	0.44	**0.50**	0.50	0.45	0.45	0.48	0.40	0.34	0.39	**0.41**	0.35	0.33
Tag	**0.70**	0.43	0.50	0.55	0.50	0.59	0.46	0.53	**0.63**	0.57	0.36	0.36	0.40	0.37	**0.47**
Des	0.55	0.59	0.45	**0.65**	0.45	0.54	0.53	0.45	**0.57**	0.51	0.36	0.40	0.41	0.42	0.44
All	0.57	0.56	0.60	**0.64**	0.58	0.50	0.62	**0.68**	0.65	0.67	0.35	0.52	**0.60**	0.40	0.37
	Dataset Dec-08(3)-C_2														
Title	0.63	0.54	0.60	0.54	**0.60**	**0.61**	0.58	0.56	0.57	0.54	0.41	0.40	**0.40**	0.35	0.35
Tag	**0.84**	0.75	0.83	0.84	0.80	**0.82**	0.80	0.75	0.75	0.68	0.44	**0.45**	0.44	0.44	0.34
Des	0.60	0.42	**0.73**	0.69	0.59	0.60	0.58	**0.65**	0.63	0.60	0.34	0.35	**0.43**	0.37	0.36
All	0.75	0.83	**0.92**	0.90	0.91	0.60	0.78	0.79	**0.88**	0.79	0.44	0.42	0.42	0.37	**0.45**

Considering the above results as base clusters, we execute three clustering ensemble algorithms, CSPA, MCLA and HGPA as shown in Table 4.

Finally, we execute the SS-CSPA algorithm with all configurations of must-link constraints, results are shown in Table 5.

We repeat the whole scheme for two other instances of dataset Dec-08(3) as well as for all three instances of data set Jan-09(6) and Feb-09(9). The overall performance is shown in Table 6.

4.3 Results Discussion

At each stage of experimental scheme, we get corresponding results being milestones for next stage.

1. Among title, tag and description data subsets, tag performs very well. If we fuse tag information as ALL, it shows better performance.
2. The M-2 configuration of must-link performs better as compared to M-1, M-3 and M-4.
3. The graphical clustering (METIS) [31] has prominently better results as compared to others two methods for this kind of sparse data.
4. The accuracy is high for dataset Dec-08(3) (small no of categories) as compared to dataset Jan-09(6) and Feb-09(9) (more categories).

Table 5. Semi-supervised cluster-based similarity partitioning (SS-CSPA) results for Dataset Dec-08(3)

Data	M-0	M-1	M-2	M-3	M-4	M-0	M-1	M-2	M-3	M-4
SubSection	Must Link - 1					Must Link - 2				
Title	0.41	0.44	0.40	**0.52**	0.42	0.47	0.39	0.42	**0.53**	0.50
Tag	**0.70**	0.44	0.43	0.55	0.49	**0.67**	0.45	0.54	0.54	0.42
Des	0.54	0.59	0.46	**0.65**	0.62	0.54	0.59	0.49	**0.66**	0.50
All	0.57	0.52	0.53	**0.65**	0.58	0.58	0.56	**0.83**	0.65	0.70
	Must Link - 3					Must Link - 4				
Title	0.46	0.45	0.43	**0.52**	0.43	0.40	0.45	0.42	**0.54**	0.42
Tag	**0.70**	0.45	0.49	0.55	0.52	**0.66**	0.45	0.43	0.55	0.40
Des	0.54	0.59	0.51	**0.65**	0.46	0.53	0.59	0.49	**0.66**	0.50
All	0.57	0.52	0.61	0.65	**0.72**	0.62	0.56	0.60	0.65	**0.72**

Table 6. Final performance of SS-CSPA

Performance	M-1	M-2	M-3	M-4	AVERAGE	M-1	M-2	M-3	M-4	AVERAGE
	Dataset Dec-08(3)-C_1					Dataset Dec-08(3)-C_2				
Increase or Same	65 %	75 %	80 %	70 %	72 %	70 %	85 %	70 %	80 %	76 %
Decreased	35 %	25 %	20 %	30 %	28 %	30 %	15 %	30 %	20 %	24 %

5. The CSPA ensembling method has better performance as compared to other two.
6. We also investigate the neutral behaviour of pairwise constraints, and find that this happens only in situations where some constraints are already implemented during the natural process of clustering ensemble.
7. The semi-supervision technique at ensembling level also performs very well.

5   Conclusions

This paper proposed a novel approach, SS-CSPA, to categorize the videos containing textual data provided by their up-loaders. Experimental results showed that the proposed approach worked well for categorization purpose. Keeping in view all the facts and statistics about the main datasets as well as must-link constraints, it is found that the available noisy text information and less dense constrains are not sufficient to fully categorize the videos data. There is a need of some more data sources like user interest videos, visual contents of corresponding videos and specifically the audio frequency patterns and bands may help us to obtain better categorization results. In our future work, while searching for more supportive information, the fusion of additional information for web video categorization is also a challenge.

Acknowledgement. This work is supported by the National Science Foundation of China (Nos. 61175047, 61170111, 61262058 and 61003142), the Fundamental Research Funds for the Central Universities (Nos. SWJTU11ZT08 and SWJTU12CX092) and the Research Fund of Traction Power State Key Laboratory, Southwest Jiaotong University (No. 2012TPL T15).

References

1. YouTube, http://www.youtube.com
2. Yang, L., Liu, J., Yang, X., Hua, X.S.: Multi-modality web video categorization. In: Multimedia Information Retrieval, pp. 265–274. ACM, New York (2007)
3. Brezeale, D., Cook, D.J.: Automatic video classification: A survey of the literature. IEEE Transaction on Systems, Man, and Cybernetics, 416–430 (2008)
4. Jiang, Y.-G., Ngo, C.-W., Yang, J.: Towards optimal bag-of-features for object categorization and semantic video retrieval. In: 6th ACM International Conference on Image and Video Retrieval, pp. 494–501. ACM, New York (2007)
5. Qi, G.J., Hua, X.-S., Rui, Y., Tang, J., Mei, T., Zhang, H.-J.: Correlative multi-label video annotation. In: 15th International Conference on Multimedia, pp. 17–26. ACM, New York (2007)
6. Fred, A., Jain, A.K.: Robust data clustering. In: IEEE Computer Society Conference on Computer Vision and Pattern Recognition, Madison, pp. 128–133 (2003)
7. Strehl, A., Ghosh, J.: Cluster ensembles a knowledge reuse framework for combining multiple partitions. Journal of Machine Learning Research 3, 583–617 (2002)
8. Topchy, A., Jain, A.K., Punch, W.: Combining multiple weak clusterings. In: IEEE International Conference on Data Mining, Washington, pp. 331–338 (2003)
9. Topchy, A., Jain, A.K., Punch, W.: A mixture model for clustering ensembles. In: SIAM Conference on Data Mining, pp. 379–390 (2004)
10. Fischer, B., Buhmann, J.M.: Bagging for path-based clustering. IEEE Transactions on Pattern Analysis and Machine Intelligence 25(11), 1411–1415 (2003)
11. Fred, A., Jain, A.K.: Data clustering using evidence accumulation. In: 16th International Conference on Pattern Recognition (ICPR), pp. 1214–1219 (2002)
12. Ghosh, J., Acharya, A.: Cluster ensembles. Wiley Interdisciplinary Reviews: Data Mining and Knowledge Discovery 1, 305–315 (2011)
13. Wagstaff, K., Cardie, C., Rogers, S., Schroedl, S.: Constrained k-means clustering with background knowledge. In: International Conference on Machine Learning, New York, pp. 577–584 (2001)
14. Wang, Z., Zhao, M., Song, Y., Kumar, S., Li, B.: YouTubeCat: learning to categorize wild web videos. In: IEEE Conference on Computer Vision and Pattern Recognition, Computer Science & Engineering, Arizona State University, Arizona, pp. 879–886 (2010)
15. Wu, X., Zhao, W.L., Ngo, C.-W.: Towards google challenge: combining contextual and social information for web video categorization. In: 17th ACM International Conference on Multimedia, pp. 1109–1110. ACM, New York (2009)
16. Ramachandran, C., Malik, R., Jin, X., Gao, J., Nahrstedt, K., Han, J.: Videomule: A consensus learning approach to multi-label classification from noisy user-generated videos. In: 17th ACM International Conference on Multimedia, pp. 721–724. ACM, New York (2009)
17. Schindler, G., Zitnick, L., Brown, M.: Internet video category recognition. In: Computer Vision and Pattern Recognition Workshops, Georgia Institute of Technology, Atlanta, pp. 1–7 (2008)

18. Zanetti, S., Zelnik-Manor, L., Perona, P.: A walk through the web's video clips. In: Proceedings of Computer Vision and Pattern Recognition Workshops, California Institute of Technology, Pasadena, pp. 1–8 (2008)

19. Liu, Y., Yu, N.: Dual linkage refinement for YouTube video topic discovery. In: IEEE International Conference on Multimedia and Expo., Singapore, pp. 1576–1581 (2010)

20. Ballan, L., Bertini, M., Del Bimbo, A., Meoni, M., Serra, G.: Tag suggestion and localization in user-generated videos based on social knowledge. In: 2nd ACM SIGMM Workshop on Social Media, pp. 3–8. ACM, New York (2010)

21. Chen, Z., Cao, J., Song, Y., Zhang, Y., Li, J.: Web video categorization based on Wikipedia categories and content-duplicated open resources. In: ACM International Conference on Multimedia, New York, pp. 1107–1110 (2010)

22. Dimitriadou, E., Weingessel, A., Hornik, K.: Voting-merging: An ensemble method for clustering. In: Dorffner, G., Bischof, H., Hornik, K. (eds.) ICANN 2001. LNCS, vol. 2130, pp. 217–224. Springer, Heidelberg (2001)

23. Topchy, A., Minaei-Bidgoli, B., Jain, A.K., Punch, W.: Adaptive clustering ensemble. In: International Conference on Pattern Recognition, vol. 1, pp. 272–275 (2004)

24. Fern, X.Z., Brodley, C.E.: Solving cluster ensemble problems by bipartite graph partitioning. In: 21st International Conference on Machine Learning, Banff, p. 36 (2004)

25. Zhou, Z.-H., Li, M.: Semi-supervised learning by disagreement. Knowledge and Information Systems, 415–439 (2010)

26. Zhou, Z.-H.: Ensemble Methods: Foundations and Algorithms. Chapman & Hall/CRC, Boca Raton, FL (2012)

27. Iqbal, A.M., Moh'd, A., Khan, Z.A.: Semi-supervised clustering ensemble by voting. In: Proc. of the International Conference on Information and Communication Systems, pp. 1–5 (2009)

28. Wang, H., et al.: Semi-Supervised Cluster Ensemble Model Based on Bayesian Network. Journal of Software 21(11), 2814–2825 (2010) (in Chinese)

29. Yang, Y., Wang, H., Lin, C., Zhang, J.: Semi-supervised clustering ensemble based on multi-ant colonies algorithm. In: Li, T., Nguyen, H.S., Wang, G., Grzymala-Busse, J., Janicki, R., Hassanien, A.E., Yu, H. (eds.) RSKT 2012. LNCS (LNAI), vol. 7414, pp. 302–309. Springer, Heidelberg (2012)

30. Frey, B.J., Dueck, D.: Clustering by passing messages between data points. Science 315(5814), 972–976 (2007)

31. Karypis, G., Kumar, V.: A fast and high quality multilevel scheme for partitioning irregular graphs. SIAM Journal on Scientific Computing, 359–392 (1998)

32. Salton, G., Buckley, C.: Term-weighing approache sin automatic text retrieval. Information Processing & Management 24(5), 513–523 (1988)

33. Cao, J., Zhang, Y.-D., Song, Y.-C., Chen, Z.-N., Zhang, X., Li, J.-T.: MCG-WEBV: A Benchmark Dataset for Web Video Analysis. Technical Report, ICT-MCG-09-001 (2009)

34. Zhou, Z., Tang, W.: Clusterer ensemble. Knowledge Based System 19(1), 77–83 (2006)

Coding Theory Tools for Improving Recognition Performance in ECOC Systems

Claudio Marrocco, Paolo Simeone, and Francesco Tortorella

DIEI, Università degli Studi di Cassino
Via G. Di Biasio 43, 03043 Cassino (FR), Italia
{c.marrocco,paolo.simeone,tortorella}@unicas.it

Abstract. Error-correcting output coding (ECOC) is nowadays an established technique to build polychotomous classification systems by aggregating highly efficient dichotomizers. This approach has exhibited good classification performance and generalization capabilities in many practical applications. In this field much work has been devoted to study new solutions both for the coding and the decoding phase, but little attention has been paid to the algebraic tools typically employed in the Coding Theory, which could provide an ECOC design approach based on robust theoretical foundations. In this paper we propose an ECOC classification system based on *Low Density Parity Check* (LDPC) Codes, a well known technique in Coding Theory. Such framework is particularly suitable to define an ECOC system that employs dichotomizers provided of a reject option. The experiments on some public data sets have demonstrated that, in this way, the ECOC system can reach good recognition rates when a suitable reject level is imposed to the dichotomizers.

Keywords: ECOC, reject option, LDPC, coding theory, ensemble learning.

1 Introduction

Among the ensemble learning algorithms, *Error Correcting Output Coding* (ECOC) has gained large popularity in the Pattern Recognition community as an effective approach for decomposing a polychotomous classification problem in several dichotomies. The reasons of this success can be attributed to the strong theoretical roots and good comprehension characterizing two-class classifiers such as Perceptrons or Support Vector Machines, the error correcting capabilities of the codes used to group classes and the good generalization capabilities due to the reduction of both bias and variance [11].

The rationale underlying the ECOC approach is to break the original M classes problem into L different binary problems. Each class is assigned a code word of length L, thus defining a $M \times L$ coding matrix in which each column specifies a particular binary problem that groups the original classes into two superclasses. For each unknown samples the outputs produced by the L dichotomizers are combined into an output word which has to be matched with the code words associated to the original classes (*decoding phase*).

Z.-H. Zhou, F. Roli, and J. Kittler (Eds.): MCS 2013, LNCS 7872, pp. 201–211, 2013.
© Springer-Verlag Berlin Heidelberg 2013

In previous papers focused on ECOC, much work has been devoted to the analysis of predefined codes (e.g., one-vs-one, one-vs-all, random codes, exhaustive codes, linear error correcting codes). Moreover, great attention has been devoted to the possibility of building efficient output codes based on the learning algorithms used [4], on the data distribution in the particular problem at hand [7], [15], [2] or on minimizing the number of dichotomizers to be trained in the ensemble [3]. In the decoding phase, the decision is typically based on the Hamming distance computed between the code words and the crisp outputs of the dichotomizers. Other decoding strategies have also been proposed, based on Euclidean distance between the soft outputs of the dichotomizers and the code words. Other distances have been also proposed, defined on the loss function used during training [1] or on a weighted loss-based distance [6].

However, the research done in this field has drifted away from the original approach proposed in [5] where the learning task was considered with a communications problem in which the identity of the correct output class for a new example is being transmitted over a channel. As a consequence, the usual setting proposed in literature for an ECOC-based classification system does not exploit all the typical features of an error correcting coding provided by Coding Theory and does not consider all the algebraic tools typically employed, which could provide an ECOC design approach based on robust theoretical foundations.

In this paper we propose an ECOC classification system based on a well known technique in Coding Theory [9]: *Low Density Parity Check* (LDPC) Codes. Besides the algebraic structure which provides a strong theoretical framework, we choose LDPC codes because they allow us to efficiently handle not only errors but also erasures, i.e., events in which the channel abstains from giving an output since there is knowledge that the symbol is likely to be in error. Since in the ECOC setting this corresponds to a dichotomizer provided with a reject option, such a framework is particularly suitable to define an ECOC system that, for each classification act, can rely on trustworthy dichotomizers and algebraically recover the outputs of the dichotomizers that do not provide a reliable response.

We have proposed a similar mechanism in [12] with the purpose of adding a reject option to the whole ECOC classification system; in that case we obtained some improvement in the error rate of the ECOC system, but at the cost of rejecting the samples for which a reliable decoding could not be reached. In this paper we introduce a new reject option for the dichotomizers that allows us to choose more accurately the operating point of the two-class classifiers. This time the possibility of rejecting a sample is limited only to the dichotomizers while the ECOC system works at zero reject. Notwithstanding, the experiments performed on some public datasets show that a significant decrease in the error rate of the ECOC system can be obtained when a suitable reject level is imposed to the dichotomizers.

2   A Glimpse at Coding Theory

The usual ECOC approach consists of representing each class label by a bit string of length L, called *code word*, with the only requirement that distinct classes are represented by distinct code words. However, this usual setting does not exploit all the typical features of an error correcting coding provided by Coding Theory. For this reason, before analyzing the proposed approach, let us introduce some basic concepts of linear codes [16] that revealed to be useful for an ECOC classification system.

Let us consider the Galois field $GF(2)$, i.e., a set of two elements $\{0, 1\}$ where a sum and a product operations, both modulo-2, are defined and let us denote with $GF^L(2)$ the vector space of all L-tuples over the field $GF(2)$. An (L, K, d) code $\mathcal{C}$ over $GF^L(2)$ is a K-dimensional vector subspace of $GF^L(2)$ where the vectors of the subspace are the code words of $\mathcal{C}$ and d is the minimum Hamming distance among them. If $\mathbf{u} = [u_0, u_1, \ldots, u_{K-1}]$ is a K-bit source message to be coded, it can be associated to a code word $\mathbf{c} = [c_0, c_1, \ldots, c_{L-1}]$ of $\mathcal{C}$ and thus, the 2^K possible source messages with length K are associated with 2^K code words with length L. The difference $L - K$ is called *redundancy*, while the ratio $r = K/L$ is the *transmission rate* of the code $\mathcal{C}$. The relation between the redundancy and d is regulated by an upper bound $d \leq L - K + 1$, which means that, for a smaller r, d can increase and thus the error correction capability.

Since $\mathcal{C}$ is a K-dimensional vector subspace, there exist K linearly independent vectors belonging to $GF^L(2)$, let us call them $\mathbf{g}_0, \ldots, \mathbf{g}_{K-1}$, which form a basis for $\mathcal{C}$. In this way, the correspondence between the source message $\mathbf{u}$ and the code word $\mathbf{c}$ can be put in terms of a linear combinations of the basis vectors through $\mathbf{u}$, i.e., $\mathbf{c} = \mathbf{u}\mathbf{G}$ where $\mathbf{G} = \left(\mathbf{g}_0 \ldots \mathbf{g}_{K-1} \right)^T$ is a $K \times L$ matrix termed the *generator matrix* of $\mathcal{C}$.

Let us now consider $\mathcal{C}^\perp$, the orthogonal complement of $\mathcal{C}$, i.e., the set of vectors belonging to $GF^L(2)$ which are orthogonal to the code words of $\mathcal{C}$. Moreover, let $\mathbf{H} = \left(\mathbf{h}_0 \ldots \mathbf{h}_{L-K-1} \right)^T$ be the $(L - K) \times L$ matrix collecting the $L - K$ vectors $\mathbf{h}_i$ of the basis of $\mathcal{C}^\perp$. In this way, each code word $\mathbf{c} = \mathbf{u}\mathbf{G}$ of the code satisfies the condition $\mathbf{H}\mathbf{c}^T = \mathbf{0}$; for this reason, $\mathbf{H}$ is called the *parity check matrix* of $\mathcal{C}$. In other words, the parity check matrix defines $L - K$ equations which allow the received word to be checked to verify if it is actually a code word of $\mathcal{C}$. In particular, if the received vector $\mathbf{o} = \mathbf{c} + \mathbf{e}$ is given by a code word $\mathbf{c}$ corrupted by an error pattern $\mathbf{e}$, we get a parity check condition given by:

$$\mathbf{H}\mathbf{o}^T = \mathbf{H}\mathbf{e}^T \neq \mathbf{0}. \tag{1}$$

According to the last equation, it is possible to show that for a linear code a decoding technique is able to recover a code word if the erroneous bits are less than $\lfloor (d - 1)/2 \rfloor$.

Finally, it is worth introducing here an useful and intuitive graphical representation of linear codes, the Tanner graph [18], that shows how each component of the output vector is involved in the parity check constraints (see Fig. 1). This is a bipartite graph with L *variable nodes*, corresponding to every component of the output vector, and $L - K$ *check nodes*, corresponding to the parity check constraints, i.e., to the rows of $\mathbf{H}$. To build the graph, every check node i is connected to a variable node j if and only if $h_{ij} = 1$ where h_{ij} is the element of

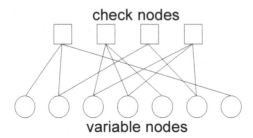

Fig. 1. An example of a Tanner graph for a generic code

the matrix $\mathbf{H}$ on the i-th row and the j-th column. The number of connections deriving from a node is usually referred as the *degree* of the node.

3   LDPC Codes in ECOC Framework

Among the techniques provided by Coding Theory we found that *Low Density Parity Check* (LDPC) codes [9] constitutes a suitable theoretical framework to be used in ECOC classification systems. LDPC codes, in fact, are characterized by a sparse pseudo-random matrix $\mathbf{H}$ that allows us to exploit the redundancy of the code words when employing decoding technique suitable with abstaining classifiers. An (a, b)-regular LDPC code is defined as a binary linear code such that in its Tanner graph every variable node has degree a and every check node has degree b. The term "low density" indicates that the number of edges in the Tanner graph is aL, where L is the length of the code. As L increases, the number of edges in the Tanner graph grows linearly in L, while for other codes it grows much faster.

To integrate LDPC codes into an ECOC framework, the first point is the choice of L and K. If M is the number of the classes in the original multiclass problem, obviously we have $K \geqslant \lceil \log_2 M \rceil$; however, for a fixed L, it is convenient to keep K as low as possible so as to decrease the transmission rate and thus, to increase the *minimum Hamming distance* (MHD) d among the code words. Once L and $K = \lceil \log_2 M \rceil$ are determined, the matrices $\mathbf{G}$ and $\mathbf{H}$ of the code $\mathcal{C}$ can be generated and thus, following the steps described in the previous section, the code words. When code words have been found a coding matrix $\mathbf{C}$ can be created assuming each row corresponding to a code word and each column to a binary subproblems as in common ECOC framework. However, in this context we do not need to choose among possible code words or to maximize the Hamming distance between the rows of the coding matrix since, according to Coding Theory, LDPC codes are already built to maximize the minimum Hamming distance d between any pair of code words.

Let us now consider the coding matrix $\mathbf{C}$ produced by the chosen code $\mathcal{C}$ which, depending on the structure of the generator matrix, could contain similar or identical columns as well as all-zeros or all-ones columns. Unlike the usual ECOC, in our approach we cannot eliminate such columns unless all the algebraic properties of the code do not hold and we cannot apply the correct decoding procedure. Actually, the all-zeros/all-ones columns are not a big problem since they can be neglected during the training of the dichotomizers so that the number of needed dichotomizers becomes lower than L. The bits corresponding to all-zeros/all-ones columns can be then inserted within the output vector before the decoding algorithm starts. Moreover, even identical columns become an issue less problematic than in the usual ECOC systems. In fact, in the decoding algorithm we should avoid the presence of dependent errors but in our approach, thanks to the sparseness of the parity check matrix, correlated outputs are likely to be forwarded to different check nodes thus making the following phases robust to such circumstance.

4   The Decoding Procedure with Abstaining Classifiers

The ECOC classification of an unknown sample $\mathbf{x}$ requires the execution of a decoding algorithm over a received code word $\mathbf{o}$ defined by the set of L binary predictions on $\mathbf{x}$. In Coding Theory received code words are considered as the result of transmitting unknown code words $\mathbf{c}$ over a noisy channel. A commonly employed model of noisy channel is the Binary Erasure Channel (BEC) in which each code word symbol is lost with a fixed constant probability timely independent of all the other symbols. The BEC can be usefully employed in ECOC framework to model classifiers with a reject option, i.e., classifiers that can abstain from a decision according to the reliability of its output. In this way, it is possible to single out the unreliable elements (binary decisions) in the output vector and process them in an appropriate way before arriving at the decoding stage. As a consequence, the decoding rules have to be modified to take their decisions only on the basis of the bits evaluated as sufficiently reliable. LDPC codes have been widely studied when applied to the BEC model and several decoding algorithm have been proposed. The most promising one that we will describe in the following has been shown to be the message-passing algorithm and its extension that has been shown to guarantee good performance when employed with codes of finite length less than 10^4 [14].

4.1   Defining the Reject Rule for Dichotomizers

Before going into details about the decoding techniques, let us consider how to design the reject option for the group of dichotomizers in the ECOC architecture. Let us assume that each dichotomizer $f_h(\mathbf{x})$ outputs a real value in the range $[-1, +1]$ and that, to take a decision about the class of $\mathbf{x}$, the value $f_h(\mathbf{x})$ is compared with a threshold τ_h. In other words, $\mathbf{x}$ is assigned the class $+1$ if $f_h(\mathbf{x}) \geq \tau_h$ otherwise, the class -1 is chosen. It is worth noting that irrespective of the value of the decision threshold τ_h, the majority of unreliable decisions correspond to the outcome values near the threshold, where the distribution of the two classes overlaps. In other words, the samples for which the output of the dichotomizer falls in this region are characterized by some ambiguity in the allocation, because their corresponding outcomes are very similar and thus quite difficult to distinguish. A way to obtain more reliable results is to employ a decision rule with two thresholds, τ_{h1} and τ_{h2} with $\tau_{h1} \leq \tau_{h2}$, such that:

$$r(f_h, \tau_{h1}, \tau_{h2}) = \begin{cases} +1 & \text{if } f_h(\mathbf{x}) > \tau_{h2}, \\ -1 & \text{if } f_h(\mathbf{x}) < \tau_{h1}, \\ reject & \text{if } f_h(\mathbf{x}) \in [\tau_{h1}, \tau_{h2}]. \end{cases} \tag{2}$$

The idea is to encapsulate the class overlap region into the *reject interval* $[\tau_{h1}, \tau_{h2}]$, so as to turn many of the errors due to the class overlap into rejects. The optimal values for the thresholds (τ_{h1}, τ_{h2}) should be chosen to satisfy two contrasting requirements: enlarging the reject region to eliminate more errors and limiting the reject region to preserve as many correct classifications as possible. In our case, we cannot choose the same pair of thresholds for all the dichotomizers because each of them has different distributions for the output score, and a unique choice would involve abnormal results for most of them. Accordingly, we imposed all dichotomizers to work at a chosen rejection rate ρ and we used the method presented by Pietraszek in [13]. This approach requires estimating the ROC curve of each dichotomizer and calculating the pair of thresholds (τ_{h1}, τ_{h2}) such that f_h abstains for no more than ρ at the lowest possible error rate. The rationale is to make all the dichotomizers work almost at the same level of reliability.

4.2   The Decoding Procedure

Once the binary reject has been applied the output vector can also contain rejected bits, i.e., $o_i \in \{0, 1, ?\}$ and thus the parity check condition in eq. (1) cannot be checked. Nevertheless, we can assume that all the bits not rejected are correct and in this way, eq. (1) becomes a system of linear equations with the rejected bits as unknowns. In particular, if we denote with E the index set of the rejected bits and with $\bar{E}$ the index set of the bits not rejected, the parity check condition $\mathbf{H}\mathbf{o}^T = \mathbf{0}$ can be written as:

$$\mathbf{H}\mathbf{o}^T = \mathbf{H}_E \mathbf{o}_E^T + \mathbf{H}_{\bar{E}} \mathbf{o}_{\bar{E}}^T = \mathbf{0} \tag{3}$$

Since we are working with the arithmetic modulo 2, this is equivalent to:

$$\mathbf{H}_E \mathbf{o}_E^T = \mathbf{H}_{\bar{E}} \mathbf{o}_{\bar{E}}^T \tag{4}$$

where $\mathbf{H}_{\bar{E}} \mathbf{o}_{\bar{E}}^T$ is a known term. This system has a unique solution if and only if the matrix $\mathbf{H}$ has a subset of $|E|$ independent rows; in this case, the solution can be found by performing Gaussian elimination and back substitution.

The parity check equations allow the rejects to be eliminated by means of an iterative procedure (*direct recovery algorithm*) that can be summarized as follows [17]:

1. Initialize the values of all check nodes to zero;
2. FOR EACH variable node, IF the node has a value in $\{0, 1\}$ THEN add this value to the values of all adjacent check nodes and remove all the edges coming from it;
3. FOR EACH check node, IF the node has degree one THEN substitute its value into the unique adjacent variable node and remove the edge;
4. IF at least a check node with degree one has been found in the previous step THEN goto step 2 ELSE exit.

It is worth noting that each check node with degree 1 singled out in step 3 can be only connected to a variable node with reject whose value is substituted in such a way to satisfy the constraint. The procedure ends when there are no more check nodes with degree 1. This means either all the check nodes have degree 0 (and thus all the rejects have been recovered) or there is some check node with degree greater than 1, i.e., a check node connected with two or more variable nodes with rejects which cannot be recovered. This case happens when $rank(\mathbf{H}_E) < |E|$. In order to have a high probability of recovering the rejected bits, the code to be chosen should have a sparse parity check matrix, as in the case of LDPC codes, with the property that $\mathbf{H}_E$ has a triangular sub-matrix with high probability when $|E|$ is not too large [16].

When $rank(\mathbf{H}_E) < |E|$ it is possible to successfully extend the previously analyzed algorithm by means of the *Guess algorithm* [14] that consists in performing several guesses of the erased bits unsolved by the recovery algorithm and can be summarized in this way:

1. Apply the recovery algorithm;
2. IF a stopping condition exists, THEN find the check nodes with degree two and guess one bit.
3. Goto step 1 until all stopping conditions have been removed.
4. Create a list of 2^g solutions where g is the number of guesses made. From the list, $\mathbf{o}_k$ with $k \in \{1, 2, ..., 2^g\}$, pick the one that satisfies $\mathbf{Ho}_k^T = \mathbf{0}$.

Obviously, compared to the recovery algorithm, the complexity of this algorithm increases with g. However, thanks to the sparsity of $\mathbf{H}$, this does not represent a problem for LDPC codes that, instead, in such situation, exhibits a strong improvement of the decoding performance [14].

It is also worth noting that the previously described decoding algorithms do not necessarily output a code word belonging to $\mathbf{C}$. In fact, in step (4) of the Guess algorithm we have implicitly made the assumption that all the bits not rejected were correctly decoded, but this is obviously not always true. The recovered code word $\tilde{\mathbf{c}}$ could thus contain some erroneous bits; moreover, such errors could propagate during the recovery of the rejected bits, even though this problem is sensibly mitigated by the sparseness of the parity check matrix. As a result, the recovery could produce a code word different from the correct one. However, a high MHD among the code words of the coding matrix increases the probability that the erroneously recovered code word does not represent another class, i.e., that $\tilde{\mathbf{c}} \in \mathcal{C}$ and $\tilde{\mathbf{c}} \notin \mathbf{C}$. In this case, an effective rule is to decide for the

Table 1. Data sets and code parameters used in the experiments

Data sets	Classes	Features	Samples	K	L	Dichot.
SatImage	6	36	6435	3	100	7
Glass	7	9	214	3	100	7
PenDigits	10	16	10992	4	100	14
Yeast	10	8	1484	4	100	14

class corresponding to the code word $\mathbf{c} \in \mathbf{C}$ with the lowest Hamming distance from the recovered code word $\tilde{\mathbf{c}}$.

5   Experiments

To evaluate the performance of the proposed ECOC classification system, some experiments have been performed on four data sets publicly available at the UCI Machine Learning Repository [8] using SVM with RBF kernel [10] as base dichotomizer. All the employed data sets have numerical input features and a variable number of classes. For each data set, 10 runs of a multiple hold out procedure have been performed to avoid any bias in the comparison. In each run, the data set has been split in three subsets: a training set (containing the 50% of the samples of each class) to train the base classifiers, a tuning set (25% of the samples of each class) to optimize the SVM parameters (γ of the RBF kernel and C) using a grid approach and a test set to evaluate the performance of the multiclass classification. The code matrix has been chosen with K dependent on the number of classes as explained in Sect. 3 and $L = 100$ so as to have a redundancy that guarantees a very high minimum Hamming distance among the code words. As stated at the end of Sect. 3, equal columns are a less problematic issue in our approach and thus, we have considered a different dichotomizer for every different column found in the matrix $\mathbf{C}$. More details on each data sets, the code parameters and the number of dichotomizers are reported in Table 1.

Our method has been compared with two classical approaches of multiclass-to-binary decomposition that are one-vs-all that discriminates one class against the others and one-vs-one that defines as many binary problems as the possible pairs of different classes. Moreover, for the sake of comparison we have also considered the Recursive ECOC (RECOC) method [19] that is the only method in literature using LDPC codes to design recursive ECOC classifiers built from a number of sparsely connected dichotomizers and employing an iterative algorithm ,i.e., the sum-product algorithm, as decoding technique. To have a fair comparison, for every approach, well optimized SVM with RBF kernel have been used as base dichotomizers.

The performance obtained with our approach has been evaluated in terms of curves reporting the error rate when varying the parameter ρ of the reject rate for the dichotomizers. The value of ρ has been varied in the interval $[0, 0.5]$ with a step of 0.05. It is worth noting that this curve is not the usual error-reject curve that plots the error rate with respect to a reject rate at the output stage.

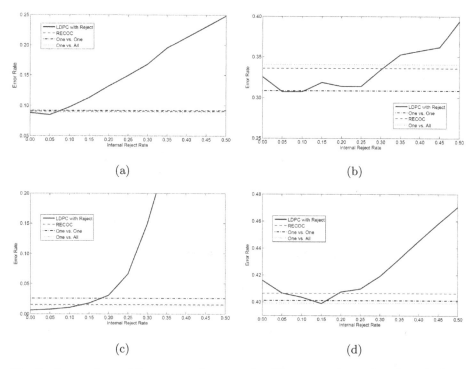

Fig. 2. Comparison of the proposed approach with commonly employed methods in literature on four data sets: (a) SatImage, (b) Glass, (c) PenDigits, (d) Yeast

Our approach, instead, uses an "internal" reject but the whole classification system works at zero reject thanks to the decoding procedure described in Sect. 4. Figs. 2 show the results of our experiments for the four datasets. Each figure represents the trend of the error rate for our approach (termed LDPC with Reject) together with lines representing the error rate obtained by the other approaches.

Two main observations can be done. The first is that employing the reject for the dichotomizers we improve the performance of the whole classification system: for three datasets (SatImage, Glass, and Yeast) there is always one point of the curve with an error rate lower than the zero reject (i.e., $\rho = 0$); for one dataset (PenDigits) the point on the y-axis has an equal performance than the following ones (i.e., $\rho = 0.05$). It is also worth noting that the error rate strongly increases for higher values of ρ. However, this is an expected behavior since increasing the number of rejects also increases the number of guesses g to be done and this produces a similar-to-random decision.

The second observation is that our approach exhibits higher performance than all the other considered methods since there is always one point of the curve with a lower error rate. A drawback of the proposed system with respect to simple approaches like one-vs-all and one-vs-one is to estimate a good value of the parameter ρ; however, it is also worth remembering that our approach does not require any selection of rows ad columns of the coding matrix since it is completely based on LDPC codes and decoding techniques deriving from the Coding Theory.

6   Conclusions and Future Works

In this paper we proposed a novel framework for an ECOC classification system founded on the strong theoretical roots of the Coding Theory techniques. The proposed approach is based on LDPC codes using code words generated without any selection of rows and columns of the coding matrix. Such codes allow us to use dichotomizers with a reject option and to define an effective algorithm for recovering the rejects so as to obtain a final decision at the output of the whole classification system. Some preliminary experiments accomplished on benchmark datasets showed that the proposed method is effective when compared with other standard approaches such as one-vs-one or one-vs-all techniques. Our approach also exhibits better performance than RECOC system that is the only method in literature using LDPC codes in the ECOC structure.

Future works will focus on the analysis of the employed coding matrix and in particular, on the properties of the LDPC code words that can be further exploited to verify to what extent we can fit the Coding Theory tools to the classification systems. Moreover, an analysis of the behavior of the Guess algorithm has to be performed to find an upper bound for the number of guessed bits so as to avoid considering high values for the parameter ρ. Finally, a deeper experimental analysis is required to extend our approach to other classifier architectures and to other decoding rules to be employed when all the bits have been recovered.

References

1. Allwein, E.L., Schapire, R.E., Singer, Y.: Reducing multiclass to binary: A unifying approach for margin classifiers. Journal of Machine Learning Research 1, 113–141 (2000)
2. Alpaydin, E., Mayoraz, E.: Learning error-correcting output codes from data. In: Ninth International Conference on Artificial Neural Networks, ICANN 1999, Conf. Publ. No. 470, vol. 2, pp. 743–748 (1999)
3. Bautista, M.Á., Escalera, S., Baró, X., Radeva, P., Vitrià, J., Pujol, O.: Minimal design of error-correcting output codes. Pattern Recognition Letters 33(6), 693–702 (2012)
4. Crammer, K., Singer, Y.: On the learnability and design of output codes for multiclass problems. In: Cesa-Bianchi, N., Goldman, S.A. (eds.) COLT, pp. 35–46. Morgan Kaufmann (2000)
5. Dietterich, T.G., Bakiri, G.: Solving multiclass learning problems via error-correcting output codes. Journal of Artificial Intelligence Research 2, 263–286 (1995)
6. Escalera, S., Pujol, O., Radeva, P.: On the decoding process in ternary error-correcting output codes. IEEE Trans. Pattern Anal. Mach. Intell. 32(1), 120–134 (2010)

7. Escalera, S., Tax, D.M.J., Pujol, O., Radeva, P., Duin, R.P.W.: Subclass problem-dependent design for error-correcting output codes. IEEE Trans. Pattern Anal. Mach. Intell. 30(6), 1041–1054 (2008)
8. Frank, A., Asuncion, A.: UCI machine learning repository (2010)
9. Gallager, R.G.: Low density parity-check codes. MIT Press (1963)
10. Joachims, T.: Making large-scale SVM learning practical. In: Schölkopf, B., Burges, C., Smola, A. (eds.) Advances in Kernel Methods - Support Vector Learning, ch. 11. MIT Press, Cambridge (1999)
11. Kong, E.B., Dietterich, T.G.: Error-correcting output coding corrects bias and variance. In: ICML, pp. 313–321 (1995)
12. Marrocco, C., Simeone, P., Tortorella, F.: Embedding reject option in ECOC through LDPC codes. In: Haindl, M., Kittler, J., Roli, F. (eds.) MCS 2007. LNCS, vol. 4472, pp. 333–343. Springer, Heidelberg (2007)
13. Pietraszek, T.: On the use of ROC analysis for the optimization of abstaining classifiers. Machine Learning 68(2), 137–169 (2007)
14. Pishro-Nik, H., Fekri, F.: On decoding of low-density parity-check codes over the binary erasure channel. IEEE Trans. Inf. Theor. 50(3), 439–454 (2006)
15. Pujol, O., Radeva, P., Vitrià, J.: Discriminant ECOC: A heuristic method for application dependent design of error correcting output codes. IEEE Trans. Pattern Anal. Mach. Intell. 28(6), 1007–1012 (2006)
16. Richardson, T.J., Urbanke, R.: Modern Coding Theory. Cambridge University Press (2008)
17. Shokrollahi, A.: An introduction to low-density parity-check codes. In: Khosrovshahi, G.B., Shokoufandeh, A., Shokrollahi, A. (eds.) Theoretical Aspects of Computer Science 2000. LNCS, vol. 2292, pp. 175–197. Springer, Heidelberg (2002)
18. Tanner, R.M.: A Recursive Approach to Low Complexity Codes. IEEE Transactions on Information Theory 27(5), 533–547 (1981)
19. Tapia, E., Bulacio, P., Angelone, L.: Recursive ECOC classification. Pattern Recognition Letters 31(3), 210–215 (2010)

Can Diversity amongst Learners Improve Online Object Tracking?

Georg Nebehay[1], Walter Chibamu[2], Peter R. Lewis[2],
Arjun Chandra[3], Roman Pflugfelder[1], and Xin Yao[2]

[1] Austrian Institute of Technology, Austria
[2] CERCIA, School of Computer Science, University of Birmingham, UK
[3] Department of Informatics, University of Oslo, Norway
{georg.nebehay.fl,roman.pflugfelder}@ait.ac.at,
{wcc081,p.r.lewis,x.yao@}cs.bham.ac.uk,
chandra@ifi.uio.no

Abstract. We present a novel analysis of the state of the art in object tracking with respect to diversity found in its main component, an ensemble classifier that is updated in an online manner. We employ established measures for diversity and performance from the rich literature on ensemble classification and online learning, and present a detailed evaluation of diversity and performance on benchmark sequences in order to gain an insight into how the tracking performance can be improved.

1 Introduction

We deal with the problem of single-target model-free object tracking in videos, meaning that a single object is to be tracked and no a priori information about the object is available. Many authors (e.g. [14,17,24,25]) formulate the task of object tracking as a binary classification problem, and use ensembles of multiple learners as binary classifiers. One of the elements required for accurate prediction in ensembles is error diversity [6]. While measures for diversity have been considered explicitly in the context of object tracking before [26], in this work, we take a different path and analyse the diversity in the state of the art object tracking method TLD (Tracking-Learning-Detection [17]) in order to gain an insight into how its performance can be improved by manipulating diversity.

As TLD consists of multiple interleaved components, we focus our analysis on its most influential component, a random fern classifier [23]. While it is not clear yet whether our findings generalize to the original TLD method, or to other object tracking methods, we do establish a baseline with the analysis of the random fern classifier, against which more involved methods can be evaluated in future. The contributions of this paper are threefold: firstly, we show how diversity can be measured in TLD. Secondly, we provide a detailed analysis with respect to diversity and performance. Thirdly, we hint at ways how performance might be improved.

This work is structured as follows. In section 2 we discuss related work in object tracking and machine learning. In section 3 we describe the state of the

Z.-H. Zhou, F. Roli, and J. Kittler (Eds.): MCS 2013, LNCS 7872, pp. 212–223, 2013.

art tracking method TLD. In section 4, we lay out our experimental setup. In section 5 we present our analysis of diversity and performance, and section 6 gives conclusions and final remarks.

2 Related Work

In this section, we first review related work in online learning for object tracking, and secondly describe existing techniques for the engineering of diversity in ensembles of learners.

2.1 Online Learning in Object Trackers

Collins et al. [8] were the first to employ binary classification in a tracking context, the two classes being the object and the immediate surrounding. They employ feature selection in order to switch to the most discriminative colour space from a set of candidates and use mean-shift for finding the mode of a likelihood surface, thereby locating the object. In a similar spirit, Grabner et al. [14] perform online boosting and Babenko et al. [1] use multiple instance learning in order to find the location of the object. All of these methods use a form of reinforcement learning, meaning that the prediction of the classifier is directly used to update the classifier. While this approach enables the use of unlabelled data for training, it typically amplifies errors made in the prediction phase, thus leading to a degradation of tracking performance. In [15], this problem is addressed by casting object tracking as a semi-supervised learning problem, where only the first appearance of the object is used for updating. Both Kalal et al. [17] and Santner et al. [25] employ an optic-flow-based mechanism for labelling the available data in order to reduce the errors made in the prediction phase and demonstrate superior results.

2.2 Diversity of Ensembles in Object Tracking

In machine learning generally, *diverse* ensembles of classifiers often provide better prediction accuracy than any of the individual members of the ensemble [6]. Visentini et al. [26] employ a combined measure of diversity and performance to select classifiers from a pool for adaptive object tracking. Additionally, diversity has been considered more generally in computer vision. Bertolami and Bunke [2] use diversity measures as indicators for the accuracy of ensemble classification for handwriting recognition. Frinken et al. [13] increase the diversity of a handwriting recognition system by combining Neural Networks, Maximum Margin Hidden Markov Models and Hidden Markov Models, and show that high diversity leads to better results. Levy et al. [19] force classifiers to learn different aspects of the data by minimizing correlation between ensemble members and show improved results on visual recognition problems.

2.3   Engineering Diversity in Online Learning

The literature is abound with methods for encouraging diversity in ensembles. Attempts at consolidating these methods into taxonomies have also been made [6,9], which can provide guidelines for encouraging diversity in different ways.

   The taxonomy by Dieterich [9] consolidates ensemble creation methods into various categories with diversity encouragement being at the heart. For the discussion in this section, we assume a standard supervised learning problem: a learning algorithm is presented with a training set $\mathcal{S}$ $\{(\mathbf{x}_1, y_1) \ldots (\mathbf{x}_N, y_N)\}$ of size N for learning some unknown function $y = f(\mathbf{x})$. The learning algorithm outputs a classifier, which is a hypothesis $h_i \in \mathcal{H}$ about the true underlying function f. The various methods found in such taxonomies have been applied mostly in the offline learning mode. They can however be adapted to the online case (e.g. [21,22]), where training instances continuously arrive one at a time as a stream of data. A brief overview of the taxonomy now follows:

Bayesian voting. In problems where it is possible to enumerate each hypothesis $h_i \in \mathcal{H}$, and calculate a prior $P(h)$, the problem of classifying a new example $\mathbf{x}$ amounts to computing $P(f(\mathbf{x}) = y | \mathcal{S}, \mathbf{x}) = \sum_{h \in \mathcal{H}} h(\mathbf{x}) P(h | \mathcal{S})$. This can be viewed as an ensemble consisting of all possible hypotheses in $\mathcal{H}$, where each hypothesis h is weighted by its posterior probability $P(h | S)$. However, Bayesian voting fails where it is not possible to enumerate all possible hypotheses and calculate the prior $P(h)$.

Manipulating training examples. L iterations of the learning algorithm are run. In each iteration a different subset of the training set $\mathcal{S}$ is used to train the classifier h_i, $i = 1 \ldots L$, thus generating multiple classifiers, each trained on a different training set. Example algorithms in this category are Bagging [3], Cross validated committees, and AdaBoost [12].

Manipulating input features. The input features are divided into feature subsets, and in each iteration i of the learning algorithm, a classifier is trained on a subset(s) of the input features. The random subspace method [16] falls into this category.

Injecting randomness. Some randomness can be induced into the learning setup, for example in a neural network ensemble by using different initial weights, or injecting noise into the input features following bootstrap sampling.

Manipulating output targets. The error-correcting output code technique [10] manipulates the y labels of the training examples in classification problems where the number of classes, k, is large. Instead of learning the problem on the original k classes, in each iteration $i = 1 \ldots L$, the k classes are divided into two subsets $\mathcal{A}$ and $\mathcal{B}$ (different in each iteration) and the input data re-labelled 0 and 1 respectively for classes in subsets $\mathcal{A}$ and $\mathcal{B}$. This results in L classifiers $h_1 \ldots h_L$. To classify a new data point $\mathbf{x}$, if $h_i(\mathbf{x}) = 0$, then each class in subset $\mathcal{A}$ receives a vote and if $h_i(\mathbf{x}) = 1$, then each class in subset $\mathcal{B}$ receives a vote. Once all L classifiers have voted, the class with the largest prediction is selected as the ensemble output.

Manipulating error functions. Diversity can be explicitly encouraged and maintained by defining and minimising a correlation term between ensemble members. Negative correlation encourages individual members to learn different parts of the training data (specialisation) allowing the ensemble to learn the entire training data better than any single or monolithic member [20]. Ensemble members are trained simultaneously allowing the members to interact and cooperate through a correlation penalty term that is introduced in the error function such that the individual error of each member is negatively correlated to the rest of ensemble errors [7].

Diversity Metrics. Several measures for a quantitative assessment of diversity in ensembles have been proposed in the literature. Kuncheva et al. [18] have conducted a wide and detailed study of various diversity measures, and conclude that there is no unique way of measuring diversity, and in general, there is no direct or distinctive relationship between the diversity of an ensemble and its accuracy. One of the most commonly used diversity measures, the *Q-statistic* [18] is calculated in a pairwise manner for any two classifiers f_i and f_j:

$$Q_{i,j} = \frac{ad - bc}{ad + bc} \tag{1}$$

The symbols a, b, c, d refer to the number of times
  a : f_i and f_j are correct,
  b : f_i is correct, f_j is incorrect,
  c : f_i is incorrect, f_j is correct,
  d : f_i and f_j are incorrect.
$Q_{i,j}$ is closer to 1 if the output of the classifiers is not diverse, and is closer to -1 if their output is diverse. An overall measure for the diversity of an ensemble of size n is then obtained by averaging all of the pairwise measurements.

3   State of the Art in Object Tracking

3.1   Tracking-Learning-Detection

Kalal et al. [17] propose a solution to the tracking problem which they call *Tracking-Learning-Detection* (TLD). TLD consists of two separate components: A **frame-to-frame tracker** that predicts the location L_j of the object in frame I_j by calculating the optical flow between frames I_{j-1} and I_j and transforming L_{j-1} accordingly. Clearly, this approach is only feasible as long as the object is visible in the scene and fails otherwise. When the object is presumably tracked correctly (according to certain criteria) the location L_j is used in order to update a **Random Fern** classifier [23] with positive training data from patches close to L_j and negative data from patches that exceed a distance. This classifier is then applied in a sliding-window manner (see figure 1) in order to re-initialize the frame-to-frame-tracker after failure. Two additional stages not described here are used for classification.

Fig. 1. In TLD, a binary ensemble classifier is used to locate the object of interest by applying it in a sliding-window manner. The ability for multi-scale detection is achieved by scaling the size of the detection window. Image is from the SPEVI[1]dataset.

3.2   Random Fern Classifier

The Random Fern classifier [23] operates on binary features $f_1 \ldots f_n$ calculated on the raw image data. These features are randomly partitioned into groups of so-called *ferns* $F_1 \ldots F_m$ of size s

$$\overbrace{f_1 \cdots f_s}^{F_1}, \overbrace{f_{s+1} \cdots f_{2s}}^{F_2} \cdots \overbrace{f_{(m-1)s+1} \cdots f_{ms}}^{F_m}. \tag{2}$$

Ferns essentially are non-hierarchical trees, meaning that the outcome of each fern is independent of the order in which features are evaluated. The main reason for favouring ferns over trees is that they can be implemented extremely efficiently, an important property for real applications.

3.3   Features

In [23], a feature vector of size s consists of s binary tests performed on gray-scaled image patches. Each test compares the brightness values of two random pixels (See figure 2). The locations of the tests are generated once at startup and remain constant throughout the rest of the processing. The same set of tests is used with appropriate scaling for all subwindows. Input images are smoothed with a Gaussian kernel to reduce the effect of noise.

3.4   Random Ferns in TLD

The posterior probability for each fern is

$$P(y = 1|F_k) = \frac{P(y = 1)P(F_k|y = 1)}{\sum_{i=0}^{1} P(y = i)P(F_k|y = i)}. \tag{3}$$

---

[1] http://www.eecs.qmul.ac.uk/~andrea/spevi.html

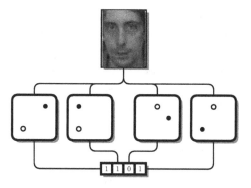

Fig. 2. Feature values depends on the brightness values of pairs of two random pixels. In this case, the outcome is the binary string 1101.

In TLD, the prior is assumed to be uniform, and the $P(F_k|y = i)$ are modelled as the absolute number of occurrences $\#p_{F_k}$ for positive training data and $\#n_{F_k}$ for negative training data. Therefore, the posterior probability becomes

$$P(y = 1|F_k) = \frac{\#p_{F_k}}{\#p_{F_k} + \#n_{F_k}}. \tag{4}$$

When $\#p_{F_k} = \#n_{F_k} = 0$, then $P(y = 1|F_k)$ is assumed to be 0 as well. Each training instance is used for training only if it was misclassified in the current frame. A decision is obtained by employing a threshold θ on the posterior probabilities combined using the mean rule

$$\frac{1}{m} \sum_{i=1}^{m} P(y = 1|F_i) \geq \theta. \tag{5}$$

4 Experimental Setup

We conduct experiments according to the following novel pattern in order to assess the diversity and the performance of the Random Fern classifier in TLD. For each frame, we closely follow the predict-update cycle of classical online learning: first we let the classifier predict labels for all subwindows. We then measure performance and diversity using the ground truth values and update the classifier according to the misclassified examples. Each experiment is run 10 times with different seeds for the random number generator. Over these runs, the mean and standard deviation of the selected metrics for performance and diversity are reported. We apply the following modifications to the original algorithm [17].

- Majority voting is used instead of the mean rule. Crisp outputs are obtained by applying the threshold θ on the posterior probabilities of the individual classifiers.

- We replace the optic-flow based tracker with manually labeled ground truth.
- We disregard the two classification stages besides the random fern classifier.

The first modification enables the use of the Q statistic. We perform the last two modifications since we are interested only in the performance limits of the classifier. The analysis of this modified version gives us a baseline against which to evaluate more involved methods in the future.

4.1   Performance Measures

We use the following statistics to measure the performance, based on the occurrences of *True Positives* (TP), *False Negatives* (FN) and *False Positives* (FP) in each frame. TPs, FNs and FPs are found by comparing algorithmic output to manually annotated ground truth. Recall, given by

$$R_j = \frac{\mathrm{TP}_j}{\mathrm{TP}_j + \mathrm{FN}_j}, \tag{6}$$

measures the fraction of positive instances that were correctly classified as positive. Precision, given by

$$P_j = \frac{\mathrm{TP}_j}{\mathrm{TP}_j + \mathrm{FP}_j}, \tag{7}$$

measures the fraction of examples classified as positive that are truly positive. The F-measure, given by

$$F_j = \frac{2R_j P_j}{R_j + P_j}, \tag{8}$$

as the harmonic mean, combines precision and recall into a single measurement. We calculate R_j, P_j and F_j for each frame and report their average values R, P, F over the whole sequence.

As the employed set of subwindows is not exhaustive, there will typically be no single subwindow of the same location and the same dimension as the manual annotation. We therefore employ the measure used in the Pascal Visual Object Challenge [11] for overlap between two bounding boxes B_1 and B_2, namely,

$$overlap = \frac{B_1 \cap B_2}{B_1 \cup B_2} = \frac{I}{(B_1 + B_2 - I)}. \tag{9}$$

If the overlap between a manual annotation and a subwindow is larger than 0.5, the subwindow is labelled positive as well.

We employ the Q-statistic (section 2.3) as a measure for diversity in each frame and report averaged values over the whole sequence. While other diversity measures are available, we chose the Q-statistic as a starting point for our analysis primarily due to its widespread use. However, we plan to investigate different measures of diversity in future work.

4.2 Sequences

We employ the following six sequences for conducting our evaluation. These sequences were used in [17,27] for evaluating object tracking methods. **David** (761 frames) shows a person walking from an initially dark setting into a bright room and undergoing various changes in appearance. **Jumping** (313 frames) shows a person jumping rope causing motion blur. **Pedestrian 1** (140 frames), **Pedestrian 2** (338 frames) and **Pedestrian 3** (184 frames) show pedestrians being filmed by an unstable camera. **Car** (945 frames) shows a moving car, exposed to low contrast recording and undergoing multiple occlusions. The appearance of the car itself stays constant over the run of the sequence.

5 Diversity Analysis of TLD

In this section we present novel analyses of diversity within TLD based object tracking. Firstly, we explore the effect of varying the parameters of the system on the selected metrics. Secondly, we artificially increase diversity in the system and analyse the resulting effects. We use the parameters $m = 30, s = 14, \theta = 0.5$ unless noted otherwise.

5.1 Effect of Parameters

The parameter m steers the number of classifiers in the ensemble. Breiman [4] proved that an ensemble of randomized decision trees does not overfit as more trees are added, meaning that performance does not decrease. However it is not clear how m affects diversity. In figure 3 we plot Q and F against m for the sequence *David*. Increasing m leads to a convergent behaviour of Q, similar to the performance metric. Q converges more quickly than the performance metrics. These findings generalize to all sequences.

The parameter θ directly influences recall and precision. High values of θ lead to an improvement of precision, as false positives are filtered out, and to a degradation of recall. Low values of θ lead to the inverse effect. Intuitively, both high and low values of θ should lead to a reduction of diversity, as the output of the individual classifiers become more similar. In table 1, θ is varied for all sequences. Surprisingly, Q decreases monotonically as θ is increased. The explanation for this effect is that high values of theta lead to many positive instances being misclassified, and therefore the set of positive training data becomes larger, causing a reduction of Q.

5.2 Increasing Diversity

In order to artificially increase diversity in the ensemble classifier, we restrict the location of the binary tests for individual classifiers to certain parts of the input image, thus decreasing the amount of information shared between them. For each classifier we randomly sample a value μ_j. We then generate the binary tests

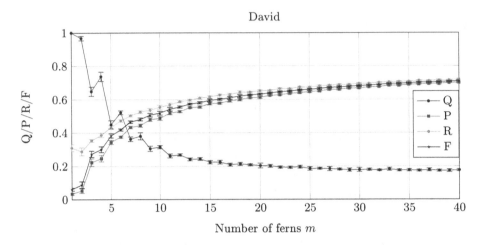

Fig. 3. Both diversity and performance exhibit a convergent behaviour when the number of ferns m is increased

Table 1. Increasing θ leads to an increase of diversity due to many positive instances being misclassified, thus increasing the size of the positive training set

| Sequence | Metric | Sensitivity threshold θ | | | | |
		0.1	0.3	0.5	0.7	0.9
car	Q	0.31±0.01	0.29±0.01	0.28±0.01	0.27±0.01	0.26±0.01
	P	0.62±0.01	0.76±0.00	0.81±0.00	0.84±0.00	0.86±0.00
	R	0.95±0.00	0.92±0.00	0.90±0.00	0.85±0.00	0.73±0.00
david	Q	0.21±0.01	0.19±0.01	0.18±0.01	0.17±0.01	0.16±0.01
	P	0.28±0.02	0.56±0.01	0.67±0.00	0.74±0.00	0.74±0.01
	R	0.82±0.00	0.76±0.00	0.70±0.00	0.58±0.01	0.34±0.01
jumping	Q	0.24±0.01	0.22±0.01	0.21±0.01	0.21±0.01	0.20±0.01
	P	0.36±0.01	0.59±0.00	0.68±0.00	0.76±0.00	0.78±0.01
	R	0.85±0.00	0.77±0.00	0.70±0.00	0.58±0.00	0.35±0.01
pedestrian1	Q	0.30±0.01	0.27±0.01	0.26±0.01	0.26±0.01	0.25±0.01
	P	0.23±0.01	0.38±0.01	0.45±0.01	0.53±0.01	0.52±0.01
	R	0.53±0.01	0.44±0.01	0.38±0.00	0.26±0.01	0.14±0.01
pedestrian2	Q	0.31±0.01	0.29±0.01	0.28±0.01	0.27±0.01	0.26±0.01
	P	0.35±0.01	0.53±0.01	0.62±0.01	0.74±0.01	0.77±0.02
	R	0.71±0.01	0.68±0.01	0.65±0.01	0.58±0.01	0.45±0.01
pedestrian3	Q	0.47±0.01	0.45±0.01	0.44±0.01	0.44±0.01	0.42±0.01
	P	0.53±0.01	0.68±0.01	0.76±0.01	0.84±0.01	0.87±0.01
	R	0.92±0.01	0.87±0.01	0.83±0.01	0.75±0.01	0.57±0.01

Table 2. Diversity increases when the locations of the binary tests become more local. Q_{bad} indicates that diversity in the classification result of misclassified instances is already very high from the start.

Sequence	Metric	Feature locality $1 - \sigma$				
		0.1	0.3	0.5	0.7	0.9
car	Q	0.28 ± 0.01	0.23 ± 0.01	0.16 ± 0.01	0.10 ± 0.00	0.08 ± 0.00
	Q_{good}	0.27 ± 0.01	0.22 ± 0.01	0.15 ± 0.01	0.10 ± 0.00	0.07 ± 0.00
	Q_{bad}	0.01 ± 0.00	0.00 ± 0.00	0.00 ± 0.00	-0.00 ± 0.00	-0.01 ± 0.00
	F	0.86 ± 0.00	0.86 ± 0.00	0.86 ± 0.00	0.85 ± 0.00	0.77 ± 0.01
david	Q	0.19 ± 0.01	0.17 ± 0.01	0.13 ± 0.01	0.11 ± 0.00	0.11 ± 0.00
	Q_{good}	0.18 ± 0.01	0.16 ± 0.01	0.13 ± 0.01	0.11 ± 0.00	0.10 ± 0.00
	Q_{bad}	0.00 ± 0.00	0.00 ± 0.00	-0.00 ± 0.00	-0.00 ± 0.00	0.00 ± 0.00
	F	0.69 ± 0.00	0.69 ± 0.00	0.69 ± 0.01	0.67 ± 0.01	0.51 ± 0.01
jumping	Q	0.22 ± 0.01	0.20 ± 0.02	0.16 ± 0.02	0.09 ± 0.01	0.07 ± 0.00
	Q_{good}	0.21 ± 0.01	0.19 ± 0.02	0.15 ± 0.02	0.09 ± 0.01	0.07 ± 0.00
	Q_{bad}	0.01 ± 0.00	0.01 ± 0.00	0.01 ± 0.00	0.01 ± 0.00	0.00 ± 0.00
	F	0.70 ± 0.00	0.70 ± 0.01	0.69 ± 0.01	0.66 ± 0.01	0.44 ± 0.02
pedestrian1	Q	0.26 ± 0.01	0.25 ± 0.01	0.21 ± 0.01	0.15 ± 0.01	0.08 ± 0.00
	Q_{good}	0.23 ± 0.01	0.22 ± 0.01	0.18 ± 0.01	0.13 ± 0.01	0.07 ± 0.00
	Q_{bad}	0.04 ± 0.00	0.04 ± 0.00	0.03 ± 0.00	0.03 ± 0.00	0.02 ± 0.00
	F	0.41 ± 0.01	0.41 ± 0.01	0.41 ± 0.02	0.40 ± 0.02	0.34 ± 0.02
pedestrian2	Q	0.27 ± 0.01	0.26 ± 0.01	0.21 ± 0.01	0.14 ± 0.01	0.08 ± 0.00
	Q_{good}	0.26 ± 0.01	0.25 ± 0.01	0.20 ± 0.01	0.13 ± 0.01	0.07 ± 0.00
	Q_{bad}	0.04 ± 0.00	0.03 ± 0.00	0.03 ± 0.00	0.02 ± 0.00	0.01 ± 0.00
	F	0.66 ± 0.01	0.67 ± 0.01	0.66 ± 0.01	0.62 ± 0.03	0.49 ± 0.04
pedestrian3	Q	0.45 ± 0.01	0.44 ± 0.02	0.38 ± 0.02	0.22 ± 0.01	0.09 ± 0.01
	Q_{good}	0.44 ± 0.01	0.42 ± 0.02	0.37 ± 0.02	0.21 ± 0.01	0.08 ± 0.00
	Q_{bad}	0.04 ± 0.00	0.04 ± 0.00	0.03 ± 0.00	0.03 ± 0.00	0.02 ± 0.00
	F	0.81 ± 0.01	0.81 ± 0.01	0.80 ± 0.01	0.77 ± 0.01	0.71 ± 0.02

from the two-dimensional uniform distribution $U(max(0, \mu_j - \sigma), min(1, \mu_j + \sigma))$. Brown and Kuncheva [5] show that the majority vote error can be decomposed into the sum of individual errors (additive term), diversity measured on correctly classified instances called good diversity (subtractive term) and diversity measured on misclassified instances called bad diversity (additive term). While the decomposition of the F measure into analogous Q terms is unknown, the notions of good and bad diversity are still helpful in our context. For this experiment, we measure Q both on correctly classified instances (Q_{good}) and on misclassified instances (Q_{bad}).

When σ is decreased, we make the following observations for all sequences in table 2: Q and Q_{good} decrease strongly. Q_{bad} starts out closely above the theoretical minimum $-\frac{1}{m}$, decreasing only slightly. Depending on the sequence, performance rapidly decreases at a certain value of σ. Increasing diversity the way we have seems to increase the error of the individual classifiers. For low values of σ, this increase is compensated for by a decreased Q_{good}, leading to a stable F. For high values of σ, the errors of the individual classifiers seem to outweigh the increased good diversity, leading to a reduction of F.

These observations suggest that we need to find a way to encourage diversity that keeps the individual classifiers from exhibiting an increased error. Since Q_{bad} is close to the theoretical minimum, increasing it can help us increase ensemble performance. Devising a training scheme that is informed by the wrongly classified instances may be one way of increasing Q_{bad}. Further analysis of the relationship between Q_{good}, Q_{bad}, and individual classifier performance, will shed more light on ways to encourage diversity that may lead to an increased overall performance.

6   Conclusions and Future Work

In this work, we presented an analysis of the state of the art in object tracking with respect to diversity and showed how it is influenced by the intrinsic parameters of its ensemble classifier. We also showed how diversity can be increased artificially and conclude that performance is reduced due to an increased error of the individual classifiers. We plan to look into methods that increase good diversity while keeping the individual accuracy stable. We also acknowledge the fact that reducing bad diversity will help increase performance.

A better understanding of the relationship between performance of individual classifiers, as well as between good and bad diversity, will help show ways on how overall performance can be increased. We also plan to explicitly reduce correlation in the system by making use of algorithms similar to minimal correlation learning [19].

As only misclassified examples are used for training, the classifier highly overfits the training data. This does not to lead to a reduction in performance as long as sequences contain sufficient training examples. When short sequences with severe changes in appearance occur, performance is affected in a negative way. The results of Minku et al. [21] suggest that an increased level of diversity could help in exactly these cases.

References

1. Babenko, B., Yang, M.H., Belongie, S.: Robust object tracking with online multiple instance learning. Pattern Analysis and Machine Intelligence 33(8) (August 2011)
2. Bertolami, R., Bunke, H.: Diversity analysis for ensembles of word sequence recognisers. In: Yeung, D.-Y., Kwok, J.T., Fred, A., Roli, F., de Ridder, D. (eds.) SSPR&SPR 2006. LNCS, vol. 4109, pp. 677–686. Springer, Heidelberg (2006)
3. Breiman, L.: Bagging Predictors. Machine Learning 24, 123–140 (1996)
4. Breiman, L.: Random forests. Machine Learning 45(1), 5–32 (2001)
5. Brown, G., Kuncheva, L.I.: "Good" and "Bad" Diversity in Majority Vote Ensembles. In: El Gayar, N., Kittler, J., Roli, F. (eds.) MCS 2010. LNCS, vol. 5997, pp. 124–133. Springer, Heidelberg (2010)
6. Brown, G., Wyatt, J., Harris, R., Yao, X.: Diversity creation methods: A survey and categorisation. Journal of Information Fusion 6, 5–20 (2005)
7. Chen, H., Yao, X.: Multiobjective neural network ensembles based on regularized negative correlation learning. Knowledge and Data Engineering 22(12) (2010)

8. Collins, R.T., Liu, Y., Leordeanu, M.: Online selection of discriminative tracking features. Pattern Analysis and Machine Intelligence 27(10), 1631–1643 (2005)
9. Dietterich, T.G.: Ensemble methods in machine learning. In: Kittler, J., Roli, F. (eds.) MCS 2000. LNCS, vol. 1857, pp. 1–15. Springer, Heidelberg (2000)
10. Dietterich, T.G., Bakiri, G.: Solving Multiclass Learning Problems via Error-Correcting Output Codes. Journal of Artificial Intelligence Research 2 (1995)
11. Everingham, M., Van Gool, L., Williams, C., Winn, J., Zisserman, A.: The pascal visual object classes (VOC) challenge. International Journal of Computer Vision 88(2), 303–338 (2010)
12. Freund, Y., Schapire, R.E.: A decision-theoretic generalization of on-line learning and an application to boosting. In: Vitányi, P.M.B. (ed.) EuroCOLT 1995. LNCS, vol. 904, pp. 23–37. Springer, Heidelberg (1995)
13. Frinken, V., Peter, T., Fischer, A., Bunke, H., Do, T.-M.-T., Artieres, T.: Improved handwriting recognition by combining two forms of hidden markov models and a recurrent neural network. In: Jiang, X., Petkov, N. (eds.) CAIP 2009. LNCS, vol. 5702, pp. 189–196. Springer, Heidelberg (2009)
14. Grabner, H., Bischof, H.: On-line boosting and vision. In: Computer Vision and Pattern Recognition, vol. 1 (2006)
15. Grabner, H., Leistner, C., Bischof, H.: Semi-supervised On-Line boosting for robust tracking. In: Forsyth, D., Torr, P., Zisserman, A. (eds.) ECCV 2008, Part I. LNCS, vol. 5302, pp. 234–247. Springer, Heidelberg (2008)
16. Ho, T.K.: The random subspace method for constructing decision forests. Pattern Analysis and Machine Intelligence 20(8), 832–844 (1998)
17. Kalal, Z., Mikolajczyk, K., Matas, J.: Tracking-Learning-detection. Pattern Analysis and Machine Intelligence 34(7), 1409–1422 (2012)
18. Kuncheva, L.I., Whitaker, C.J.: Measures of Diversity in Classifier Ensembles and Their Relationship with the Ensemble Accuracy. Machine Learning 51(2), 181–207 (2003)
19. Levy, N., Wolf, L.: Minimal correlation classification. In: Fitzgibbon, A., Lazebnik, S., Perona, P., Sato, Y., Schmid, C. (eds.) ECCV 2012, Part VI. LNCS, vol. 7577, pp. 29–42. Springer, Heidelberg (2012)
20. Liu, Y., Yao, X., Higuchi, T.: Evolutionary ensembles with negative correlation learning. Evolutionary Computation 4(4), 380–387 (2000)
21. Minku, L.L., White, A.P., Yao, X.: The Impact of Diversity on Online Ensemble Learning in the Presence of Concept Drift. Knowledge and Data Engineering 22(5), 730–742 (2010)
22. Oza, N.C.: Online Bagging and Boosting. Systems, Man and Cybernetics (2005)
23. Ozuysal, M., Calonder, M., Lepetit, V., Fua, P.: Fast keypoint recognition using random ferns. Pattern Analysis and Machine Intelligence 32(3), 448–461 (2010)
24. Saffari, A., Leistner, C., Santner, J., Godec, M., Bischof, H.: On-line random forests. In: International Conference on Computer Vision Workshops (2009)
25. Santner, J., Leistner, C., Saffari, A., Pock, T., Bischof, H.: PROST: Parallel robust online simple tracking. In: Computer Vision and Pattern Recognition (2010)
26. Visentini, I., Kittler, J., Foresti, G.L.: Diversity-based classifier selection for adaptive object tracking. In: Benediktsson, J.A., Kittler, J., Roli, F. (eds.) MCS 2009. LNCS, vol. 5519, pp. 438–447. Springer, Heidelberg (2009)
27. Yu, Q., Dinh, T.B., Medioni, G.: Online tracking and reacquisition using co-trained generative and discriminative trackers. In: Forsyth, D., Torr, P., Zisserman, A. (eds.) ECCV 2008, Part II. LNCS, vol. 5303, pp. 678–691. Springer, Heidelberg (2008)

A New Perspective of Support Vector Clustering with Boundary Patterns

Yuan Ping, Huina Li, Yong Zhang, and Zhili Zhang

Department of Computer Science and Technology
Xuchang University, 461000 Xuchang, China
pyuan.lhn@gmail.com

Abstract. To overcome the pricey computation required by redundant kernel function matrix and poor label performance, in a novel perspective, we present support vector clustering with boundary patterns (BPSVC for abbreviation) for efficiency. For the first phase, the conventional method of estimating the support vector function with the whole data is altered by only essential boundary patterns. Thence, BPSVC only need to solve a much simpler optimization problem. For the second phase of cluster labeling, both convex decomposition and cone cluster labeling method are employed by an ensemble labeling strategies for further improvements on accuracy and efficiency. Both theoretical analysis and experimental results show its superiorities in comparison of the state-of-the-art methods, especially for large-scale data analysis.

Keywords: data analysis, support vector clustering, convex decomposition, boundary pattern, cluster labeling.

1 Introduction

With the advantage of generating cluster boundaries of arbitrary shape, support vector clustering (SVC)[12, 15] has attracted many researchers and been extensively applied to wide variety of domains, e.g., instance-based learning, pattern denoising and medical information processing etc[6, 13, 14].

However, the literatures show that training to estimate a support function and cluster labeling are two major bottlenecks which might degrade its popularity. As a quadratic programming problem, the prior can be solved by many classic algorithms, such as sequential minimal optimization and entropy-based algorithms[16], in approximately $O(N^2)$ kernel evaluations, here N is the number of data points. In addition, data block based methods [1, 17] may be a good choice even though persistent parameter tuning is generally required. Correspondingly, the time complexity of the latter is $O(N^2 m)$ with $m \ll N$ which is the sample rate on each edge. Naturally, the studies argue that cluster labeling takes most of the computation time in the entire SVC clustering process. For efficiency, especially on large-scale data, some insightful methods have been designed to replace the complete graph (CG)[15], such as support vector graph (SVG)[15], proximity graph of delaunay (DD)[9], minimum spanning tree (MST),

Z.-H. Zhou, F. Roli, and J. Kittler (Eds.): MCS 2013, LNCS 7872, pp. 224–235, 2013.
© Springer-Verlag Berlin Heidelberg 2013

k-nearest neighbor (kNN)[10], divide and conquer-based[11, 17], cone cluster labeling (CCL)[8], equilibrium based approaches[18–21], fast support vector clustering (FSVC) [14], double centroids (DBC) labeling [4] and position regularized support vector clustering (PSVC)[2], etc.

However, we find that many of them reach lower time consumption with higher error or improve accuracy at the cost of efficiency. We consider to make improvements in terms of decreasing both the number of points N and the sample rate m. Three works are included in the proposed support vector clustering with boundary patterns (BPSVC) method, i.e., selecting critical points on cluster boundaries, constructing a so-called minimum hypersphere in feature space by the selected points, and integrating our convex decomposition based clustering labeling (CDCL)[3] and CCL to complete labeling under an ensemble labeling strategy. Benchmarks depict the main contributions of this paper including:

- The proposal of constructing hypersphere by only the boundary points at much lower cost of time and space to estimate the support function.
- The ensemble labeling strategy, especially for transferring connectivity checks between all pair-wise points (or SVs in [15], or stable equilibrium points (SEPs) in [18]) into between neighboring convex hulls with significantly reduced sample rate to avoid redundant checks.

2   Preliminaries

2.1   Estimating a Trained Support Function

Following [14, 13], the support function is defined as a positive scalar function $f : \mathbb{R}^n \to \mathbb{R}^+$ where a level set of f estimates a support of a data distribution and which can be decomposed into several disjoint connected components corresponding to different clusters. In support vector domain description (SVDD)[12], estimating a support function is to find the exact SVs by solving the dual problem in Eq.(1) where C is a constant for penalty and $\mathbf{x}_i$ corresponds to coefficient $\beta_i (i = 1, \ldots, N)$ if its $0 < \beta_i < C$ is a support vector.

$$\max_{\beta_j} \sum_j K(\mathbf{x}_j, \mathbf{x}_j)\beta_j - \sum_{i,j} \beta_i \beta_j K(\mathbf{x}_i, \mathbf{x}_j)$$

$$\text{s.t.} \quad \sum_j \beta_j = 1, \quad 0 \le \beta_j \le C, \quad j = 1, \ldots, N \tag{1}$$

By optimizing Eq.(1) with Gaussian kernel $K(\mathbf{x}_i, \mathbf{x}_j) = e^{-q\|\mathbf{x}_i - \mathbf{x}_j\|^2}$, the objective trained support function can be formulated by a squared radial distance of the image of $\mathbf{x}$ from the sphere center $\boldsymbol{\alpha}$ given by

$$f(\mathbf{x}) = K(\mathbf{x}, \mathbf{x}) - 2\sum_j \beta_j K(\mathbf{x}_j, \mathbf{x}) + \sum_{i,j} \beta_i \beta_j K(\mathbf{x}_i, \mathbf{x}_j) \tag{2}$$

Theoretically, the squared radius R^2 is usually defined by the value of $f(\mathbf{x}_i)$ while $\mathbf{x}_i$ is one of SVs.

2.2   Cluster Assignments

Since SVs locate on the border of clusters in data space, a simple graphical connected-component method can be used for labeling. For any two points, $\mathbf{x}_i$ and $\mathbf{x}_j$, we can check m segmers sampled on the line segment connecting them by traveling its image in the hypersphere. According to Eq.(2), $\mathbf{x}_i$ and $\mathbf{x}_j$ should be labeled the same cluster index while all the m segmers are always lying in the hypershpere, i.e., $f(\mathbf{x}_{\tilde{m}}) \leq R^2$ for $\tilde{m} = 1, \cdots, m$.

3   Support Vector Clustering with Boundary Patterns

Notice that the hypersphere is determined by SVs which are a subset of boundaries. Obviously, either SVs or boundaries are sufficient for constructing the hypersphere. Thus, we prefer a transferred strategy which collects a candidate set of them from cluster boundaries.

3.1   Obtaining Boundary Patterns

To select the most informative points, a border-edge pattern selection method presented by Li and Maguire [22] is preferred in this study. It confirms that, on the cluster boundaries, every point actually has all or most of its nearest neighbors sitting on one side of the tangent plane passing through it. Therefore, boundaries identification is to count the ratio of a point's nearest neighbors on two sides. Following [22], for a given point $\mathbf{x}_i$ with its k nearest neighbors $\mathbf{x}_j (j = 1, 2, \ldots, k)$, we can reformulate the procedure as follows:

- Setting a threshold γ $(0 < \gamma < 1)$ to control the curvature of the aforementioned surface.
- Generating normal vector $\boldsymbol{n}_i = \sum_{j=1}^{k} \boldsymbol{v}_{ij}$, where $\boldsymbol{v}_{ij} = \mathbf{x}_j - \mathbf{x}_i$.
- Calculating $l_i = \frac{1}{k} \sum_{j=1}^{k} g(\boldsymbol{n}_i^T \cdot \boldsymbol{v}_{ij})$, where the function $g(\mathbf{x})$ returns 1 if $\mathbf{x} \geq 0$, otherwise it returns 0.
- Cluster boundary identification. If $l_i \geq 1 - \gamma$, then $\mathbf{x}_i$ is considered as one of the boundary points.

3.2   Constructing Hypersphere for Support Function

Given a data set $\mathcal{X} = \{\mathbf{x}_1, \mathbf{x}_2, \ldots, \mathbf{x}_N\}$ in $\mathbb{R}^d$, selecting the cluster boundaries will return a subset $\mathcal{Z}_S = \{\mathbf{x}_{s_1}, \mathbf{x}_{s_2}, \ldots, \mathbf{x}_{s_M}\} \subseteq \mathcal{X}$ which contains the most informative points for constructing the hypersphere. That is, each point $\mathbf{x}_{s_i}$ $(i = 1, \ldots, M)$ can be approximately expected as one of SVs.

Estimating Coefficients for Boundaries. Consider the definition of SVs[15], we have

$$f(\mathbf{x}_{s_1}) = f(\mathbf{x}_{s_2}) = \cdots = f(\mathbf{x}_{s_M}) \tag{3}$$

Since $K(\mathbf{x}_{s_i}, \mathbf{x}_{s_i})$ with Gaussian kernel and $\sum_{i,j} \beta_i \beta_j K(\mathbf{x}_{s_i}, \mathbf{x}_{s_j})$ are respectively equal in Eq.(3), it is easy to check that Eq.(3) has the following expression:

$$
\begin{cases}
\sum_j \beta_j [K(\mathbf{x}_{s_j}, \mathbf{x}_{s_1}) - K(\mathbf{x}_{s_j}, \mathbf{x}_{s_2})] = 0 \\
\sum_j \beta_j [K(\mathbf{x}_{s_j}, \mathbf{x}_{s_1}) - K(\mathbf{x}_{s_j}, \mathbf{x}_{s_3})] = 0 \\
\cdots \\
\sum_j \beta_j [K(\mathbf{x}_{s_j}, \mathbf{x}_{s_1}) - K(\mathbf{x}_{s_j}, \mathbf{x}_{s_M})] = 0
\end{cases}
\tag{4}
$$

where $j \in [1, M]$ and $\sum_j \beta_j = 1$. Let $\boldsymbol{\beta} = [\beta_1, \beta_2, \cdots, \beta_M]^T$, $\mathbf{0} = [0, 0, \cdots, 0]^T$ and $\boldsymbol{Q} = [Q_1, Q_2, \cdots, Q_{M-1}]^T$ where

$$
\begin{aligned}
Q_j = [1 - K(\mathbf{x}_{s_1}, \mathbf{x}_{s_{j+1}}), K(\mathbf{x}_{s_2}, \mathbf{x}_{s_1}) - K(\mathbf{x}_{s_2}, \mathbf{x}_{s_{j+1}}), \cdots, \\
K(\mathbf{x}_{s_M}, \mathbf{x}_{s_1}) - K(\mathbf{x}_{s_M}, \mathbf{x}_{s_{j+1}})]
\end{aligned}
\tag{5}
$$

then the Eq.(4) can be further written as

$$
\boldsymbol{Q}\boldsymbol{\beta} = \mathbf{0}
$$
$$
\text{s.t.} \quad \sum_i \beta_i = 1, \quad \beta_i \geq 0
\tag{6}
$$

Since $\boldsymbol{Q}$ is determined by the cluster boundaries, $\boldsymbol{\beta}$ can be found by solving this linear system of equations with inequality constraint. Using $\boldsymbol{\beta}$ and Eq.(2), the required hypersphere with radius R can be constructed. However, the Eq.(6) can hardly be solved directly. Thus an alternative method is constructed by converting Eq.(6) into a quadratic programming problem.

Consider Eq.(6), we get

$$
\sum_j (Q_j \boldsymbol{\beta})^2 = \begin{bmatrix} Q_1\boldsymbol{\beta} \\ Q_2\boldsymbol{\beta} \\ \cdots \\ Q_{M-1}\boldsymbol{\beta} \end{bmatrix}^T \times \begin{bmatrix} Q_1\boldsymbol{\beta} \\ Q_2\boldsymbol{\beta} \\ \cdots \\ Q_{M-1}\boldsymbol{\beta} \end{bmatrix} = 0
\tag{7}
$$

where Q_j $(j = 1, \cdots, M - 1)$ is either positive or negative and each element of $Q_j\boldsymbol{\beta}$ or $(Q_j\boldsymbol{\beta})^2$ is 0. Naturally, it can be approximately reformulated by

$$
\min \quad \boldsymbol{\beta}^T \boldsymbol{H} \boldsymbol{\beta}
$$
$$
\text{s.t.} \quad \sum_j \beta_j = 1, \beta_j \geq 0, j = 1, \cdots, M
\tag{8}
$$

where $\boldsymbol{H} = \boldsymbol{Q}^T\boldsymbol{Q}$ is a Hessian matrix in $\mathbb{R}^{M \times M}$. Note that it is a standard convex quadratic program, its global optimal solution can be obtained effectively and the value of the object function can be guaranteed very close to 0 for $(Q_j\boldsymbol{\beta})^2 \geq 0$. Obviously, the penalty factor C is no longer existing.

Removal of Less Informative Points. For real problem, it is poetical that all the boundary patterns are expected to be SVs. One aspect is that none of the boundary pattern selection methods can guarantee a hundred percent correct. On the other hand, taking two nearest neighboring patterns to constructing the hypersphere is unnecessary. As depicted by Fig.1, compared with CG, in spite of a correct result achieved by BPSVC (see Fig.1b) which estimates an acceptable support function by solving the convex quadratic program (8), too many points lying on the cluster boundaries are recognized as SVs. Obviously, some of these data points are useless.

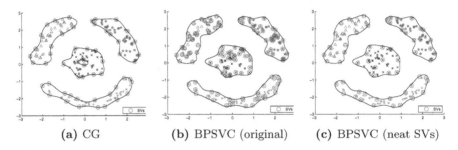

| (a) CG | (b) BPSVC (original) | (c) BPSVC (neat SVs) |

Fig. 1. Comparison of cluster boundaries and clustering results on **ring**[14]. (a) CG[15] ($q = 2, C = 1$). (b) BPSVC ($k = 10, \gamma = 0.2, q = 3.125$) before removing any boundary point. (c) BPSVC ($k = 10, \gamma = 0.2, q = 3.125$) after removing boundary points whose corresponding coefficient lower than 10^{-3}.

Our intuitive solution is quite simple: since the importance of a data point $\mathbf{x}_{s_j}$ relates to its corresponding coefficient β_j ($j = 1, \cdots, M$) directly in constructing center $\boldsymbol{\alpha} = \sum_j \beta_j \Phi(\mathbf{x}_j)$, where $\Phi(\cdot)$ is a nonlinear map function. To further obtain a neat data set, those points with coefficients lower than a predefined threshold β_s should be removed for uselessness or little information. Obviously, a smaller β_s allows more redundant data to profile cluster boundaries accurately, but sometimes go along with overfitting; whereas a smoother profile would be generated by a greater β_s at the risk of much more overlapped regions between clusters. Following the principle of SVC, a large number of experiments suggest that an appropriate threshold β_s for removing the less informative points should be $10^{-\lceil \lg N \rceil}$ more or less, e.g., $\beta_s = 10^{-3}$ for **ring** in Fig.1c.

Estimating the Radius of the Hypersphere. Notice that programming (8) is a compromise of programming (1). Therefore, the strict zero can hardly be achieved though we expected it should be. After the removal of useless boundary points, in reality, we get a reduced set $\mathcal{Z}_R = \{\mathbf{x}_{r_1}, \mathbf{x}_{r_2}, \ldots, \mathbf{x}_{r_L}\} \subseteq \mathcal{Z}_S$ with $L \leq M$, which are the exact SVs. However, the relation of their distances to the center of the hypersphere is

$$f(\mathbf{x}_{r_1}) \approx f(\mathbf{x}_{r_2}) \approx \cdots \approx f(\mathbf{x}_{r_L}) \tag{9}$$

Finally, consider the numerical problem in practical[1], we construct the hyperpshere whose radius is the maximum distance from $\mathbf{x}_{r_l}$ ($l \in [1, L]$) to its center, i.e., $R^2 = \max_{l=1}^{L} f(\mathbf{x}_{r_l})$.

3.3   Ensemble Labeling Strategy for Efficiency

Following the aforementioned works, a subset of boundary points are selected as exemplars for cluster assignments. Intuitively, three conventional strategies can be employed, i.e., checking the full pairs of SVs like SVG, directly calculating the distance between each pair of SVs to verify if their cones are intersected (like CCL), and traveling the line segments connecting each pair of SEPs converged from the SVs while the solution of finding the minimal hypersphere, $\partial \mathbf{x}/\partial t = -\nabla f(\mathbf{x})$, is considered as a gradient dynamical system (e.g., reduced complete graph (R-CG) [4, 18]). However, as the previously stated, these strategies suffer from some obvious drawbacks. For the first, it is time-consuming while processing large-scale or high dimension data[4, 14]. Although the adjacency matrix of SVs can be calculated very fast, a radius lower than 1 is essential for the second strategy to find connected components. However, due to numerical problem, the radius which should be lower than 1 cannot be guaranteed. Therefore, CCL cannot be employed directly. Finally, as noted in Ref.[20] and detailed by Ref.[4], only SEPs employed by the third strategy to represent data for connectivity checking usually lead to relatively high error on irregular shaped data set. Therefore, to achieve improvements on both efficiency and clustering quality, we prefer a simple but effective ensemble labeling strategies of CDCL and CCL. As depicted in lines 4~8 of Algorithm 1, it prefers CDCL since the constructed hypersphere's radius is greater than 1; otherwise, CCL is employed.

4   Implementation

In this section, we give description of the proposed BPSVC method as well as some remarks distinguishing from the others. For the given γ and k, line 2 of Algorithm 1 collects cluster boundaries for constructing the objective hypersphere by line 3. Three essential elements of the support function, i.e., the radius of the hypersphere R, the final set of SVs $\mathcal{Z}_R$ with respect to their coefficient β are obtained by ConstructHypersphere($\mathcal{Z}_s, q, \beta_s$). To get a nest set of SVs, the threshold β_s for removing useless boundary points is set in line 1 of the Algorithm 1. Notice that although in Algorithm 1 we start from selecting cluster boundaries by measuring the full data set, the computation is significantly reduced as the calculation is repeated in data space and a rather lower size of data ($M, L \ll N$) remained for both line 3, and ConnectivityAnalysisofConvexHulls($\mathcal{Z}_R, q$) in line 5 or ConnectivityAnalysisofSVs($\mathcal{Z}_R, R$) in line 7. Specifically, taking this tidy data into the rather simple convex quadratic program (8) makes the proposed BPSVC handle large-scale problem efficiently.

Actually, the ensemble labeling strategies is implemented by line 4~8. Since $R \geq 1$, CDCL is employed to decompose $\mathcal{Z}_R$ into N_c groups for constructing convex hulls (detailed in Ref.[3]). Then the connectivity analysis of clusters can be done between convex hulls. It believes that the far from associate external locations to convex hulls is, the greater probability that the corresponding local region to be sparse distribution with data points is. Practically, it

---

Algorithm 1. BPSVC($\mathcal{X}, \gamma, k, q$)

Input: the data set $\mathcal{X}$, number of neighbors k
        threshold γ and Gaussian kernel width q
Output: clustering labels for all the data points
1 set $\beta_s = 1/|\mathcal{X}|$
2 $\mathcal{Z}_s \leftarrow$ SelectClusterBoundaries($\mathcal{X}, \gamma, k$)
3 $\{\mathcal{Z}_R, R, \boldsymbol{\beta}\} \leftarrow$ ConstructHypersphere($\mathcal{Z}_s, q, \beta_s$)
4 **if** $R \geq 1$ **then**
5     $\boldsymbol{A} \leftarrow$ ConnectivityAnalysisofConvexHulls($\mathcal{Z}_R, q$)
6 **else**
7     $\boldsymbol{A} \leftarrow$ ConnectivityAnalysisofSVs($\mathcal{Z}_R, R$)
8 **end**
9 Labels $\leftarrow$ FindConnComponents($\boldsymbol{A}$)
10 **for each** $\mathbf{x} \in \mathcal{X} \backslash \mathcal{Z}_R$
11    $inx \leftarrow$ find the nearest SV from $\mathbf{x}$
12   Labels[$\mathbf{x}$] $\leftarrow$ Labels[$\boldsymbol{v}_{inx}$]
13 **end**
14 **return** Labels

---

does reduce the average sample rate m significantly. All of these tasks are completed by function ConnectivityAnalysisofConvexHulls(H) in line 5. After that, the adjacency matrix $\boldsymbol{A}$ is obtained for connectivity analysis by means of any standard algorithm. Otherwise, while R is lower than 1, the invoked function ConnectivityAnalysisofSVs($\mathcal{Z}_R, R$) will check the connectivity among SVs following the CCL method, which could be explained by Ref.[8]. By now, the output of FindConnComponents($\boldsymbol{A}$) in line 9 is an array with size N_c which contains the cluster labels. Finally, the remaining data points are separately assigned with the labels of their nearest SVs.

5 Experiments

5.1 Datasets and Experimental Settings

To demonstrate the performance of the proposed BPSVC, in this section, we conduct comparisons among ten state-of-the-art methods, i.e., CG, DD, k-NN ($k = 4$), MST, R-CG, E-SVC, CCL, FSVC, PSVC and CDCL. The employed data sets (described in Table 1) include: five-Gaussians, twocircles and D31 from Refs.[14, 13, 24] and iris, wisconsin, zoo, movement_libras and shuttle from UCI repository[25]. For fair comparisons, all the simulations are carried out in MATLAB 2011b on system with Intel Dual Core 2.66 GHz and 3GB RAM, and all of the data sets are employed without any preprocessing.

Table 1. Description of the benchmark data sets

Data sets	dims	size	# of classes
twocircles	2	300	2
iris	4	150	3
wisconsin	9	683	2
zoo	16	101	7
movement_libras	90	360	15
five-Gaussians	2	1000	5
D31	2	3100	31
shuttle	9	43500	7

To measure the clustering accuracy, we use adjusted rand index (ARI)[6, 23], rand index (Rand), jaccard coefficient (Jaccard)[5], and normalized mutual information (NMI)[7], which are a widely used similarity measure between two data partitions where both true labels and predicted cluster labels are given.

5.2 Benchmark Results

Table 2 shows the performance achieved by the evaluated algorithms. Notice that the time cost is an average value of ten times of the execution for each data. Rank of each algorithm is given depending on its performance measure followed by corresponding rank (from 1 to 3). In particular, the value of rank 1 for each test item is highlighted by boldface.

Table 2. Benchmark results on data sets with different sizes

Data	Methods	(C,q)	ARI	Rand	Jaccard	NMI	Time(sec.)
twocircles	CG	0.5,0.125	**1.00000**[a]	**1.00000**[a]	**1.00000**[a]	**1.00000**[a]	10.01
	DD	0.3,0.0638	**1.00000**[a]	**1.00000**[a]	**1.00000**[a]	**1.00000**[a]	11.66
	kNN	0.32,0.125	0.69679	0.84854	0.69612	0.76529[c]	2.38
	MST	0.3,0.3252	0.59935	0.79541	0.58953	0.64044	17.56
	R-CG	0.3, 0.1072	0.67695	0.83864	0.67625	0.75411	2.93
	E-SVC	0.2,0.074	0.73547[c]	0.86785[c]	0.73486[c]	0.77801	28.62
	CCL	0.1, 0.0633	0.76193[b]	0.88581[b]	0.77423[b]	0.80800[b]	9.71
	FSVC	0.1, 50	0.14592	0.57411	0.14552	0.50433	0.75[b]
	CDCL	0.1,0.1385	**1.00000**[a]	**1.00000**[a]	**1.00000**[a]	**1.00000**[a]	0.77[c]
	PSVC	—,0.1385	**1.00000**[a]	**1.00000**[a]	**1.00000**[a]	**1.00000**[a]	14.78
	BPSVC	—,11.3379	**1.00000**[a]	**1.00000**[a]	**1.00000**[a]	**1.00000**[a]	**0.22**[a]
iris	CG	0.46,15.4321	0.61780	0.84886	0.55187	0.67392	1.13
	DD	0.5, 13.8504	0.58334	0.83696	0.51645	0.65589	2.87
	kNN	0.45, 13.8504	0.64143	0.85718	0.57655	0.68600	0.36[b]
	MST	0.45, 4.0816	0.79457	0.91257	0.74981	0.79331	1.08
	R-CG	0.29,12.5	0.73737	0.89208	0.68095	0.75000	3.07
	E-SVC	0.19,1.3889	0.56812	0.77629	0.59514	0.76117	4.74
	CCL	0.03, 0.7436	0.88579[b]	0.94953[b]	0.85776[b]	0.87052[b]	0.76
	FSVC	0.33, 2.4691	0.56196	0.77271	0.57842	0.71256	1.07
	CDCL	0.19,9.4518	**0.92218**[a]	**0.96564**[a]	**0.90075**[a]	**0.90112**[a]	0.67[c]
	PSVC	—,15.4321	0.61467	0.84904	0.54504	0.68354	1.50
	BPSVC	—, 4.0816	0.85089[c]	0.93548[c]	0.8148[c]	0.82681[c]	**0.15**[a]

Table 2. *(Continued)*

Data	Methods	(C, q)	ARI	Rand	Jaccard	NMI	Time(sec.)
wisconsin	CG	0.4, 0.3472	0.77930	0.88971	0.80951	0.69747	66.45
	DD	—	—	—	—	—	—
	kNN	0.4, 0.3472	0.76243	0.88110	0.79463	0.68049	31.65
	MST	0.4, 0.3472	0.66311	0.82986	0.70612	0.61597	85.72
	R-CG	0.1, 0.0868	0.80345	0.90241	0.83473	0.66434	22.25
	E-SVC	0.2,0.1134	0.13441	0.59395	0.54846	0.14341	443.23
	CCL	0.1,0.005	0.90763[b]	0.94861[b]	0.90957[b]	0.81521[b]	124.35
	FSVC	0.1,1.3889	0.66871	0.83192	0.70242	0.45672	13.06[c]
	CDCL	0.105,0.0595	0.86850[c]	0.93482[c]	0.88747[c]	0.77555[c]	**1.31**[a]
	PSVC	—,2.8345	0.2574	0.63714	0.52731	0.22633	192.95
	BPSVC	—,4.8828	**0.91712**[a]	**0.95882**[a]	**0.92658**[a]	**0.80295**[a]	4.41[b]
zoo	CG	0.49, 0.4287	0.93421[c]	0.97663[c]	0.90367[c]	0.90763[c]	0.62
	DD	—	—	—	—	—	—
	kNN	0.49, 0.4287	0.93421[c]	0.97663[c]	0.90367[c]	0.90763[c]	**0.26**[a]
	MST	0.49, 0.4287	0.93421[c]	0.97663[c]	0.90367[c]	0.90763[c]	0.38[c]
	R-CG	0.27, 0.3916	**0.95702**[a]	**0.98455**[a]	**0.93633**[a]	**0.92036**[a]	3.82
	E-SVC	0.27, 0.3916	**0.95702**[a]	**0.98455**[a]	**0.93633**[a]	**0.92036**[a]	19.85
	CCL	0.1,2.5826	0.83426	0.89861	0.79016	0.84893	1.03
	FSVC	0.1,0.2551	0.84625	0.94416	0.79033	0.85331	2.67
	CDCL	0.39,0.5	0.94691[b]	0.98079[b]	0.92215[b]	0.90934[b]	2.83
	PSVC	—,0.4058	0.7441	0.91723	0.65878	0.85325	0.73
	BPSVC	—,50	0.93421[c]	0.97663[c]	0.90367[c]	0.90763[c]	0.30[b]
movement_libras	CG	0.5,5.5556	0.24218	0.93013	0.15922	0.70155	15.14
	DD	—	—	—	—	—	—
	kNN	0.5, 3.8580	0.26661[c]	0.91360[c]	0.18532[c]	0.66459[c]	7.26[b]
	MST	0.5, 3.8580	0.24872	0.91102	0.17385	0.65660	41.07
	R-CG	0.5, 5.5556	0.23559	0.93375	0.15194	0.70258	252.89
	E-SVC	—	—	—	—	—	—
	CCL	0.5, 5.5556	0.08987	0.93873	0.04556	0.70874	26.91
	FSVC	0.3,0.4132	0.14205	0.93861	0.04478	0.70352	226.09
	CDCL	0.32,4.8828	0.33195[b]	0.92098[b]	0.23010[b]	0.68084[b]	78.57
	PSVC	—,4.3253	0.25407	0.91882	0.17412	0.67012	12.18[b]
	BPSVC	—, 2.9744	**0.37034**[a]	**0.92103**[a]	**0.25995**[a]	**0.68956**[a]	4.20[a]
five-Gaussians	CG	0.15,22.8269	0.47118	0.83982	0.39755	0.63436	89.01
	DD	0.2,19.5313	0.61487	0.89230	0.51115	0.69568	25.63
	kNN	0.21,15.4321	0.26661	0.91360	0.18532	0.66459	7.26[c]
	MST	0.14,0.5	0.67807	0.90912	0.57554	0.72340	52.84
	R-CG	0.14,17.3010	0.86934[c]	0.95987[c]	0.80832[c]	0.85852[c]	12.90
	E-SVC	0.14,17.3010	0.85854	0.95707	0.79335	0.84993	775.69
	CCL	0.1,0.005	0.00211	0.31746	0.19109	0.091794	650.39
	FSVC	0.1,50	0.71373	0.91753	0.59742	0.77843	2.39[b]
	CDCL	0.14,17.3010	**0.88074**[a]	**0.96344**[a]	**0.82348**[a]	**0.86874**[a]	7.67
	PSVC	—,29.5858	0.00124	0.30446	0.19180	0.13823	750.21
	BPSVC	—,2.0406	0.87043[b]	0.96020[b]	0.80979[b]	0.86845[b]	1.19[a]
D31	CG	0.15, 0.5	0.23448	0.87570	0.15893	0.63159	3594.75
	DD	0.5, 2.9744	0.54946	0.95681	0.39769	0.8488	500.37
	kNN	0.5, 2.9744	0.72032	0.98334	0.57344	0.88154	242.70
	MST	—	—	—	—	—	—
	R-CG	0.1, 5.2029	0.86532[c]	0.99160[c]	0.76938[c]	0.88985[c]	53.27
	E-SVC	—	—	—	—	—	—
	CCL	—	—	—	—	—	—

Table 2. *(Continued)*

Data	Methods	(C, q)	ARI	Rand	Jaccard	NMI	Time(sec.)
	FSVC	1,200000000	0.56109	0.96430	0.40702	0.82011	4.18[a]
	CDCL	0.1,5.5556	0.90199[a]	0.99420[a]	0.82643[a]	0.94355[a]	19.79[c]
	PSVC	—,12.5	0.45178	0.94391	0.31265	0.80943	7041.01
	BPSVC	—,1.3889	0.87670[b]	0.99224[b]	0.78685[b]	0.90216[b]	6.72[b]
	CG	—	—	—	—	—	—
	DD	—	—	—	—	—	—
	kNN	—	—	—	—	—	—
	MST	—	—	—	—	—	—
shuttle	R-CG	—	—	—	—	—	—
	E-SVC	unknown	0.59[15]	—	—	—	—
	CCL	—	—	—	—	—	—
	FSVC	unknown	0.58[15]	—	—	—	—
	CDCL	—	—	—	—	—	—
	PSVC	—	—	—	—	—	—
	BPSVC	—,0.0078	0.68574[a]	0.86416[a]	0.82084[a]	0.62654[a]	878.19[a]

Note: [a]Rank 1, [b]Rank 2, [c]Rank 3; "—" means not available.

In terms of accuracy, it is apparent that BPSVC is better for most of data sets (namely `twocicles`, `wisconsin`, `movement_libras` and `shuttle`). Furthermore, it achieves first three ranks consistently on the other data sets. With the help of hypersphere construction and convex decomposition, BPSVC reaches global optimal solutions consistently. For time consumption, BPSVC employs much fewer points to work out the support function, while the others keep solving the same quadratic programming problem with different parameters to achieve their best performance. Thus BPSVC finishes the clustering works fastest on five out of eight data sets. Its advantage is obvious on relative large-scale data, e.g., `shuttle`. Due to memory limitation, we cannot afford the requirement of kernel matrix from FSVC[14], thus a direct citation of experiment result is given.

6   Concluding Remarks

This paper develops a support vector clustering with boundary method namely BPSVC from a new perspective. It gives an optimal solution for these known problems, i.e., a requirement of huge memory for kernel matrix, too many redundant point pairs and a great sample rate.

Even though BPSVC gives consistent results for various cases, how to shrink the cluster boundaries to leave a number of outliers while obtains high quality profiles for clusters in input space, might be an open issue for further improvements on both efficiency and accuracy. And how to redefine the coefficient β_j for the remaining patterns after removing unless data needs to be further investigated as well.

Acknowledgements. This work is supported by the grant from the National Natural Science Foundation of China under Grant No. 60972077, partially under

Grant No. 70921061, the Foundation of He'nan Educational Committee under Grant No.13A413750, 13A413747, the Natural Science Foundation of He'nan Province of China under Grant No. 122102210543, and Xuchang Municipal Natural Science Foundation under Grant No. 5018.

References

1. Wang, C.-D., Lai, J.-H., Huang, D., Zheng, W.-S.: SVStream: A Support Vector Based Algorithm for Clustering Data Streams. IEEE Transactions on Knowledge and Data Engineering, 1–14 (2011) (in press), doi:10.1109/TKDE.2011.263
2. Wang, C.-D., Lai, J.-H.: Position Regularized Support Vector Domain Description. Pattern Recognition 46(3), 875–884 (2013)
3. Ping, Y., Tian, Y.J., Zhou, Y.J., Yang, Y.X.: Convex Decomposition based Cluster Labeling Method for Support Vector Clustering. Journal of Computer Science and Technology 27(2), 428–442 (2012)
4. Ping, Y., Zhou, Y.J., Yang, Y.X.: A Novel Scheme for Acclereating Support Vector Clustering. Computing and Infomatics 31(6), 613–638 (2012)
5. Tan, P.-N., Steinbach, M., Kumar, V.: Introduction to Data Mining. Addison Wesley (2006)
6. Xu, R., Wunsch, D.C.: Clustering. A John Wiley&Sons (2008)
7. Strehl, A., Ghosh, J.: Cluster Ensembles - A Knowledge Reuse Framework for Combining Multiple Partitions. Journal of Machine Learning Research (3), 583–617 (2002)
8. Lee, S.-H., Daniels, K.M.: Cone Cluster Labeling for Support Vector Clustering. In: Proceedings of 6th SIAM Conference on Data Mining, pp. 484–488. SIAM, Bethesda (2006)
9. Yang, J.H., Estivill-Castro, V., Chalup, S.K.: Support Vector Clustering Through Proximity Graph Modelling. In: Proceedings of 9th International Conference on Neural Information Processing (ICONIP 2002), pp. 898–903. Orchid Country Club, Singapore (2002)
10. Estivill-Castro, V., Lee, I., Murray, A.T.: Criteria on Proximity Graphs for Boundary Extraction and Spatial Clustering. In: Cheung, D., Williams, G.J., Li, Q. (eds.) PAKDD 2001. LNCS (LNAI), vol. 2035, pp. 348–357. Springer, Heidelberg (2001)
11. Puma-Villanueva, W.J., Bezerra, G.B., Lima, C.A.M., Zuben, F.J.V.: Improving Support Vector Clustering with Ensembles. In: Proceedings of International Joint Conference on Neural Networks, Montreal, Quebec, Canada, pp. 13–15 (2005)
12. Tax, D.M.J., Duin, P.R.W.: Support Vector Domain Description. Pattern Recognition Letters 20(11-13), 1191–1199 (1999)
13. Jung, K.-H., Kim, N., Lee, J.: Dynamic Pattern Denoising Method using Multibasin System with Kernels. Pattern Recognition 44(8), 1698–1707 (2011)
14. Jung, K.-H., Lee, D., Lee, J.: Fast Support-based Clustering Method for Large-scale Problems. Pattern Recognition 43(5), 1975–1983 (2010)
15. Ben-Hur, A., Horn, D., Siegelmann, H.T., Vapnik, V.N.: Support Vector Clustering. Journal of Machine Learning Research (2), 125–137 (2001)
16. Guo, C.H., Li, F.: An Improved Algorithm for Support Vector Clustering based on Maximum Entropy Principle and Kernel Matrix. Expert Systems with Applications 38(7), 8138–8143 (2011)
17. Ling, P., Zhou, C.-G., Zhou, X.: Improved Support Vector Clustering. Engineering Applications of Artificial Intelligence 23(4), 552–559 (2010)

18. Lee, J., Lee, D.: An Improved Cluster Labeling Method for Support Vector Clustering. IEEE Transactions on Pattern Analysis and Machine Intelligence 27(3), 461–464 (2005)
19. Lee, J., Lee, D.: Dynamic Characterization of Cluster Structures for Robust and Inductive Support Vector Clustering. IEEE Transactions on Pattern Analysis and Machine Intelligence 28(11), 1869–1874 (2006)
20. Lee, J., Lee, D.: Equilibrium-based Support Vector Machine for Semisupervised Classification. IEEE Transactions on Neural Networks 18(2), 578–583 (2007)
21. Lee, J., Jung, K.-H., Lee, D.: Constructing Sparse Kernel Machines Using Attractors. IEEE Transactions on Pattern Analysis and Machine Intelligence 20(4), 721–729 (2009)
22. Li, Y.H., Maguire, L.: Selecting Critical Patterns Based on Local Geometrical and Statistical Information. IEEE Transactions on Pattern Analysis and Machine Intelligence 33(6), 1189–1201 (2011)
23. Hubert, P.A.L.: Compariring partitions. Journal of Classification 2, 193–218 (1985)
24. Veenman, C.J., Reinders, M.J.T., Backer, E.: A Maximum Variance Cluster Algorithm 24(9), 1273–1280 (2002)
25. Frank, A., Asuncion, A.: UCI Machine Learning Repository, http://archive.ics.uci.edu/ml

Ensembles of Optimum-Path Forest Classifiers Using Input Data Manipulation and Undersampling

Moacir P. Ponti Jr. and Isadora Rossi*

Institute of Mathematical and Computer Sciences, University of São Paulo
(ICMC/USP)
13560-970 São Carlos, SP, Brazil
moacir@icmc.usp.br, isarossi@grad.icmc.usp.br
http://www.icmc.usp.br/~moacir

Abstract. The combination of multiple classifiers was proven to be useful in many applications to improve the classification task and stabilize
results. In this paper we used the Optimum-Path Forest (OPF) classifier to investigate input data manipulation techniques in order to use
less data from the training set without hampering the classification accuracy. The data undersampling can be useful to speed-up the classification
task, and could be specially useful with large datasets. The results indicate that the OPF-based ensemble methods allow a significant reduction
on the size of the training set, while maintaining or slightly improving
accuracy. We provide intuition for a case of failure and report the results
of synthetic and real datasets.

Keywords: Optimum-Path Forest, undersampling, pasting of small votes.

1 Introduction

Ensemble learning techniques include methods to create multiple classifiers and
to combine their decisions. There are many approaches designed to produce
ensembles of classifiers such as input data manipulation, e.g. Bagging [1] input
feature manipulation, e.g. random subspace method [9], among others. These
method are based on classifiers trained with different training sets obtained
by manipulation of samples or features. The decision fusion is often obtained
by majority voting (when only class labels are available) or by averaging (for
support/confidence output for each class).

Although the most often cited advantage of ensemble learning is to **improve
the recognition accuracy by adding complexity** to the system [17], the
same framework can be also used to **reduce the computational cost** while
maintaining or even slightly improving the accuracy. This kind of effect is specially interesting for large datasets, for which it is not feasible to use all data
available to create the classification model, due to the high computational cost.

---

* Supplementary material for this paper such as the code and datasets can be found
  at http://www.icmc.usp.br/~moacir/project/MCS

Z.-H. Zhou, F. Roli, and J. Kittler (Eds.): MCS 2013, LNCS 7872, pp. 236–246, 2013.

It is clearly important to investigate methods that uses smaller training sets or undersamples the data. However, it is also challenging to design a successful pattern recognition system with a single classifier trained with a reduced number of samples, since they often require sufficient and relevant data in order to perform well [5]. Therefore, multiple classifiers can be a way out, providing that the base classifier is able to produce suficient diversity and accuracy with undersampled training sets [3], and that its training computational cost is not deterrent.

Questions such as what exact sampling methodology to use, and the right sample size are difficult to answer. However, some of those questions are addressed by [13]. Our claim is that ensemble methods can help by reducing the variance of the results obtained by several classifiers trained with **reduced randomly sampled** data, without the need to fine tune sample size. An example of ensemble approach developed in this context is pasting of small votes, proposed by Breiman [2]. It builds classifiers from "bites", bags of small size, of data. Two main algorithms were originally proposed: Rvote and Ivote. Rvote selects random objects from the training set in order to build the classifiers, whereas Ivote selects the samples based on their importance. Distributed versions of both algorithms were proposed afterwards by Chawla et al. [4].

Related works performed undersampling mainly to deal with imbalanced data. For example, the IRUS method, proposed in 2009, severely undersamples the majority class, and creates a large number of distinct training sets [20]. Balance cascade learners were also proposed in the same context [11]. Another study used ensembles with undersampling in order to classify websites reputation [8]. The Ensembles on Random Patches tackles larger datasets by using random selection of both samples and features [12]. Also, ensemble methods based on undersampling in terms of the bias-variance decomposition of the error were developed before [21].

The Optimum-Path Forest (OPF) classifier [15] was proposed in 2009 and two studies on multiple OPF classifiers were published. The first reported the use of disjoint and distributed partitions of the training set, with good results [18], the second used OPF outputs in a graph modeled through a Markov Random Field in order to obtain a hierarchy of the decisions [19]. Since this classifier appears to have produced diversity under input data manipulation according to the first study [18], and to handle well small sizes of data as shown in the second one [19], it is likely to have the potential to be investigated as a base classifier for an ensemble using both input data manipulation and undersampling.

In this study, we propose input sample manipulation to construct ensembles with a reduced number of samples to train each OPF base classifier. The outputs are combined by majority voting, since OPF outputs only class labels. As far as we know there are no papers on the creation of ensembles of OPF classifiers based on pasting of small votes with an undersampled training set.

The methods were implemented to be included on the free library LibOPF [14] and are available at the project webpage. The behavior and the accuracy improvement is compared to the OPF single classifier using datasets with different characteristics.

2   The OPF Classifier

This method interprets samples as vertices of a graph. The training step connects the samples from the same class in order to produce a tree for each class, using a specified distance space adjacency relation. The set of trees is called the optimum path forest (OPF). Each tree has a prototype vertex (obtained by minimizing the distance from the vertex of the current class to a vertex of another class). This prototype is considered the root of the tree. A new sample is classified by connecting it to the tree that offers the optimum cost path to its root. This classifier was proposed in 2009 [14] and showed good performance on different applications. It handles multi-class problems natively and has no parameters to adjust.

Using an input training set S of size N, a subset S is selected with N' samples, from which M samples are used to train each of the L base classifiers and construct the ensemble.

3   Combining OPF Classifiers

All algorithms were developed based on pasting of small votes. The aim was to explore the OPF characteristics in order to create classifiers that produce diverse outputs under data perturbation. The algorithms are given a previously undersampled training set. The bite size, M, is then computed by the algorithms. In this study we used $M = N'/L$, allowing the methods to be tested even with scarce data.

OPF–Rvote uses small training sets (bites), obtained at random from the original training set [2]. The procedure can be done with or without replacement. The Algorithm 1 shows the complete procedure.

OPF–Ivote uses smaller training sets, but they are obtained by an algorithm that computes the importance of each sample, inserting mainly samples that are misclassified by the combination of the previously trained classifiers [2]. It uses an evaluation set in order to obtain the errors. The Algorithm 2 shows the complete procedure.

4   Method and Experiments

Six ensembles were created with each algorithm, with $L = 3, 5, 9, 11, 13$ and 15 base classifiers. Since one of our objectives was to investigate the data undersampling, we used only half the samples in the training set, i.e. $N' = N/2$, so half the data was used to form the reduced training set N'.

Algorithm 1. OPF-Rvote

Require: Ensemble size L, training set S' of size N', size of each bite M, sampling with replacement R (true or false)

1: **for** $i = 1$ to L **do**
2:    **if** $R = $ true **then**
3:       $S_i \leftarrow M$ sampled items from S', with replacement.
4:    **else**
5:       $S_i \leftarrow M$ sampled items from S', without replacement.
6:    **end if**
7:    Train classifier h_i using S_i.
8: **end for**
9: **for** each new pattern **do**
10:    Compute the majority voting of h_i, $i = 1, ..., L$.
11: **end for**

Algorithm 2. OPF-Ivote

Require: Ensemble size L, training set S' of size N', size of each bite M, evaluation set A.

1: $S_1 \leftarrow M$ sampled items from S', with replacement.
2: Train classifier h_1 using S_1.
3: Compute the error E_1 of the classifier h_1 using A
4: **for** $i = 2$ to L **do**
5:    **while** training set S_i has less than M samples **do**
6:       Select one sample x from S
7:       Classify x using the previous classifiers h_j $j = 1, ..., i - 1$
8:       $label \leftarrow$ majority voting of $h_j(x)$, $j = 1, ..., i - 1$.
9:       **if** $label$ was misclassified **then**
10:          $S_i \leftarrow x$
11:       **else**
12:          $S_i \leftarrow x$ with probability $p = E_{i-1}/(1 - E_{i-1})$
13:       **end if**
14:    **end while**
15:    Train classifier h_i using S_i.
16:    Compute the error E_i of the classifier h_i using A
17: **end for**
18: **for** each new pattern **do**
19:    Compute the majority voting of h_i, $i = 1, ..., L$.
20: **end for**

Moreover, the bite size is $M = N'/L$, so that larger ensembles contains base classifiers trained with less samples. This choice was made in order to study the behavior of the OPF classifiers when dealing with less data. It also allow the methods to be tested with datasets with fewer samples and different ensemble sizes.

We used two hold-out settings for the experiments:

- Configuration 1 (Single OPF, OPF-Rvote): 30% train, and 70% test.
- Configuration 2 (OPF-Ivote): 25% train, 5% evaluation and 70% test.

The OPF-Ivote uses the evaluation set A in order to compute the errors E_i, as an estimate of the out-of-bag error (see Algorithm 2).

Each experiment was repeated 100 times. The average and standard deviation were computed by these repetitions. The evaluation was based on an accuracy value that takes into account the balance between the classes:

$$\text{Acc} = 1 - \frac{\sum_{i=1}^{c} E(i)}{2c},$$

where c is the number of classes, and $E(i) = e_{i,1} + e_{i,2}$ is the partial error of c, computed by:

$$e_{i,1} = \frac{FP(i)}{N - N(i)} \text{ and } e_{i,2} = \frac{FN(i)}{N(i)}, i = 1, ..., c,$$

where $FN(i)$ (false negatives) is the number of samples belonging to i incorrectly classified as belonging to other classes, and $FP(i)$ (false positives) the samples $j \neq i$ that were assigned to i.

Detailed information about the datasets used in the experiments are shown in Table 1, including synthetic and real data:

- **Lithuanian**: "Lithuanian" type classes as proposed by Raudys [6].
- **Banana**: "banana-shaped" distributed classes [6].
- **Gaussian**: Gaussian distributed classes of different covariance matrices [6].
- **Corel-GCH**: global histogram color features from 1000 images obtained from the Corel image dataset [10].
- **Pterygium**: identification of a common ophthalmic disease [16].
- **Skin**: classification of skin and non-skin image pixels, obtained from face images of various age, ethnicities and genders from FERET database and PAL database [7].
- **Spambase**: identification of spam emails (advertisements for products/web sites, make money fast schemes, chain letters, pornography, etc.) from non-spam emails [7].

5   Results and Discussion

The average accuracy and standard deviation values for the single classifier and the best ensemble result are shown in Table 2, including the number of base classifiers L for the best ensemble result. The value of M, the actual size of the training set for each base classifier is also shown in the same table. The accuracy

Table 1. Dataset characteristics

Dataset	#Samples	#Classes	#Features
Lithuanian	1,000	4	3
Banana	1,000	2	4
Gaussian	100,000	4	3
Corel-GCH	1,000	10	64
Spambase	4,601	2	57
Pterigyum	7,651	2	89
Skin	245,057	2	4

Table 2. Results

Dataset	Single-OPF	N	Ensemble methods	L	M
Lithuanian	76.2±0.6	300	RVote: 78.6±0.3	13	
			RVote-WR: 79.2±0.0	9	16
			IVote: 78.9±0.0	9	
Banana	80.5±0.3	300	RVote: 81.6±0.1	13	
			RVote-WR: 82.3±0.2	9	16
			IVote: 81.5±0.0	7	
Gaussian	96.1±0.2	30,000	RVote: 96.3±0.1	7	
			RVote-WR: 96.4±0.0	5	3,000
			IVote: 96.4±0.0	9	
Corel-GCH	77.6±0.2	300	RVote: 74.4±0.1	3	
			RVote-WR: 74.6±0.1	3	50
			IVote: 75.0±0.0	3	
Spambase	86.5±0.1	1,380	RVote: 84.7±0.1	3	
			RVote-WR: 85.1±1.2	3	230
			IVote: 84.9±0.1	3	
Pterygium	94.1±2.5	2,550	RVote: 98.9±0.1	15	
			RVote-WR: 98.7±0.0	15	115
			IVote: 98.8±0.0	15	
Skin	99.2±0.1	73,517	RVote: 99.7±0.0	11	
			RVote-WR: 99.7±0.1	13	2,827
			IVote: 99.8±0.1	13	

curves for different ensemble sizes of OPF-Rvote, OPF-Rvote-WR (withouth replacement) and OPF-Ivote are shown in Figure 1 for the synthetic datasets and in Figure 2 for real datasets.

The combination of multiple OPF classifiers showed slight improvement on the classification accuracy. We observed a significant improvement for the Pterygium, a larger dataset with 7651 available samples and 89 features. The creation

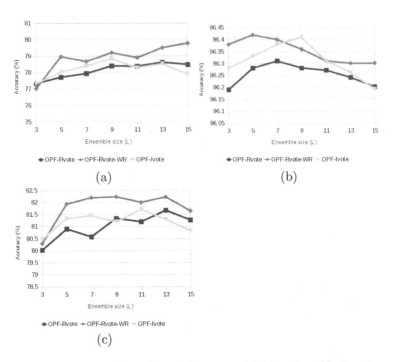

Fig. 1. Average accuracy of synthetical datasets: a) Lithuanian, b) Gaussian and c) Banana-shaped

of ensembles with data undesampling was successful, achieving an improvement of ≈ 5% with a final result of 99% accuracy for the ensemble with 15 base classifiers. The experiments showed a small improvement: ≈ 3% in Lithuanian, ≈ 3% Gaussian and ≈ 2% Banana-shaped dataset. It is interesting to see, however, that the improvement could be achieved for a reduced training set size and adapting the bite size M as the ensemble size, L, grow larger, since M was defined as a function of L. Also, where there is redundancy, such as in Skin and Pterygium datasets, the method is able to make use of fewer training instances, improve speed, while increasing the accuracy.

The exception was the dataset Corel-GCH that obtained worst results with ensemble methods. This example show when the method fails: under a low number of examples per class. Since the training set has 30 samples/class, when the ensemble size is 3, each base classifier has to be trained using just $M = \lfloor (30/2)/3 \rfloor = 5$ images/class. Moreover, this dataset presented a low variance characteristic (between the samples), not desirable when combining classifiers created by Bagging and pasting of small votes [17].

The potential of the OPF for ensemble learning by input data manipulation and undersampling can be seen in Figure 3, where are shown results of using from 50% (as in the experiments) to 20% of the Pterygium original training set.

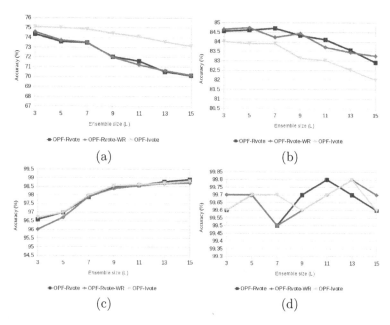

Fig. 2. Average accuracy of real datasets: a) COREL-GCH, b) Spambas, c) Pterigyum and d) Skin

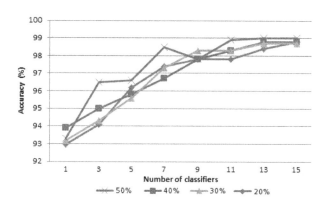

Fig. 3. OPF-Rvote for Pterygium dataset with different sizes of N'

5.1   Running Time Analysis

The platform used to run the experiment was a 2.1GHz Intel®dual-core portable computer with 2GB RAM memory and a 64 bit operating system Linux 3.2.0-33 kernel version. The running time were computed using the same 10 repetitions experiment. The average was obtained and the standard deviation was insignificant (below 0.01. For the datasets with 1000 samples the running time of both single and ensemble methods to train and classify was very similar, around 0.05 sec. However, for the 100,000 samples dataset, we observed a speed improvement, as shown in Figure 4.

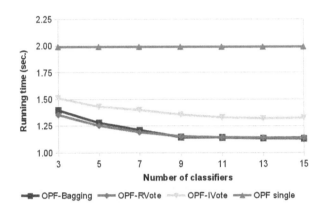

Fig. 4. Running time of training and classification for the Gaussian 100k samples dataset

As expected, the OPF-Bagging and the OPF-Rvote were faster, since the OPF-Ivote has a error computing step that uses an evaluation set in order to choose the samples. This running time improvement was possible due to the reduced training set and bites calculated based on the ensemble size. Since OPF training is $O(N^2)$, it is faster to train several classifiers with $(N/2)/L$ samples than to train a single one with N.

6   Conclusions

This paper reports results of ensembles of OPF using input data manipulation and the undersampling of the data. The results indicate that it is possible to maintain or even improve the accuracy with ensemble of OPF classifiers while using less data to train the base classifiers.

We observed the expected behavior of Bagging and pasting of small votes, that is, for higher variance cases, the input manipulation methods are able to reduce the variance and consequently improve the accuracy. More importantly, this effect was achieved using half the available data and a bite size inversely proportional to the ensemble size, an important result for large dataset applications. However, for the dataset with lower variance and also few examples per class, the proposed methods degraded the results.

As an important contribution, we showed that OPF classifier is a promising base classifier for parallel combination using a reduced number of training examples. Besides, the running time of the ensembles is compatible with large datasets. Further studies should investigate diversity and bias/variance of OPF ensembles to a deeper understanding of the present results.

Acknowledgment. This work was supported by FAPESP (grants 2011/16411-4 and 2011/10282-4), and Conselho Nacional de Desenvolvimento Científico e Tecnológico – CNPq – Brasil (grant 482760/2012-5).

References

1. Breiman, L.: Bagging predictors. Machine Learning Journal 24(2), 123–140 (1996)
2. Breiman, L.: Pasting small votes for classification in large databases and on-line. Machine Learning 36, 85–103 (1999)
3. Brown, G., Wyatt, J., Harris, R., Yao, X.: Diversity creation methods: a survey and categorisation. J. Information Fusion 6(1), 1–28 (2005)
4. Chawla, N.V., Hall, L.O., Bowyer, K.W., Moore Jr., T.E., Kegelmeyer, W.P.: Distributed pasting of small votes. In: Roli, F., Kittler, J. (eds.) MCS 2002. LNCS, vol. 2364, pp. 52–61. Springer, Heidelberg (2002)
5. Domingos, P.: A few useful things to know about machine learning. Communications of the ACM 55(10), 78–87 (2012)
6. Duin, R.P.W.: Prtools v.3 - a matlab toolbox for pattern recognition. In: Proc. of SPIE, p. 1331 (2000)
7. Frank, A., Asuncion, A.: UCI machine learning repository (2010), http://archive.ics.uci.edu/ml
8. Geng, G.G., Wang, C.H., Li, Q.D., Xu, L., Jin, X.B.: Boosting the performance of web spam detection with ensemble under-sampling classification. In: Fourth International Conference on Fuzzy Systems and Knowledge Discovery, FSKD 2007, vol. 4, pp. 583–587 (2007)
9. Ho, T.: The random subspace method for constructing decision forests. IEEE Trans. Pattern Analysis and Machine Intelligence 20(8), 832–844 (1998)
10. Li, J., Wang, J.Z.: Automatic linguistic indexing of pictures by a statistical modeling approach. IEEE Trans. Pattern Analysis and Machine Intelligence 25(9), 1075–1088 (2003)
11. Liu, X.Y., Wu, J., Zhou, Z.H.: Exploratory undersampling for class-imbalance learning. IEEE Transactions on Systems, Man, and Cybernetics, Part B: Cybernetics 39(2), 539–550 (2009)
12. Louppe, G., Geurts, P.: Ensembles on random patches. In: Flach, P.A., De Bie, T., Cristianini, N. (eds.) ECML PKDD 2012, Part I. LNCS, vol. 7523, pp. 346–361. Springer, Heidelberg (2012)
13. Martinez-Munoz, G., Suarez, A.: Out-of-bag estimation of the optimal sample size in bagging. Pattern Recognition 43, 143–152 (2010)
14. Papa, J.P., Falcao, A.X., Suzuki, C.T.N.: LibOPF: a library for optimum-path forest (OPF) classifiers (2009), http://www.ic.unicamp.br/~afalcao/libopf/
15. Papa, J., Falcao, A.X., Suzuki, C.T.N.: Supervised pattern classification based on optimum-path forest. Int. J. Imaging Systems and Technology 19(2), 120–131 (2009)
16. Papa, J., Pagnin, A., Schellini, S., Ponti Jr., M., Spadotto, A., Guido, R.C., Chiachia, G., Falcao, A.X.: Feature selection through gravitational search algorithm. In: 36th Int. Conf. Acoustics, Speech and Signal Processing (ICASSP), pp. 2052–2055. IEEE, Prague (2011)

17. Ponti Jr., M.P.: Classifier combination: from the creation of ensembles to the decision fusion. In: IEEE Proceedings of the 24th SIBGRAPI Conference on Graphics, Patterns and Images Tutorials (SIBGRAPI-T), pp. 1–10. IEEE, Alagoas (2011)
18. Ponti Jr., M.P., Papa, J.P.: Improving accuracy and speed of Optimum-Path Forest classifier using combination of disjoint training subsets. In: Sansone, C., Kittler, J., Roli, F. (eds.) MCS 2011. LNCS, vol. 6713, pp. 237–248. Springer, Heidelberg (2011)
19. Ponti Jr., M.P., Papa, J.P., Levada, A.L.M.: A Markov Random Field model for combining Optimum-Path Forest classifiers using decision graphs and Game Strategy Approach. In: San Martin, C., Kim, S.-W. (eds.) CIARP 2011. LNCS, vol. 7042, pp. 581–590. Springer, Heidelberg (2011)
20. Tahir, M.A., Kittler, J., Mikolajczyk, K., Yan, F.: A multiple expert approach to the class imbalance problem using inverse random under sampling. In: Benediktsson, J.A., Kittler, J., Roli, F. (eds.) MCS 2009. LNCS, vol. 5519, pp. 82–91. Springer, Heidelberg (2009)
21. Valentini, G.: An experimental bias-variance analysis of svm ensembles based on resampling techniques. IEEE Trans. Systems, Man and Cybernetics — Part B 35(6) (2005)

Random Oracle Ensembles for Imbalanced Data

Juan J. Rodríguez, José-Francisco Díez-Pastor, and César García-Osorio

University of Burgos
{jjrodriguez,jfdpastor,cgosorio}@ubu.es

Abstract. In the Random Oracle ensemble method, each base classifier is a mini-ensemble of two classifiers and a randomly generated oracle that selects one of the two classifiers. The performance of this method have been previously studied, but not for imbalanced data sets. This work studies its performance for this kind of data. As the Random Oracle ensemble method can be combined with any other ensemble method, this work considers its combination with four ensemble methods: Bagging, SMOTEBoost, SMOTEBagging and RUSBoost. The last three methods combine classical, not specific for imbalance, ensemble methods (i.e., Bagging, Boosting), with pre-processing approaches designed for imbalance (i.e., random undersampling, SMOTE). The results show that Random Oracles improves all these methods.

1 Introduction

The classification of imbalanced data sets may require specific techniques. In addition, special performance measures are required, since with imbalanced data, conventional classification measures (for example, accuracy) are not useful. A measure used for the imbalance is the AUROC: area under the ROC curve (Receiver Operating Characteristics) [1]. Often when these curves are only considered, this area is called AUC. Another useful curve for imbalance is the Precision-Recall curve [2], the area under this curve is called AUPRC. These two curves (and their corresponding measures) are generated from the confidences given by the classifiers to their predictions. Another measure used for imbalance is the F-measure, the harmonic mean of precision and recall.

Various approaches for dealing with imbalanced data sets have been proposed [3]. In the data level approached the data is pre-processed, altering the classes distribution. For example, SMOTE [4] generates artificial instances of the minority class. Other approaches are based on ensemble methods. A common approach is to combine a method of ensembles that imbalance is not specific to a pre-processing technique.

A combination of Boosting and SMOTE is SMOTEBoost [5]. For each Boosting iteration, SMOTE is applied to the data, generating artificial instances of the minority class. SMOTEBagging [6] used Bagging and SMOTE. Each base classifiers is trained with a balanced data set: the classes have the same number of instances. These training data sets have instances that are obtained from resampling the original training data and artificial instances generated with SMOTE.

Z.-H. Zhou, F. Roli, and J. Kittler (Eds.): MCS 2013, LNCS 7872, pp. 247–258, 2013.
© Springer-Verlag Berlin Heidelberg 2013

In order to increase the diversity, the percentage of instances from resampling and SMOTE is variable. RUSBoost [7] is a method based on SMOTEBoost, but it used random undersampling of the majority class instead of SMOTE of the minority class.

This paper studies the performance of the Random Oracle ensemble method [8,9,10] in imbalanced data. As this method can be used in conjunction with other ensemble methods, it will be combined with these ensemble methods for imbalance.

The rest of the paper is organized as follows: Sect. 2 describes the Random Oracles ensemble method. The experiments and results are presented in Sect. 3. Diversity and error of the base classifiers is considered in Sect. 4. Finally, Sect. 5 shows the conclusions.

2   Random Oracles

A Random Oracle classifier is formed by two classifiers and a randomly generated oracle. The oracle selects one of the two classifiers. It can be seen as a random discriminant function, it splits the data into two subsets without taking into consideration the class labels or cluster structure. Moreover, a Random Oracle classifier can be used as the base classifier of any other ensemble method.

Given a base method, the process for training a Random Oracle classifier is: 1) Select randomly the Random Oracle. 2) Split the training data in two subsets using the Random Oracle. 3) For each subset of the training data, build a classifier.

Given a test instance, the prediction is obtained in the following way: 1) Use the Random Oracle to select one of the two classifiers. 2) Return the prediction (and its confidence) given by the selected classifier.

For oracles with small computational complexity (both in training and prediction), the computational complexity of a Random Oracle classifier is very similar to the complexity of the base classifier. In the prediction phase, only one of the two classifiers is used. In the training phase, two classifiers are built. Nevertheless, they are trained with a disjoint partition of the training examples and the training time of any method depends, usually at least linearly, on the number of training examples.

Different types of Oracles can be considered. In this work two are considered: Linear and Spherical Random Oracles. The linear oracle divides the space into two subspaces using a hyperplane. To build the oracle, two different training objects are selected at random, the points that are at the same distance from the two training objects define the hyperplane. Each remaining training object is assigned to the subspace of the selected training object for which is closer. Algorithms 1 and 2 show the training and testing phases of this method.

The spherical oracle also divides the space into two subsets: inside and outside of a sphere. A training object is selected randomly as the center of the sphere. Another seven training examples are selected randomly, the distances from the center to these examples is calculated, the radius is the median of these seven

---

Input: Training dataset D; base learning method $\mathbb{L}$
Output: Random Oracle Model RO
RO.instance $[1] \leftarrow \{\mathbf{x} \mid (\mathbf{x}, y)$ is a random instance from $D\}$
RO.instance $[2] \leftarrow \{\mathbf{x} \mid (\mathbf{x}, y)$ is a random instance from $D\}$
$D_1 \leftarrow \emptyset$ // The training dataset for the 1st sub-model
$D_2 \leftarrow \emptyset$ // The training dataset for the 2nd sub-model
foreach *instance* $(\mathbf{x}, y) \in D$ **do**
   if distance $(RO$.instance$[1], \mathbf{x}) <$ distance$(RO$.instance$[2], \mathbf{x})$ **then**
     | $D_1 \leftarrow D_1 \cup \{(\mathbf{x}, y)\}$ // Add the instance to the 1st subset
   else
     | $D_2 \leftarrow D_2 \cup \{(\mathbf{x}, y)\}$ // Add the instance to the 2nd subset
RO.model $[1] \leftarrow \mathbb{L}(D_1)$ // Train the 1st sub-model
RO.model $[2] \leftarrow \mathbb{L}(D_2)$ // Train the 2nd sub-model

Algorithm 1. Random Linear Oracle method: training phase

Input: Trained Random Oracle RO; instance $\mathbf{x}$
Output: Predicted value
if distance $(RO$.instance$[1], \mathbf{x}) <$ distance$(RO$.instance$[2], \mathbf{x})$ **then**
   | **return** RO.model $[1]$.predict$(\mathbf{x})$ // Predict with the 1st sub-model
else
   | **return** RO.model $[2]$.predict$(\mathbf{x})$ // Predict with the 2nd sub-model

Algorithm 2. Random Linear Oracle method: prediction phase

---

distances. This is done with the purpose of having some guarantee that there will be training examples inside and outside of the sphere. As an additional source of diversity, the distances are calculated in a random subspace (a subset of the features). This random subspace is only used for defining the sphere, the two classifiers are trained using all the features.

In Random Subspaces [11] and Bagging each classifier is trained with a data set randomly obtained from the original training data. In these methods the used data sets are different because some information of the original training data is lost, diversity is obtained at the potential cost of decreased accuracy of the base classifiers. The objective of Random Oracles is to have diversity without losing information. The Random Oracle classifier is trained using all the attributes and instances, the diversity is obtained because the two classifiers in the Random Oracle are trained with different partitions of the training data. These ensemble methods can be used together.

It is also possible to use Random Oracles with more than two sub-regions, but in our experiments they do not improve the performance.

In this work, the distances are calculated according to the Euclidean distance, numerical attributes are scaled within [0,1]. For nominal attributes we consider that the distance is 0 or 1 depending if the two values are different or equal.

3   Experiments

3.1   Data Sets

Two collections of datasets were used. The HDDT collection[1] contains the binary imbalanced datasets used in [12]. The KEEL collection[2] contains the binary imbalanced datasets from the KEEL repository of [13].

Table 1 shows the characteristics of the 20 data sets in the HDDT collection and Table 2 the 66 data sets in the KEEL collection. Many data sets in these two collections are available or are modifications of data sets in the UCI Repository [14].

Table 1. Characteristics of the data sets from the HDDT collection.(#E: number of examples, #A: number of attributes (numeric/nominal), IR: imbalance ratio).

Data set	Examples	Attributes Numeric	Nominal	Imbalance Ratio
boundary	3505	0	175	27.50
breast-y	286	0	9	2.36
cam	18916	0	132	19.08
compustat	13657	20	0	25.26
covtype	38500	10	0	13.02
credit-g	1000	7	13	2.33
estate	5322	12	0	7.37
german-numer	1000	24	0	2.33
heart-v	200	5	8	2.92
hypo	3163	7	18	19.95
ism	11180	6	0	42.00
letter	20000	16	0	24.35
oil	937	49	0	21.85
optdigits	5620	64	0	9.14
page	5473	10	0	8.77
pendigits	10992	16	0	8.63
phoneme	5404	5	0	2.41
PhosS	11411	480	0	17.62
satimage	6430	36	0	9.29
segment	2310	19	0	6.00

3.2   Settings

Weka [15] was used for the experiments. Unless explicitly specified, the parameters for the different methods take the default values given by Weka.

The decision tree method used for constructing the base classifiers was J48 (Weka's re-implementation of C4.5 [16]). As recommended for imbalanced data [12], it was used without pruning and collapsing but with Laplace smoothing at the leaves. C4.5 with this options is called C4.4 [17].

Several ensemble methods were considered. The first ensemble method is Bagging [18]. It was the best method for imbalanced data in [12] and [19].

---

[1] Available at http://www.nd.edu/~dial/hddt/
[2] Available at http://sci2s.ugr.es/keel/imbalanced.php

Table 2. Characteristics of the data sets from the KEEL collection. #E: number of examples, #N: number of numeric attributes, #D: number of discrete attributes, IR: imbalance ratio.

data set	#E	#N	#D	IR	data set	#E	#N	#D	IR
abalone19	4174	7	1	129.44	glass5	214	9	0	22.78
abalone9-18	731	7	1	16.40	glass6	214	9	0	6.38
cleveland-0_vs_4	177	13	0	12.62	haberman	306	3	0	2.78
ecoli-0-1-3-7_vs_2-6	281	7	0	39.14	iris0	150	4	0	2.00
ecoli-0-1-4-6_vs_5	280	6	0	13.00	led7digit-0-2-4-5-6-7-8-9_vs_1	443	7	0	10.97
ecoli-0-1-4-7_vs_2-3-5-6	336	7	0	10.59	new-thyroid1	215	5	0	5.14
ecoli-0-1-4-7_vs_5-6	332	6	0	12.28	new-thyroid2	215	5	0	5.14
ecoli-0-1_vs_2-3-5	244	7	0	9.17	page-blocks-1-3_vs_4	472	10	0	15.86
ecoli-0-1_vs_5	240	6	0	11.00	page-blocks0	5472	10	0	8.79
ecoli-0-2-3-4_vs_5	202	7	0	9.10	pima	768	8	0	1.87
ecoli-0-2-6-7_vs_3-5	224	7	0	9.18	segment0	2308	19	0	6.02
ecoli-0-3-4-6_vs_5	205	7	0	9.25	shuttle-c0-vs-c4	1829	9	0	13.87
ecoli-0-3-4-7_vs_5-6	257	7	0	9.28	shuttle-c2-vs-c4	129	9	0	20.50
ecoli-0-3-4_vs_5	200	7	0	9.00	vehicle0	846	18	0	3.25
ecoli-0-4-6_vs_5	203	6	0	9.15	vehicle1	846	18	0	2.90
ecoli-0-6-7_vs_3-5	222	7	0	9.09	vehicle2	846	18	0	2.88
ecoli-0-6-7_vs_5	220	6	0	10.00	vehicle3	846	18	0	2.99
ecoli-0_vs_1	220	7	0	1.86	vowel0	988	13	0	9.98
ecoli1	336	7	0	3.36	wisconsin	683	9	0	1.86
ecoli2	336	7	0	5.46	yeast-0-2-5-6_vs_3-7-8-9	1004	8	0	9.14
ecoli3	336	7	0	8.60	yeast-0-2-5-7-9_vs_3-6-8	1004	8	0	9.14
ecoli4	336	7	0	15.80	yeast-0-3-5-9_vs_7-8	506	8	0	9.12
glass-0-1-2-3_vs_4-5-6	214	9	0	3.20	yeast-0-5-6-7-9_vs_4	528	8	0	9.35
glass-0-1-4-6_vs_2	205	9	0	11.06	yeast-1-2-8-9_vs_7	947	8	0	30.57
glass-0-1-5_vs_2	172	9	0	9.12	yeast-1-4-5-8_vs_7	693	8	0	22.10
glass-0-1-6_vs_2	192	9	0	10.29	yeast-1_vs_7	459	7	0	14.30
glass-0-1-6_vs_5	184	9	0	19.44	yeast-2_vs_4	514	8	0	9.08
glass-0-4_vs_5	92	9	0	9.22	yeast-2_vs_8	482	8	0	23.10
glass-0-6_vs_5	108	9	0	11.00	yeast1	1484	8	0	2.46
glass0	214	9	0	2.06	yeast3	1484	8	0	8.10
glass1	214	9	0	1.82	yeast4	1484	8	0	28.10
glass2	214	9	0	11.59	yeast5	1484	8	0	32.73
glass4	214	9	0	15.46	yeast6	1484	8	0	41.40

As ensemble methods for imbalance the following were used: SMOTEBoost, SMOTEBagging and RUSBoost. SMOTEBoost has a parameter, the number of artificial instances to generate. Three values were considered: 100%, 200% and 500% of the number of instances in the minority class.

For each considered ensemble method there are three versions: one without Random Oracles, and two with Random Oracles: Linear and Spherical For all the ensembles the number of classifiers was 100.

The results were obtained with 5 × 2-fold cross validation [20]. Average ranks [21,22] were used for comparing several methods across several data sets.

3.3   Results

Table 3 shows the comparison of methods with and without Random Oracles, according to the AUROC. Each entry in the table shows the number of wins (W), ties (T) and losses (L) when comparing the method with the oracle (linear or spherical) with the original method. For 20 data sets, according to a two-tailed sign test at $\alpha = 0.05$ a classifier is significantly better than another if the number of wins plus half the number of ties is at least 15 [21]. For the HDDT collection in all the considered pairs the number of wins is at least 15, with the only exception of RUSBoost and the Linear Oracle. For 66 data sets, the required number of wins (plus half the number of ties) is 41. For the KEEL collection the number of wins is at least 41.

Table 3. Comparison of methods with and without Random Oracles, according to the AUROC

Method	Linear Oracle		Spherical Oracle	
	HDDT	KEEL	HDDT	KEEL
	W/T/L	W/T/L	W/T/L	W/T/L
Bagging	20/0/0	57/1/8	20/0/0	59/1/6
SMOTEBagging	20/0/0	61/2/3	20/0/0	61/2/3
SMOTEBoost (100%)	19/0/1	50/2/14	19/0/1	52/2/12
SMOTEBoost (200%)	16/0/4	53/2/11	18/0/2	56/2/8
SMOTEBoost (500%)	17/0/3	49/2/15	18/0/2	56/2/8
RUSBoost	14/0/6	41/2/23	16/0/4	44/2/20

Table 4 shows the comparison of methods with and without Random Oracles, according to the AUPRC. In all the cases the balance if favorable for the method with Random Oracles, in the great majority of the cases the balances are significant according to the sign test.

Table 4. Comparison of methods with and without Random Oracles, according to the AUPRC

Method	Linear Oracle		Spherical Oracle	
	HDDT	KEEL	HDDT	KEEL
	W/T/L	W/T/L	W/T/L	W/T/L
Bagging	19/0/1	56/1/9	20/0/0	62/1/3
SMOTEBagging	19/0/1	55/2/9	20/0/0	59/2/5
SMOTEBoost (100%)	19/0/1	59/2/5	16/0/4	58/2/6
SMOTEBoost (200%)	14/0/6	56/2/8	16/0/4	60/2/4
SMOTEBoost (500%)	15/0/5	57/2/7	17/0/3	57/2/7
RUSBoost	16/0/4	39/2/25	15/0/5	41/2/23

Table 5 shows the comparison of methods with and without Random Oracles, according to the F-measure. In this case the advantage for the Random Oracles, compared with the other considered measures, is reduced. Nevertheless, the balance is never favorable for the method without Random Oracle.

Table 5. Comparison of methods with and without Random Oracles, according to the F-measure

Method	Linear Oracle		Spherical Oracle	
	HDDT	KEEL	HDDT	KEEL
	W/T/L	W/T/L	W/T/L	W/T/L
Bagging	11/0/9	33/4/29	10/0/10	36/3/27
SMOTEBagging	15/0/5	51/4/11	17/0/3	50/4/12
SMOTEBoost (100%)	13/0/7	51/3/12	14/0/6	48/4/14
SMOTEBoost (200%)	12/0/8	52/3/11	15/0/5	51/4/11
SMOTEBoost (500%)	14/0/6	52/4/10	14/0/6	55/4/7
RUSBoost	12/0/8	42/1/23	13/0/7	43/3/20

Until now the results in this section show the comparison of a method and the corresponding method augmented with Random Oracles. They show clearly the advantage of using Oracles, but another interesting issue is what are the best methods among all the considered. Average ranks [21,22] are used to compare all these methods. For each data set all the methods are sorted, from best to worst, according to the considered performance measure. The best method has rank 1, the second rank two and so on. If several methods have the same performance they are assigned an average rank (e.g., if two methods have the best result, both have a rank of 1.5). The average rank of a method is calculated as the mean across all the data sets.

The Iman and Davenport Test is used to determine the presence of differences among all the compared methods. As a post-hoc procedure, using the best method as the control, the Hochberg procedure is used [22]. Table 6 shows the average ranks according to the AUROC. For the two collections of data sets, all the methods with Random Oracles are above all the methods without random oracles. The top positions are for SMOTEBoost with Spherical Random Oracles. In these tables a double horizontal line is used to indicate for which methods the Hochberg procedure rejects ($\alpha = 0.05$) the hypotheses.

Table 7 shows the average ranks according to the AUPRC. For the KEEL collection all the methods with Random Oracles have better ranks than all the methods without Random Oracles. This is not the case for the HDDT collection, due to the positions of RUSBoost with Random Oracles. Nevertheless, all the methods with Random Oracles have a better rank than the corresponding method without Random Oracles. The top positions are for SMOTEBoost with Random Oracles.

Table 8 shows the average ranks according to the F-measure. The advantage of Random Oracles is less clear than for the other measures, but still there is an advantage. For all the methods with Random Oracle the rank is better than the corresponding method without Random Oracles. The top positions are, for the two collections, for SMOTEBoost with Spherical Oracle.

Table 6. Average ranks according to the AUROC. The prefix "L-" is used for methods with Linear Random Oracle and "S-" for methods with Spherical Random Oracle.

(a) HDDT

Method	Ranking
S-SMOTEBoost (100%)	5.950
S-SMOTEBoost (200%)	6.350
L-SMOTEBoost (100%)	7.025
S-SMOTEBagging	7.400
L-SMOTEBoost (200%)	7.400
S-SMOTEBoost (500%)	7.425
L-SMOTEBoost (500%)	8.200
L-SMOTEBagging	8.650
S-Bagging	9.450
S-RUSBoost	9.450
L-Bagging	9.900
L-RUSBoost	10.400
SMOTEBoost (100%)	11.100
SMOTEBoost (200%)	11.300
RUSBoost	11.300
SMOTEBoost (500%)	11.950
SMOTEBagging	13.500
Bagging	14.250

(b) KEEL

Method	Ranking
S-SMOTEBoost (500%)	6.9242
S-SMOTEBoost (200%)	7.2803
S-SMOTEBagging	7.5303
L-SMOTEBoost (200%)	7.5758
S-RUSBoost	7.9470
S-Bagging	7.9848
L-SMOTEBagging	8.0833
L-SMOTEBoost (500%)	8.1742
S-SMOTEBoost (100%)	8.4470
L-Bagging	8.5303
L-SMOTEBoost (100%)	8.5606
L-RUSBoost	8.5909
RUSBoost	9.9621
SMOTEBoost (500%)	12.1061
SMOTEBoost (200%)	12.3485
SMOTEBoost (100%)	13.0758
SMOTEBagging	13.7727
Bagging	14.1061

Table 7. Average ranks according to the AUPRC

(a) HDDT

Method	Ranking
L-SMOTEBoost (100%)	6.05
S-SMOTEBoost (100%)	6.45
L-SMOTEBoost (200%)	6.85
S-SMOTEBagging	7.20
S-SMOTEBoost (200%)	7.25
S-SMOTEBoost (500%)	8.15
S-Bagging	8.70
L-SMOTEBoost (500%)	9.05
L-SMOTEBagging	9.10
L-Bagging	9.50
SMOTEBoost (200%)	10.00
SMOTEBoost (100%)	10.50
S-RUSBoost	10.85
L-RUSBoost	11.05
SMOTEBoost (500%)	12.05
RUSBoost	12.40
SMOTEBagging	12.55
Bagging	13.30

(b) KEEL

Method	Ranking
S-SMOTEBoost (500%)	7.2121
L-SMOTEBoost (200%)	7.2348
S-Bagging	7.2803
S-SMOTEBoost (200%)	7.3182
S-SMOTEBoost (100%)	7.5152
L-SMOTEBoost (100%)	7.5303
L-SMOTEBoost (500%)	7.8561
S-SMOTEBagging	7.9394
L-Bagging	8.2576
S-RUSBoost	8.4242
L-SMOTEBagging	8.7121
L-RUSBoost	8.9773
RUSBoost	10.3636
SMOTEBagging	13.2424
Bagging	13.1667
SMOTEBoost (200%)	13.1970
SMOTEBoost (500%)	13.2879
SMOTEBoost (100%)	13.4848

Table 8. Average ranks according to the F-measure

(a) HDDT		(b) KEEL	
Method	Ranking	Method	Ranking
S-SMOTEBoost (500%)	6.200	S-SMOTEBoost (500%)	6.3864
S-SMOTEBoost (200%)	6.500	S-RUSBoost	7.1894
L-SMOTEBoost (500%)	6.950	L-RUSBoost	7.5682
L-SMOTEBoost (200%)	7.600	S-SMOTEBagging	7.6591
S-SMOTEBagging	8.050	L-SMOTEBoost (200%)	7.7121
L-SMOTEBoost (100%)	8.050	L-SMOTEBoost (500%)	7.2955
S-SMOTEBoost (100%)	8.750	S-SMOTEBoost (200%)	8.1818
S-RUSBoost	8.800	L-SMOTEBoost (100%)	8.6061
SMOTEBoost (500%)	8.900	L-SMOTEBagging	8.7045
SMOTEBoost (200%)	9.250	S-SMOTEBoost (100%)	9.1894
L-SMOTEBagging	9.350	RUSBoost	9.6288
L-RUSBoost	9.550	SMOTEBagging	10.8788
SMOTEBoost (100%)	9.650	SMOTEBoost (500%)	10.9318
RUSBoost	9.900	L-Bagging	11.4091
SMOTEBagging	11.200	S-Bagging	11.7424
L-Bagging	13.825	SMOTEBoost (200%)	12.0833
S-Bagging	14.125	Bagging	12.8712
Bagging	14.350	SMOTEBoost (100%)	12.9621

4   Diversity Study

One possible explanation for the improvements obtained with the use of Random Oracles is that they can introduce additional diversity in the base classifiers. This diversity is a necessary ingredient of successful ensembles. The other is the accuracy of the base classifiers, but if the classifiers are very accurate they cannot be very diverse.

Kappa-error diagrams [23] are used to represent the diversity (measured with κ) and error of the classifiers in an ensemble. Each pair of classifiers is represented as a point, the x axis is the diversity between the two classifiers, the y-axis is the average error of the two classifiers.

Figure 1 shows for two data sets these diagrams, for Bagging and L-Bagging. They also show the average diversity and error across all the pairs of classifiers.

Each κ-error diagram shows a single data set. For several data sets, their information can be summarized with κ-error relative movement diagram [24]. Figure 2 shows these diagrams for Bagging and L-Bagging. Each arrow represents a data set, the head coordinates are given by the differences between the average diversity and error of the two considered classifiers.

The number in the corners (e.g., 74, 1) indicate how many arrows go to the corresponding quadrant (e.g., top-left, right-bottom). From 85 data sets, in 84 cases the arrows go to the left, that is, the diversity is increased when using Bagging with linear oracles instead of Bagging without oracles. As expected, the

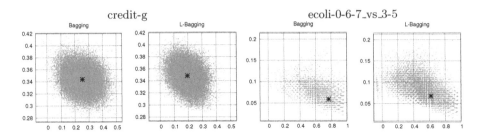

Fig. 1. Kappa error diagrams

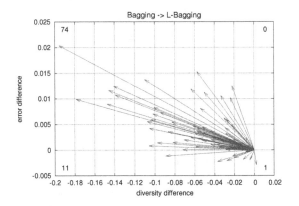

Fig. 2. Kappa error relative movement diagram

increased diversity usually (in 74 data sets) comes with increased error in the base classifiers.

5   Conclusions

The performance of Random Oracles have been studied when included in ensemble methods for imbalanced data. Four ensemble methods have been considered: Bagging, SMOTEBagging, SMOTEBoost and RUSBoost. According to the AUROC, the AUPRC and the F-Measure in two sets of data sets (HDDT and KEEL) including Random Oracles improves the results.

From the four ensemble methods, the best global results are for SMOTEBoost. This method has a parameter, the number (or percentage) of artificial instances of the minority class to generate. In this work three values have been considered, the results could improve if the parameter were adjusted for each data set.

The diversity of the base classifiers has been studied, Random Oracle usually improves this diversity. This can be the cause of the better performance of ensembles with Random Oracles.

As base classifiers, only decision trees have been used. Other base classifiers could give better results or affect in different ways the behaviour of the ensemble methods. The performance of the different ensemble methods with other base classifiers can be studied in future work.

Acknowledgements. This work was supported by the Projects TIN2011-24046 and IPT-2011-1265-020000 of the Spanish Ministry of Economy and Competitiveness.

We wish to thank the developers of Weka [15]. We also express our gratitude to the donors of the different data sets.

References

1. Flach, P., Hernandez-Orallo, J., Ferri, C.: A coherent interpretation of AUC as a measure of aggregated classification performance. In: 28th International Conference on Machine Learning (ICML 2011), pp. 657–664. ACM (June 2011)
2. Davis, J., Goadrich, M.: The relationship between Precision-Recall and ROC curves. In: Proceedings of the 23rd International Conference on Machine Learning, ICML 2006, pp. 233–240. ACM, New York (2006)
3. Galar, M., Fernandez, A., Barrenechea, E., Bustince, H., Herrera, F.: A review on ensembles for the class imbalance problem: Bagging-, boosting-, and hybrid-based approaches. IEEE Transactions on Systems, Man, and Cybernetics, Part C: Applications and Reviews 42(4), 463–484 (2012)
4. Chawla, N.V., Bowyer, K.W., Hall, L.O., Kegelmeyer, W.P.: SMOTE: Synthetic minority over-sampling technique. Journal of Artificial Intelligence Research 16, 321–357 (2002)
5. Chawla, N.V., Lazarevic, A., Hall, L.O., Bowyer, K.W.: SMOTEBoost: Improving prediction of the minority class in boosting. In: Lavrač, N., Gamberger, D., Todorovski, L., Blockeel, H. (eds.) PKDD 2003. LNCS (LNAI), vol. 2838, pp. 107–119. Springer, Heidelberg (2003)
6. Wang, S., Yao, X.: Diversity analysis on imbalanced data sets by using ensemble models. In: IEEE Symposium Series on Computational Intelligence and Data Mining (IEEE CIDM 2009), pp. 324–331 (2009)
7. Seiffert, C., Khoshgoftaar, T., Van Hulse, J., Napolitano, A.: Rusboost: A hybrid approach to alleviating class imbalance. IEEE Transactions on Systems, Man and Cybernetics, Part A: Systems and Humans 40(1), 185–197 (2010)
8. Kuncheva, L.I., Rodríguez, J.J.: Classifier ensembles with a random linear oracle. IEEE Transactions on Knowledge and Data Engineering 19(4), 500–508 (2007)
9. Rodríguez, J.J., Kuncheva, L.I.: Naïve bayes ensembles with a random oracle. In: Haindl, M., Kittler, J., Roli, F. (eds.) MCS 2007. LNCS, vol. 4472, pp. 450–458. Springer, Heidelberg (2007)
10. Pardo, C., Rodríguez, J.J., Díez-Pastor, J.F., García-Osorio, C.: Random oracles for regression ensembles. In: Okun, O., Valentini, G., Re, M. (eds.) Ensembles in Machine Learning Applications. SCI, vol. 373, pp. 181–199. Springer, Heidelberg (2011)
11. Ho, T.K.: The random subspace method for constructing decision forests. IEEE Transactions on Pattern Analysis and Machine Intelligence 20(8), 832–844 (1998)
12. Cieslak, D., Hoens, T., Chawla, N., Kegelmeyer, W.: Hellinger distance decision trees are robust and skew-insensitive. Data Mining and Knowledge Discovery 24(1), 136–158 (2012)

13. Alcalá-Fdez, J., Fernández, A., Luengo, J., Derrac, J., García, S.: KEEL data-mining software tool: Data set repository, integration of algorithms and experimental analysis framework. Multiple-Valued Logic and Soft Computing 17(2-3), 255–287 (2011)
14. Frank, A., Asuncion, A.: UCI machine learning repository (2010), http://archive.ics.uci.edu/ml
15. Hall, M., Frank, E., Holmes, G., Pfahringer, B., Reutemann, P., Witten, I.H.: The WEKA data mining software: An update. SIGKDD Explorations 11(1) (2009)
16. Quinlan, J.R.: C4.5: Programs for Machine Learning. Machine Learning. Morgan Kaufmann, San Mateo (1993)
17. Provost, F., Domingos, P.: Tree induction for Probability-Based ranking. Machine Learning 52(3), 199–215 (2003)
18. Breiman, L.: Bagging predictors. Machine Learning 24(2), 123–140 (1996)
19. Rodríguez, J.J., Díez-Pastor, J.F., García-Osorio, C., Santos, P.: Using model trees and their ensembles for imbalanced data. In: Lozano, J.A., Gámez, J.A., Moreno, J.A. (eds.) CAEPIA 2011. LNCS, vol. 7023, pp. 94–103. Springer, Heidelberg (2011)
20. Dietterich, T.G.: Approximate statistical test for comparing supervised classification learning algorithms. Neural Computation 10(7), 1895–1923 (1998)
21. Demšar, J.: Statistical comparisons of classifiers over multiple data sets. Journal of Machine Learning Research 7, 1–30 (2006)
22. García, S., Fernández, A., Luengo, J., Herrera, F.: Advanced nonparametric tests for multiple comparisons in the design of experiments in computational intelligence and data mining: Experimental analysis of power. Information Sciences 180(10), 2044–2064 (2010)
23. Margineantu, D.D., Dietterich, T.G.: Pruning adaptive boosting. In: Proc. 14th International Conference on Machine Learning, pp. 211–218. Morgan Kaufmann (1997)
24. Maudes, J., Rodríguez, J.J., García-Osorio, C.: Disturbing neighbors diversity for decision forests. In: Okun, O., Valentini, G. (eds.) Applications of Supervised and Unsupervised Ensemble Methods. SCI, vol. 245, pp. 113–133. Springer, Heidelberg (2009)

Towards a Framework for Designing Full Model Selection and Optimization Systems

Quan Sun, Bernhard Pfahringer, and Michael Mayo

Department of Computer Science
The University of Waikato
Hamilton, New Zealand
{qs12,bernhard,mmayo}@cs.waikato.ac.nz

Abstract. People from a variety of industrial domains are beginning to realise that appropriate use of machine learning techniques for their data mining projects could bring great benefits. End-users now have to face the new problem of how to choose a combination of data processing tools and algorithms for a given dataset. This problem is usually termed the Full Model Selection (FMS) problem. Extended from our previous work [10], in this paper, we introduce a framework for designing FMS algorithms. Under this framework, we propose a novel algorithm combining both genetic algorithms (GA) and particle swarm optimization (PSO) named GPS (which stands for **G**A-**P**SO-FM**S**), in which a GA is used for searching the optimal structure for a data mining solution, and PSO is used for searching optimal parameters for a particular structure instance. Given a classification dataset, GPS outputs a FMS solution as a directed acyclic graph consisting of diverse data mining operators that are available to the problem. Experimental results demonstrate the benefit of the algorithm. We also present, with detailed analysis, two model-tree-based variants for speeding up the GPS algorithm.

1 Introduction

Machine learning users now have to face the new problem of how to choose a combination of data processing tools and algorithms. The goal is usually defined as maximizing or minimizing a quantitative measure. In classification problems, the goal could be optimising the classification accuracy, the Lift score or the ROC area (AUC); in regression problems the goal could be optimising RMSE (root mean squared error), MAE (mean absolute error), or any proper loss function.

Sometimes the final goal might be a combination of multiple goals. Traditionally, these problems are addressed separately in the feature selection, model or parameter selection and the meta-learning fields. A practical data mining problem consists of many sub-problems which presents an extremely large search space that could be a very time-consuming task for humans to explore manually. Therefore, strategies and methods that can help people to choose, or that could automatically suggest, an optimised data mining solution is useful. In this

Z.-H. Zhou, F. Roli, and J. Kittler (Eds.): MCS 2013, LNCS 7872, pp. 259–270, 2013.
© Springer-Verlag Berlin Heidelberg 2013

paper, we propose a framework which can be used for designing new FMS algorithms, and we also present a novel FMS algorithm which is a realization and an application of the proposed framework.

1.1 Data Mining in the DMO Space and Framework

We first define the DMO space and discuss potential approaches for searching the space. We here attempt to define a search space that consists of all data mining actions (operators) that are available to a given dataset for a user-specified goal, such as a set of outlier filters, a set of feature selection methods, a set of data transformation techniques and a set of base learning algorithms. In this sense, we call the subject of interest "the space of data mining operators (DMO)", or simply "the DMO space" [10].

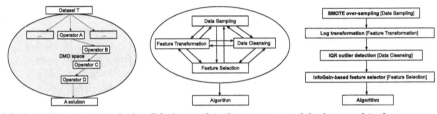

(a) An illustration of the DMO space

(b) A graphical representation of the DMO template used by GPS

(c) A graphical representation of a DMO solution template instance

Fig. 1. A full model defined by the GPS algorithm

In this search space, a data mining solution is abstracted as a directed acyclic graph (DAG) consisting of DMOs that are connected based on some relations: see Figure 1 (a) for an illustration. For simplicity, in Figure 1 (a) we consider that an optimal data mining solution is given by a DAG defined by four DMOs (A, B, C and D) for dataset T. The DMO space is represented by the largest oval, which consists of all DMOs applicable to T. The directed arrows represent the relationships (action rules) in the DAG. If Operator A is an outlier filter, Operator B is a feature reduction method, Operator C is a decision tree algorithm, and Operator D is a post-processing method, the DAG can be interpreted as follows: given a dataset T, in an optimal solution we first use the outlier detection method (DMO A) to remove outliers, and then we employ the feature selection method (DMO B) to remove useless features, and then build a decision tree model (DMO C), and finally, we use a probability calibration method (DMO D) to calculate the model outputs. This is a very large search space because in theory there exists an arbitrary number of DMOs (including an arbitrary number of link directions, node orders and arrangements). Therefore, the next question is how to search in this space?

In practice, due to the resources at hand, usually we do not search in an infinite DMO space, and, moreover, we can make the DMO space a finite space by defining the DMOs that are to be included. For example, given a dataset T, and given we have one outlier detection algorithm, two feature selection methods, three classification algorithms and that the goal is to build a model that gives the lowest classification error on T, typically, we can define the following *node* type DMO objects:

$DMO_{filter}, DMO_{no-filter}, DMO_{feature-selection-1},$
$DMO_{feature-selection-2}, DMO_{no-feature-selection},$
$DMO_{algorithm-1}, DMO_{algorithm-2}, DMO_{algorithm-3}.$

Given these DMOs, if we want to preprocess the data, we can define some *function* type DMOs that output a new data object. For example:

$data \Longleftarrow DMO_{preprocessing-1}(DMO_{filter}, DMO_{feature-selection-1})$
$data \Longleftarrow DMO_{preprocessing-2}(DMO_{no-filter}, DMO_{feature-selection-1})$
$data \Longleftarrow DMO_{preprocessing-3}(DMO_{filter}, DMO_{feature-selection-2})$
...

where $DMO_{preprocessing-1,2,3}$ are *function* type DMOs. We can also define more complex *function* type DMOs which take *function* and *node* type DMOs as inputs and output a solution. For example:

$solution \Longleftarrow DMO_{build-model}(DMO_{preprocessing-1}, DMO_{algorithm-1})$
$solution \Longleftarrow DMO_{build-model}(DMO_{preprocessing-2}, DMO_{algorithm-2})$
$solution \Longleftarrow DMO_{build-model}(DMO_{preprocessing-3}, DMO_{algorithm-1})$
...

where $DMO_{build-model}$, and $DMO_{preprocessing-1,2,3}$ are all *function* type DMOs. In this way, we are free to define which, and what kind of, DMOs are to be added to the DMO space.

To meet the data mining goal, we could simply search all the DMO function-object relations (paths) in the space. Therefore, the solution which has the lowest classification or regression error could be the output of a grid-search-like exhaustive search. One advantage of an exhaustive search in a finite DMO space is that the user controls the search complexity. Another advantage is that the DMO relations can be designed by a data mining expert and then shared and reused. For example, if an expert designed a good DMO search space for an unbalanced binary classification problem, she can probably share it with her colleagues or reuse it for a new project.

However, the disadvantage is also obvious because the search complexity grows dramatically as the number of DMOs increases, with the result that if the search space is too large, due to computational and time costs the user may have to terminate the search before all DMOs are explored. To overcome this problem, we may need to think about questions such as how promising DMOs can be automatically defined/generated for a given dataset.

In the previous examples, we have defined some DMOs by hand. One could generate DMOs simply by generating all possible DMO combinations of different types, but doing so would create an extremely large (even infinite) search space, and the problem becomes intractable. We here propose a semi-automatic method to solve this problem.

Firstly, we define some DMO functions, and add these functions to the DMO search space as we did on previous page. Secondly, we define some templates (rules) for searching. Here are two examples:

$$solution \Longleftarrow$$
$$DMO_{chain-search}(DMO_{[filter]}, DMO_{[feature-selection]},$$
$$DMO_{[tree-model]}) \tag{1}$$
$$solution \Longleftarrow$$
$$DMO_{chain-search}(DMO_{random-topology-search}(DMO_{[filter]},$$
$$DMO_{[feature-selection]}), DMO_{[tree-model]}) \tag{2}$$

Template (1) is a chain solution. Here a chain solution means the order (such as from left to right) of each DMO does matter. A "[...]" is a placeholder for a certain type of DMO object: in this example, the [filter] placeholder can be substituted by any filter-type DMO. The [feature-selection] placeholder follows the same rule, and the [tree-model] placeholder can be substituted only by a "tree" type model. In template (2), we can see a new DMO function called "random-topology-search", which means that the order of the DMOs will be changed automatically during the search. So we can see that template (1) is actually a subset of template (2). Once we have a set of DMO objects added, and a DMO template defined, then we have a finite DMO space.

So far, we have defined a DMO space that consists of *node* type DMOs, *function* type DMOs and DMO templates. In the template part of the search space, we will have to make a decision on what kind of search strategy to use when searching for substitution DMOs for placeholders. We here consider only cases where an exhaustive search (in the case of too many DMOs permutations) is not feasible, and we are particularly interested in a search method that optimises a problem by iteratively trying to improve a candidate DMO with regard to a given measure of quality (the goal metric). These methods are usually referred to as a "heuristic search", such as the best-first search, the local search (using neighborhood relation) and the population-based evolutionary algorithms.

1.2   Related Work

The PSMS system proposed in [3], is an application of Particle Swarm Optimization (PSO) to the problem of full model selection for binary classification problems. In total, 3 feature transformation objects, 13 feature selection objects and 10 classifier objects are used in the PSMS system. A full PSMS model is defined as a 16-dimensional particle position. For the details of the PSMS system, we refer the reader to [3]. Based on the experimental results in [3], the PSMS system shows promising results when it is compared with the Pattern Search (PS) strategy [9] for the FMS problem. The system also showed competitive performance compared with other search strategies in a model selection competition.

From the system architecture point of view, PSMS assumes a full model has three components: feature transformation, feature selection, and learning algorithm. In the DMO framework, we can define the following DMO template for the search space covered by the PSMS system:

$solution \Longleftarrow$

    $DMO_{chain-search}($

        $DMO_{random-topology-search}(DMO_{[feature-transformation]},$

        $DMO_{[feature-selection]}),$

    $DMO_{[algorithm]})$

We can see that the search space covered by the above DMO template is a simplified presentation of a full model, because a full model may have other components, such as data cleansing and data sampling. Extended from our previous work [10], in the next section, we introduce a novel search strategy for the FMS problem, which covers five data mining components, namely, data cleansing, data sampling, feature transformation, feature selection and algorithm DMOs.

2   The GPS Search Strategy

In this section, we propose a novel algorithm for searching a FMS solution in the DMO space. The algorithm combines both genetic algorithm (GA) [6] and particle swarm optimization (PSO) [7], in which GA is used for searching the optimal template instance of a DMO template, and PSO is used for searching the optimal parameter set for a particular template instance. The motivation is that GA is usually considered a good strategy for combinational optimization problems, whereas PSO is usually considered good at numerical optimization.

The proposed algorithm is named as GPS (**GA-PSO FMS**). It can be seen as a realization and an application of the DMO framework. Before introducing the GPS algorithm, we first define a DMO template. Here, we assume a FMS solution consists of five DMOs:

    $DMO_{[data-cleansing]},$

    $DMO_{[data-sampling]},$

    $DMO_{[feature-transformation]},$

    $DMO_{[feature-selection]},$ and

    $DMO_{[algorithm]}.$

Then, a DMO template for the FMS problem covered by GPS is defined as:

$solution \Longleftarrow$

    $DMO_{chain-search}($

        $DMO_{random-topology-search}($

        $DMO_{[data-cleansing]}, DMO_{[data-sampling]},$

        $DMO_{[feature-transformation]}, DMO_{[feature-selection]}),$

    $DMO_{[algorithm]})$ (3)

Graphically, this template can be represented as Figure 1 (b). The four DMOs at the top can be performed in any order, then followed by an *Algorithm* DMO. Figure 1 (c) shows a solution instance of the DMO template, which can be interpreted as: given a dataset, we firstly apply the data sampling technique,

---

Algorithm 1. Pseudocode of the GPS strategy for searching a FMS solution

---

  procedure GPS(T,P,M,W,G)

    Input:

    T (number of generations for GA), P (population size for GA), M (number of evolutions for PSO), W (swarm size for PSO), G (goal metric)

    Get P random template instances based on template (3).

    Populate template instances with objects in the DMO pools (Table 2)

    for $i \leftarrow 1$ to T **do**

        Use a standard PSO procedure **PSO**(M,W,G,I) to search for the optimal parameters for each template instance I (optimising the goal metric G), and assign an evaluation score to each template instance I. This procedure is similar to the PSMS system [3].

        Do *crossover* // single point crossover among the top 20% template instances.

        Do *mutation* // randomly choose 30% template instances from the population, and randomly change one DMO in each template instance.

        Replace the worst N template instances with the N new template instances generated in above two steps, here we use $N = (20\% + 30\%) \times P$.

        solution$_{best}$ $\leftarrow$ population$_{best}$

    end for

    return *solution$_{best}$*

  end procedure

SMOTE [2], followed by applying log-transformation, then, we do IQR outlier detection, and then use information gain based feature selection; finally, an AdaBoost.M1 [4] model is built based on the transformed data. We call such a solution a "DMO solution template instance", shortened to "template instance".

For each of the five DMOs we have defined in template (3), we have a pool of data mining tools available. For this research, the filters and algorithms in the WEKA [5] machine learning package are used. Table 2 shows the tools that are included in the GPS system.

Algorithm 1 shows the pseudocode of the GPS algorithm. The basic steps of the system are: firstly a initial population of DMO template instances is randomly generated based on a predefined template (e.g., template (3) and Figure 1 (b)), the placeholders of each template instance are randomly populated with the objects in the pools of DMOs (e.g., Figure 1 (c)). Then for each GA iteration (generation), PSO is used for searching an optimal parameters for each template instance (similar to the PSMS system). The population of template instances is then sorted by their PSO-based evaluation scores. After the PSO optimization procedures are done, typical GA operators, such as crossover and mutation, can be applied for generating new template instances which are used for replacing the template instances with relatively low evaluation scores. The above procedure is repeated T times, where T is the number of GA generations. Finally, the template instance with the best evaluation score is returned as the GPS solution.

Table 1. Data sets: basic characteristics

Original data sets			Final binary data sets
Data set with release year	#Insts	Atts:Classes	Class distribution (#Insts)
Adult 96	48,842	14:2	23% vs 77% (10,000)
Chess 94	28,056	6:18	48% vs 52% (8,747)
Connect-4 95	67,557	42:3	26% vs 74% (10,000)
Covtype 98	581,012	54:7	43% vs 57% (10,000)
KDD09 Customer Churn 09	50,000	190:2	8% vs 92% (10,000)
Localization Person Activity 10	164,860	8:11	37% vs 63% (10,000)
MAGIC Gamma Telescope 07	19,020	11:2	35% vs 65% (10,000)
MiniBooNE Particle 10	130,065	50:2	28% vs 72% (10,000)
Poker Hand 07	1,025,010	11:10	45% vs 55% (10,000)
UCSD FICO Contest 10	130,475	334:2	9% vs 91% (10,000)

3 Comparing GPS to PSMS and Other Learning Systems

We experiment with ten classification problems. All of them are real-world datasets which can be downloaded from the UCI repository, the UCSD data mining contest repository and the KDD Cup repository. These data sets were selected because they are large and come from different research and industrial areas. To speed up the experiments, all five multi-class datasets were converted to binary problems by retaining only the two largest classes from each. After this conversion to binary problems, for datasets that are larger than 10,000 instances, a subset of 10,000 instances is randomly selected for experiments. Table 1 shows the basic properties of the original and the final datasets.

To test the performance of the GPS algorithm, we implemented a variant[1] of the PSMS system proposed in [3] with the DMO pools defined in Table 2. The two systems are set to optimise the AUC performance[2] and are tested under 30 configurations (3 experiments per dataset): for GPS, the population size for GA and the swarm size for PSO are both set to 10, and the number of PSO evolutions is set to 10; for PSMS, the swarm size is set to 10.

For each dataset, three experiments were conducted. Let g be the number of GA generations for GPS; when $g=10$, the number of PSO evolutions for PSMS is set to 1000; when $g=20$, the number of PSO evolutions for PSMS is set to 2000; when $g=30$, the number of PSO evolutions for PSMS is set to 3000. So, for each experiment, the training cost for both systems is roughly the same. The objective functions of both GPS and PSMS are based on the respective training set AUC performance obtained from 3-fold cross validation of a particular template instance. The AUC performance of two popular ensemble learning algorithms, AdaBoost.M1 [4] with 1,000 decision stumps, and Random Forest [1] with 1,000 unpruned random trees are also reported as baseline performance.

Figure 2 (a) to Figure 2 (j) show the comparison results based on the AUC performance obtained from 5 times 3-fold cross validation. Figure 2 (k) gives a summary in terms of number of wins. Overall, on the 10 datasets, the GPS algorithm wins 83% (25 wins) of the 30 experiments. The results demonstrate the benefit of combining GA and PSO for the FMS problem. Also, we can see

---

[1] In our implementation, the dimensionality of each particle is adapted automatically based on the number of parameters of a particular DMO.

[2] The balanced error rate (BER) was used in the original PSMS system.

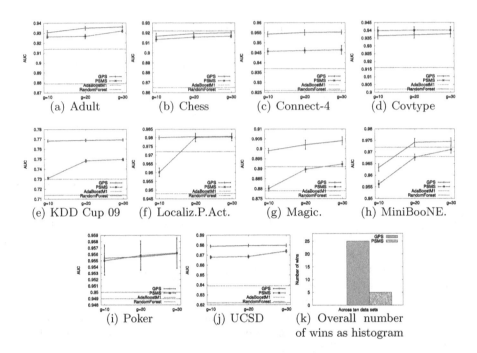

Fig. 2. A comparison of AUC performance between GPS and PSMS under 30 different configurations; the number of PSO evolutions for GPS is set to 10; x-axis g is the number of GA generations for GPS; when $g{=}10$, the number of PSO evolutions for PSMS is set to 1000; when $g{=}20$, the number of PSO evolutions for PSMS is set to 2000; when $g{=}30$, the number of PSO evolutions for PSMS is set to 3000

that the best performance of both GPS and PSMS outperform AdaBoost.M1 and Random Forest on 9 out of the 10 datasets, which indicates the advantage of a full model over the single algorithm model. Another interesting pattern is that the GPS algorithm outperforms the baseline algorithms with big margin on datasets with a relatively imbalanced class distribution.

4   Speeding Up the GPS System

The training complexity of the GPS algorithm depends on the base learners found and evaluated during the search. The main cost for GPS is the cost for estimating a base learner's performance (e.g., cross validation). The algorithm searches for a full model consisting of many data mining operators. Therefore, although GPS is powerful in modeling, the user may have to wait for several hours, or even days on relatively large data. For example, on the reduced version of the KDD Cup 2009 data (with 50,000 data points and 190 numeric attributes), the GPS system took about six hours to complete on an AMD 2.8G PC with 16G RAM (number of GA generations, number of PSO evolutions, GA's population size and PSO's swarm size were all set to 10, and 3-fold cross validation was

Table 2. WEKA algorithms and filters that are used as the DMO objects

Data Sampling	Data Cleansing	Feature Trans.	Feature Sel.
SMOTE oversampling	NumericCleaner	Normalize	CfsSubsetEval
Resample with replacement	RemoveUseless	Standardize	InfoGainAttributeEval
Resample no replacement	ReplaceMissingValues	Center	GainRatioAttributeEval
Do nothing	Do nothing	AddNoise	OneRAttributeEval
		Discretize	PrincipalComponents
		NominalToBinary	ChiSquaredAtt.Eval
		NumericTransform	Do nothing
		Do nothing	

Algorithm	HyperParameters
Bagging with Random Tree	num.Bagging.Iterations, num.Atts., depth.Tree
Bagging with REPTree	num.Bagging.Iterations, num.Folds., depth.Tree
AdaBoost.M1 with DecisionStump	num.Boosting.Iterations , useResample
LogitBoost with DecisionStump	num.Boosting.Iterations , useResample
Bagging with J48 Decision Tree	num.Bagging.Iterations , prune , conf.
RotationForest with REPTree	num.Iterations, Percentage.removed, projection

used in the objective function). Therefore, in this section, we present a strategy for speeding up the GPS algorithm. Before introducing the new algorithm, we first review the model tree idea.

A model tree [11] is a decision tree system that uses linear models at the leaves instead of using discrete class labels for classification tree or mean as the prediction for regression tree. Model trees inherit the advantageous scalable feature of decision tree systems since the training data is stored in a tree structure. Some variants that have been designed based on the model tree idea show promising results, such as the logistic model tree [8].

We here propose a novel GPS-based model tree algorithm named the Full Model Tree, because GPS builds a full model on a given dataset. The idea is that instead of training the GPS algorithm on the full training data, we build GPS models at the leaves of a tree structure. In the second set of experiments in this paper, we compare GPS to Full Model Tree with two different tree structures, namely, the perfect binary tree and the random binary tree based on the following definitions.

Definition 1: A perfect binary tree is a binary tree with all leaf nodes at the same depth. All internal nodes have degree 2.

Definition 2: A random binary tree is a binary tree formed by inserting nodes one at a time according to a random mechanism.

Next, we show that theoretically when the above two binary tree structures are used, and if the tree height is greater than zero and the training complexity of GPS is worse than linear, then GPS-based Full Model Tree is faster than GPS when training on the same data.

Assume the running time of the normal GPS algorithm (GPS-0) for training its model on a dataset of n data points is $O(f(n))$, and that for the GPS-based Full Model Tree is $O(g(n))$. Based on our preliminary experiments, we found that the *empirical* training complexity of GPS is worse than linear on most of the datasets we have tested, so here we consider the case for $f(n) > n^1, n > 1$.

Theorem 1. *For a perfect-binary-tree-based GPS Full Model Tree T with height $h \geq 1$. If GPS-0's empirical training complexity is worse than linear, such as $f(n) > n^1, n > 1$, then we have $g(n) < f(n)$.*

Proof. Let l be the number of leaf nodes of T, we have $l = 2^h$, ($l \geq 2$). Let k be the number of data points at each leaf of T, we have $k = n/l$. Then, we have $g(n) = l \times f(k)$ and $f(n) = f(k \times l)$. Let $f(n) = n^x$, so we have $x > 1$.

$$f(n) - g(n) = f(k \times l) - l \times f(k) = (k \times l)^x - l \times k^x$$
$$= k^x \times l^x - k^x \times l = k^x \times (l^x - l) = k^x \times l^{x-1} > 0.$$

Theorem 2. *For a random-binary-tree-based GPS Full Model Tree T with height $h \geq 1$. If GPS-0's empirical training complexity is worse than linear, such as $f(n) > n^1, n > 1$, then we have $g(n) < f(n)$.*

Proof. Let l be the number of leaf nodes of T, $h \geq 1$ so we have $l \geq 2$. Let k_i be the number of data points at leaf i of T, we have $\sum_{i=1}^{l} k_i = n$. Let $f(n) = n^x$, we have $x > 1$. Then, we have $g(n) = \sum_{i=1}^{l} g(k_i) = \sum_{i=1}^{l} k_i^x$ and $f(n) = n^x = (\sum_{i=1}^{l} k_i)^x$. Therefore, $f(n) - g(n) = \{(\sum_{i=1}^{l} k_i)^x - \sum_{i=1}^{l} k_i^x\} > 0$.

The two theorems state that theoretically the two Full Model Tree variants are faster than GPS in the case that the training complexity of GPS is worse than linear. Results above are also applicable to memory consumption stating that the two Full Model Tree variants are supposed to be more memory efficient than GPS if the training complexity in terms of memory of GPS is worse than linear. The results also imply that if GPS's training complexity is linear or better, then theoretically the Full Model Tree variants will not speed up the original GPS algorithm. Next, we describe how the GPS-based Full Model Trees are built.

When growing a perfect binary tree, firstly the algorithm checks if the tree height is equal to a user-specified value h. If tree height $= h$ or the current data contains only one class, then make a leaf node and build a GPS model, else the best variable is selected for splitting. Here, the *best* is based on the information gain measure of a variable. For numeric variables, we examine information gain using the median of a variable as the splitting point; for nominal variables, we balance the number of data points from distinct categorical values. For instance, imagine a nominal variable has two distinct categorical values A and B; if the data we are to split has 100 data points, where 80 of them belong to A, and 20 of them belong to B, then we randomly select 30 data points from A, and put them into B. If the nominal variable has three distinct categorical values, say A with 60 data points, B with 30 data points, and C with 10 data points, then we merge B and C first, and then balance A and BC by randomly moving 10 data points from A to BC. The same balancing strategy is also applicable to a nominal variable having more than three distinct categorical values. In this way the amount of data from the current node is roughly equally split for its two child nodes.

When growing a random binary tree, firstly the algorithm checks if the tree height is equal to a user-specified value h. If tree height $= h$ or the current data contains only one class, then make a leaf node and build a GPS model, else the algorithm randomly chooses one of the best K variables for splitting. Here, the *best* is based on the information gain measure of a variable, where K is a user-specified value. For numeric variables, the best splitting point is found by trying

Table 3. Performance and runtime of the GPS and the Full Model Tree algorithms; A "⊖" indicates that in terms of AUC, the GPS algorithm is significantly better than the respective algorithm; A "◊" indicates that in terms of runtime, the GPS algorithm is significantly slower than the respective algorithm; level of significance 0.05

Dataset	GPS	FMT-perfect	FMT-random	GPS	FMT-perf.	FMT-rand.
		AUC			Runtime (mins)	
Adult	0.94 ± 0.002	0.93 ± 0.003	0.93 ± 0.002 ⊖	45 ± 6	37 ± 4 ◊	48 ± 11
Connect-4.	0.95 ± 0.002	0.95 ± 0.002	0.95 ± 0.003 ⊖	91 ± 5	77 ± 9 ◊	74 ± 14 ◊
KDD Cup.	0.77 ± 0.002	0.77 ± 0.002	0.76 ± 0.003 ⊖	178 ± 9	157 ± 11 ◊	189 ± 8
Mini.B.E.	0.98 ± 0.002	0.98 ± 0.002 ⊖	0.97 ± 0.003 ⊖	124 ± 7	123 ± 9	135 ± 12
UCSD.	0.68 ± 0.003	0.68 ± 0.002 ⊖	0.67 ± 0.002 ⊖	487 ± 16	417 ± 19 ◊	476 ± 17

all possible splitting points between two neighbored numbers (the splitting point with the highest gain will be selected); for nominal variables, the data is split between the majority categorical value and the other categorical values.

Next, we examine both the predictive performance and the runtime of the two Full Model Tree variants (one uses the perfect binary tree structure, the other uses the random binary tree structure, namely, FMT-perfect and FMT-random, respectively) to the original GPS algorithm.

We use five medium size datasets for this experiment. Table 1 shows the properties of these datasets. The original *KDDCup09* dataset has 50,000 data points, 190 numeric variables and 40 categorical variables. To speed up the experiment, the 40 categorical variables were removed from the data because some variables have thousands of distinct values. We set the height of both FMT-perfect and FMT-random to 3. So, for FMT-perfect, there will be $2^3 = 8$ leaf GPS models to be built, and each leaf will have $n/8 \pm 1$ data points where n is the total number of training data points. The K value for FMT-random is set to $log(M)+1$, where M is the number of variables. For the GPS algorithm, the number of generations for GA, the population size for GA, the number of evolutions for PSO, and the swarm size for PSO are all set to 10. The objective function of GPS is based on 2-fold cross validation.

Table 3 shows the comparison results based on 5 times 3-fold cross validation. The AUC performance and the runtime are reported. For the AUC performance, we can see that GPS significantly outperforms FMT-random on all datasets, indicating that FMT-random is not good enough to be used as a GPS alternative. The GPS algorithm significantly outperforms FMT-perfect on two datasets; for the other three datasets, the performance of GPS and FMT-perfect has no significant difference. This indicates that for these three datasets, FMT-perfect can be used as a GPS alternative. In terms of runtime, the FMT-random algorithm is significantly faster than GPS only on the *Connect-4.* dataset. One reason could be that the number of data points at the leaf nodes of FMT-random are not the same, so the empirical training complexity of FMT-random varies at each leaf. We can see that FMT-perfect is faster than FMT-random on all datasets because usually the number of leaf nodes of FMT-random is less than that for FMT-perfect. The results show that FMT-perfect is significantly faster then GPS on 4 out of 5 datasets, indicating that FMT-perfect is a viable approach for speeding up GPS. Overall, on 3 out of 5 datasets, namely, *Adult, Connect-4.*, and

KDDCup09, the perfect-binary-tree-based Full Model Tree could significantly speed up the GPS algorithm without sacrificing GPS's predictive power.

5   Conclusions

We proposed a framework (in the DMO space setting) which can be used for designing new FMS (full model selection) algorithms, and a novel FMS algorithm which can be seen as a realization and an application of the framework. Our experiments on ten real-world problems show that the GPS algorithm performs very competitively with PSMS, the state-of-the-art PSO-based FMS algorithm. We also examined the feasibility of using the model tree idea for speeding up the GPS algorithm. Our experimental results suggest that using the perfect binary tree as the internal tree structure for GPS-based Full Model Tree is a viable approach when the empirical training complexity of GPS is worse than linear. The techniques described in this paper could probably also be applied to regression and label ranking problems, but this needs to be verified in a future study. Another future work direction is to compare the performance of the GPS systems to fine-tuned base-level ensemble algorithms. The 5-DMO template (3) defined in Section 2 is only one of many possible templates for practical data mining solutions, in future research we will also investigate methods for optimizing alternative templates simultaneously in a cloud environment.

References

1. Breiman, L.: Random forests. Machine Learning 45(1), 5–32 (2001)
2. Chawla, N.V., Bowyer, K.W., Hall, L.O., Kegelmeyer, W.P.: Smote: synthetic minority over-sampling technique. J. Artif. Int. Res. 16, 321–357 (2002)
3. Escalante, H.J., Montes, M., Sucar, L.E.: Particle swarm model selection. Journal of Machine Learning Research 10, 405–440 (2009)
4. Freund, Y., Schapire, R.: Experiments with a new boosting algorithm. In: Thirteenth International Conference on Machine Learning, pp. 148–156 (1996)
5. Hall, M., Frank, E., Holmes, G., Pfahringer, B., Reutemann, P., Witten, I.: The weka data mining software: An update. SIGKDD Explorations 11(1) (2009)
6. Holland, J.H.: Adaptation in Natural and Artificial Systems. MIT Press, Cambridge (1992)
7. Kennedy, J., Eberhart, R.: Particle swarm optimization. In: Proceedings of the IEEE International Conference on Neural Networks, vol. 4, pp. 1942–1948. IEEE (1995)
8. Landwehr, N., Hall, M., Frank, E.: Logistic model trees. Mach. Learn. 59, 161–205 (2005)
9. Momma, M., Bennett, K.P.: A pattern search method for model selection of support vector regression. In: Proceedings of the SIAM International Conference on Data Mining. SIAM (2002)
10. Sun, Q., Pfahringer, B., Mayo, M.: Full model selection in the space of data mining operators. In: Proceedings of the Fourteenth International Conference on Genetic and Evolutionary Computation Conference Companion, pp. 1503–1504. ACM (2012)
11. Wang, Y., Witten, I.H.: Induction of model trees for predicting continuous classes. Poster Papers of the 9th European Conference on Machine Learning. Springer (1997)

Transfer Learning with Part-Based Ensembles

Shiliang Sun, Zhijie Xu, and Mo Yang

Department of Computer Science and Technology, East China Normal University
500 Dongchuan Road, Shanghai 200241, P.R. China
slsun@cs.ecnu.edu.cn, momo.yang12@gmail.com

Abstract. Transfer learning is one of the most important directions in current machine learning research. In this paper, we propose a new learning framework called Multi-source part-based Transfer Learning (Ms-pbTL), which is one kind of parameter transfer with multiple related source tasks. Dissimilar to many traditional works, we consider how to transfer information from one task to another in the form of integrating transferred information between parts. We regard all the complex tasks as a collection of several constituent parts respectively. It means that transfer learning between two complex tasks can be accomplished by sub-transfer learning between their parts. Then, after completing the above information transfer between the source and target tasks, we integrate the models of all the parts in the target task into a whole. Experiments on some real data sets with support vector machines (SVMs) validate the effectiveness of our proposed learning frameworks.

Keywords: Transfer learning, part-based model, multi-source learning, support vector machine.

1 Introduction

Traditional machine learning usually depends on the availability of a large number of data from a single task to train an effective model. However, researchers often confront the situations that there are not enough data available and they have to resort to data from other tasks (source tasks) to aid the learning of the target task. In some cases, even though there are many training examples for the target task, integrating information from other tasks or data sets can still be helpful to improve the performance. Due to the above reasons, some machine learning strategies have been investigated, including multi-task learning [1–3], multi-view learning [4, 5], lifelong learning [6, 7] and transfer learning [8–10]. In this paper, we would like to develop new methods for effective transfer learning.

It should be noted that the target task and source task often have different data distributions in real applications. For example, Wall Street firms often hire physicists to solve finance problems even though there is nothing in common superficially between these two problems [11]. It is easy to see in this example that humans can deal with some problems through applying knowledge learned in one domain to an entirely different one. The reason humans can do this is that they have the ability to choose the essence or the most related part of knowledge

Z.-H. Zhou, F. Roli, and J. Kittler (Eds.): MCS 2013, LNCS 7872, pp. 271–282, 2013.
© Springer-Verlag Berlin Heidelberg 2013

which is useful to learn the new task. Nevertheless, computers cannot directly distinguish whether one part in the source task is good or bad to transfer to the target task. As a result, it is significant for us to teach the computer to judge the importance of every part in one complex task. One solution to this problem is to weight different parts differently by their contributions to the task.

Many collections of data exhibit a common underlying structure: they consist of a number of identical parts, each with a range of possible states [12]. Sometimes, although one part of a source task is unsuitable to help learn the target task, there may still exist another part which is useful and helpful for this work. In our paper, to find such parts, we utilize a part-based model. This approach has been used in many fields, especially image processing and computer vision. For instance, Bar-Hillel et al. [13] used this principle to establish the model for the task of object recognition. However, they have just made use of the part-based model in one single task. In this paper, we extend the part-based model to transfer learning. Moreover, for the purpose of avoiding negative transfer [14], which is a situation where knowledge from a source task unexpected deteriorates the performance of target task, the principle of multi-source learning [15] is used in our learning framework as well. Different from the usage of multi-source learning in transfer learning like [16], we will analyze the contribution of every part in every source task to help judge the importance of different parts in the target task. With the help of these principles, we can not only reduce the problem of negative transfer, but also improve the effectiveness of transfer learning.

In this paper, we propose a framework named Multi-source part-based Transfer Learning (Ms-pbTL). It is an extension of pbTL [17], which is a process of parameter transfer using one source task. In pbTL, all the complex tasks are regarded as a collection of constituent parts, and every task can be divided into several parts respectively. This means transfer learning between two complex tasks can be accomplished by sub-transfer learning tasks between their parts. This method is also used in Ms-pbTL.

Some main functions of the frameworks of the Ms-pbTL are described in the following points. Firstly, due to the usage of the part-based model, one task can be divided into a number of parts so as to exploit different latent knowledge. Secondly, the multi-source principle lets us have opportunities to obtain more sets of parameters from different source tasks synchronously, which can be combined in the target task. Finally, we can make a difference not only between different parts in one single source task, but also between the corresponding parts in all the source tasks. This step makes it possible to focus more on the parts which can contribute much more to the target task. Besides these points, in our frameworks, from every part in all the source tasks, we can obtain a set of parameters which can be transferred to its corresponding part in the target task to construct an ensemble of classifiers with support vector machines (SVMs) [18, 19]. At the same time, notice that the parameters about a certain part of one source task can only be transferred into the corresponding part of the target task. Depending on this rule, after reusing all the sets of parameters from the source tasks to help train the ensemble classifiers on their corresponding parts in the target task, we combine these classifiers into a final classifier with weights determined by their accuracy rates. The effectiveness of our proposed learning frameworks is supported by experiments on multiple real data sets.

The remainder of this paper is organized as follows. In Section 2, we describe our framework Ms-pbTL in detail. Then, experiments with our proposed methods are provided in Section 3. Section 4 concludes the whole paper and gives future work directions.

2   Multi-source Part-Based Transfer Learning

In this section, we present our transfer learning framework, Ms-pbTL. It is essential for us to pay attention to one special characteristic of this part-based model. According to the part-based principle, the whole will be divided into several parts to learn separately. Different from multi-view learning, some latent relationships exist between every pair of two adjacent parts in the part-based model. For example, one picture of human can be divided into three parts such as head, the upper part of the body and the lower part of the body. Obviously, every part contains many particular features which only exist in the part itself. Nevertheless, serving as the joint of the head and the upper part of the body, the neck is the part which can belong to both of them. As a result, all the features about the neck can be contained in both of these two parts. Consequently, we can summarize the characteristics of the part-based model as follows. On one hand, every part of the whole contains a number of distinctive features which will not belong to other parts. On the other hand, there exist a few features which are used to describe the intersection between two adjacent parts and may appear in both of them.

In order to fully use the benefits of the part-based model in parameter transfer learning, a basic learning framework named part-based Transfer Learning (pbTL) which utilizes one source task was proposed in [17]. For the purpose of avoiding negative transfer and improving the effectiveness of transfer learning, we propose its extended version, a new learning framework, Multi-source part-based Transfer Learning (Ms-pbTL). In the first step, we use SVM to train classifiers and learn a set of optimal parameters for every part of each source task. Then, these sets of parameters need to be transferred to the parts of the target task correspondingly. This is the process of parameter transfer learning. Following this, we can learn several better classifiers trained on the basis of these parameters about every part in the target task and combine them into a final classifier in a weighted fashion at last. The function of this step is to determine which part can contribute much more. Before presenting our learning framework Ms-pbTL, we need to state some key considerations here.

Firstly, we use RBF as the kernel function in the SVM. The detailed formula can be written as:

$$k(x_i, x_j) = e^{-\|x_i - x_j\|^2 / 2\sigma^2}. \tag{1}$$

Therefore, due to the characteristics of RBF and SVM, the core elements in our parameter transfer learning are the usual regularization coefficient C in the SVM formulation and the width parameter σ in the above (1).

Secondly, we divide the source tasks and target task into several parts in terms of features correspondingly. In this step, we need not only to divide these features into different parts averagely, but also to consider in a characteristic of the part-based model that different parts may be related to each other and contain some common features. For example, according to the supposition that $dimension = 10$ (the dimension of the data set) and $N = 3$ (the number of parts to be divided into), now we can split the first nine features into three parts averagely and supply the last feature for every part so as to create three interdependent four-dimensional parts.

Thirdly, although samples of source and target tasks come from different distributions, we suppose they can be mapped into the same class label set.

The detailed process of Ms-pbTL is given in Table 1.

2.1   Remarks on Ms-pbTL

Through getting the optimal set of parameters of every part in the source task and transferring them to the target task part by part, the merits of different parts can be clearly shown. Moreover, the goal of treating different parts differently can be reached by defining weights as well. The weights can be calculated as:

$$W_{T_i} = \frac{Accuracy_{T_i}}{\sum_{i=1}^{N} Accuracy_{T_i}}. \tag{2}$$

What is more, in (2), to further distinguish the importance of different parts in the target task, we calculate the weights of the classifiers of different parts by the distribution of their accuracy rates on the training data set. These accuracy rates show the percent of samples which are predicted correctly by f_{T_N}. Furthermore, in the output step, we compose the final classifier by the sum of the product of every classifier and its weight. The detailed formula can be written as:

$$h_f(x) = \begin{cases} 1 & \text{if} \quad \sum_{i=1}^{N} W_{T_i} \times f_{T_i}(x) \geq 0 \\ -1 & \text{otherwise}. \end{cases} \tag{3}$$

In Ms-pbTL, we use several source tasks $S_1, \cdots, S_n$ simultaneously to learn the target task. From step 1 to 3, we divide all the source tasks and the target task into N parts. In step 4, due to the fact that now we have n source tasks, we can learn n sets of optimal parameters from them to help every part in the target task to come to n sub-classifiers, respectively. Then, with the help of the accuracy vector acquired in step 5, we can combine all the sub-classifiers about one part into a final sub-classifier in line with step 6. After that, we calculate the weights of each classifier to obtain a final one.

Table 1. Framework of multi-source part-based transfer learning

---

Input:
set of n source tasks $S_1, \cdots, S_n$ (now each S_i here is a source task) and one
target task T, where $S_1, \cdots, S_n$ and T belong to different distributions, but
contain the same class label set $Y = \{1, -1\}$ as Os-pbTL.
Initialize the number of parts to be divided. : N
Initialize the parameter set (C, σ) in SVM with RBF.

---

1. Divide every source task into N parts by their features under the same
   rule: $\{(S_{11}, \cdots, S_{1N}), \cdots, (S_{n1}, \cdots, S_{nN})\}$.
2. Get N classifiers of every source task on the basis of its parts:
   $\{(f_{S_{11}}, \cdots, f_{S_{1N}}), \cdots, (f_{S_{n1}}, \cdots, f_{S_{nN}})\}$,
   and their optimal parameter vectors:
   $\{[(C_{S_{11}}, \sigma_{S_{11}}), \cdots, (C_{S_{1N}}, \sigma_{S_{1N}})], \cdots, [(C_{S_{n1}}, \sigma_{S_{n1}}), \cdots, (C_{S_{nN}}, \sigma_{S_{nN}})]\}$.
   (In this paper, we will use cross validation to learn these optimal
   parameter vectors)
3. Divide the target task into N parts corresponding to the source tasks:
   $T_1, \cdots, T_N$.
4. According to every part in the target task, we will come to n sub-
   classifiers: $\{[f_{T_{11}}, \cdots, f_{T_{1n}}], \cdots, [f_{T_{N1}}, \cdots, f_{T_{Nn}}]\}$ by the optimal sets of
   parameters about the corresponding parts of n source tasks acquired in
   step 2.
5. Calculate the accuracy rate about every classifier obtained in step 4:
   $\{[Accuracy_{T_{11}}, \cdots, Accuracy_{T_{1n}}], \cdots, [Accuracy_{T_{N1}}, \cdots, Accuracy_{T_{Nn}}]\}$.
6. Calculate the final classifier of every part of the target task, respec-
   tively:for $i = 1, \cdots, N$
   $f_{T_i} = \sum_{j=1}^{n} Accuracy_{T_{ij}} \times f_{T_{ij}}$
   end
7. Calculate the accuracy rate $\{Accuracy_{T_1}, \cdots, Accuracy_{T_N}\}$ of every
   classifier $\{f_{T_1}, \cdots, f_{T_N}\}$ in the target task.
8. Calculate the weights of $\{f_{T_1}, \cdots, f_{T_N}\}$:
   for $i = 1, \cdots, N$
   $W_{T_i} = \dfrac{Accuracy_{T_i}}{\sum_{i=1}^{N} Accuracy_{T_i}}$
   end

Output the hypothesis:
$$h_f(x) = \begin{cases} 1 & \text{if } \sum_{i=1}^{N} W_{T_i} \times f_{T_i}(x) \geq 0 \\ -1 & \text{otherwise} \end{cases}$$

In addition to all the description above, it is also important for us to discuss
more here. In our paper, we have not considered the problem caused by the
diversity between the corresponding parts of the source and target tasks. For
example, in step 4, we train the sub-classifiers of every part in the target task on
the basis of the optimal sets of parameters learned in the source tasks directly
and make no changes. However, sometimes, the sets of parameters learned in the

source tasks are not suitable enough to be reused in the target task because of the diversity mentioned above. As a result, if we want to deal with this problem and come to the sets of parameters which are more suitable for the learning of the target task, we can actually just initialize the parameter vectors of every part in the target task by the optimal sets of parameters obtained in the corresponding parts of the source tasks. After that, the work is to update them through continuous iterations with some other processors such as neural networks until coming to the satisfied ones.

3   Experiments

In this section, we implement two groups of experiments. We start with a basic group of experiments with real data sets so as to illustrate the effectiveness of our learning framework. In this group of experiments, for the purpose of implementing our method, Ms-pbTL, we employ two source tasks simultaneously to help learning a target task. Moreover, we do a further study about the influence caused by varying the number of source tasks in Ms-pbTL.

In all these experiments, we set the parameter $N = 3$, which represents the number of parts to be used in the target task and source tasks. Certainly, in practical applications, this number of different tasks can be different and needs to be decided by the characteristics of different tasks. In addition, we compare our method, Ms-pbTL, with basic SVM, transfer learning with basic SVM (Transfer SVM) and pbTL in all the experiments.

3.1   Learning with Two Source Tasks

In this section, we run some experiments on real data sets from UCI repository. Note that all the data sets used here are transformed into binary-class problems. Then, due to the characteristics of different data sets, we use different ways to generate the target task and source tasks and run five sets of experiments on four data sets.

On one hand, data sets $\frac{\text{Segmentation}}{\text{path:cement}}$ and $\frac{\text{Digit}}{5:8}$ are multi-class problems. They are divided into several binary-class sub-data sets by their labels to generate the target task and source tasks. On the other hand, data sets $\frac{\text{Digit}}{3:8}$, German and $\frac{\text{WQ}}{\text{level 5: 7}}$ are binary-class problems, as a result, we need to divide them into several sub-data sets by one specific rule to generate the target task and source tasks. Table 2 provides the summary of the used real data sets.

For each data set in Table 2, we use a specific rule to divide it into the target task and source tasks.

Segmentation is one seven-class data set. We divide the whole data set into several binary-class sub-data sets by their labels to generate the target task and source tasks. We use all the data with label *sky* and *window* as the source task A, the data with label *grass* and *foliage* as the source task B and the data with label *path* and *cement* as the target task.

Table 2. Summary of data sets

Real data set	Segmentation path:cement	Digit $\frac{}{5:8}$	Digit $\frac{}{3:8}$	German	WQ $\frac{}{\text{level } 5: 7}$
Total number of examples	1980	3361	1126	1000	2337
Size of the source task A	$\frac{660}{\text{sky:window}}$	$\frac{1115}{6:2}$	488	230	877
Size of the source task B	$\frac{660}{\text{grass:foliage}}$	$\frac{1134}{3:9}$	330	411	648
Target training set	$\frac{330}{\text{path:cement}}$	$\frac{500}{5:8}$	150	159	500
Target testing set	$\frac{330}{\text{path:cement}}$	$\frac{618}{5:8}$	158	200	312
Dimensions	19	64	64	24	11
Number of classes	6	6	2	2	2

Handwritten Digit is one ten-class data set and here we use two different ways to generate the target task and source tasks. Firstly, similar to data set Segmentation, in $\frac{\text{Digit}}{5:8}$, we use all the data with label 6 and 2 as the source task A, the data with label 3 and 9 as the source task B and the data with label 5 and 8 as the target task.

After that, we get all the data with label 3 and 8 to generate one binary-class data set, $\frac{\text{Digit}}{3:8}$, to run another set of experiments. According to this data set, we divide it into the target task and source tasks on the basis of the value of *dimension six*. All the data according with the rule *dimension six* < 5 belong to the source task A, $5 \le$ *dimension six* < 10 belong to the source task B and *dimension six* ≥ 10 for the target task.

German Credit Data is one binary-class data set. We split the data set on the basis of the feature *Duration of month*. The source task A consists of all the data following the rule *Duration* > 24 while the source task B consists of all the data following the rule $12 <$ *Duration* ≤ 24 and *Duration* ≤ 12 for the target task.

Wine Quality (WQ) is one eleven-class data set and the assignment of it is to grade the wine quality between 0 to 10. Because the data of different classes are not balanced, we select all the data with label *level* 5 and 7 to generate one binary-class data set which contains 2337 samples. Then, we divide this data set into the target task and source tasks on the basis of the value of feature *Residual sugar*. The source task A consists of all the data following the rule *Residual sugar* < 3 while the source task B consists of all the data following the rule $3 \le$ *Residual sugar* < 8 and *Residual sugar* ≥ 8 for the target task.

Finally, note that we only make use of the source task A to run the experiments of Transfer SVM and pbTL. Furthermore, in our framework Ms-pbTL, due to the fact that we divide all the data sets into three parts by their features randomly, we run the experiments of every data set for ten times and get the mean of them as the final scores. Certainly, standard deviation (Std) will be also calculated synchronously. Moreover, it is significant to demonstrate here that, in order to keep the characteristics of the part-based model, we make most of features be owned by only one part and a few features be shared among all the parts in one task. For example, in our experiments, we make use of the Handwritten Digit data set which contains 64 features. We realign the features randomly at the beginning. Then we divide the first 60 features into three parts averagely and share the rest 4 features for all parts. As a result, we create three interdependent 24-dimensional parts. Table 3 shows the classification results.

Table 3. Accuracy rates of different methods (%)

	SVM	Transfer SVM	pbTL	Ms-pbTL
$\frac{\text{Segmentation}}{\text{path:cement}}$	83.33	87.88	$91.61_{\pm2.66}$	$94.09_{\pm2.05}$
$\frac{\text{Digit}}{5:8}$	58.99	49.02	$88.35_{\pm6.23}$	$93.46_{\pm1.67}$
$\frac{\text{Digit}}{3:8}$	51.27	56.33	$66.52_{\pm8.19}$	$80.57_{\pm7.42}$
German	72.00	74.00	$74.70_{\pm0.42}$	$75.95_{\pm0.93}$
$\frac{\text{WQ}}{\text{level 5: 7}}$	69.55	70.19	$74.39_{\pm2.33}$	$76.47_{\pm2.54}$

Table 3 shows that pbTL and Ms-pbTL outperform the standard SVM and Transfer SVM in every data set and the results of Ms-pbTL are better than pbTL. In data sets $\frac{\text{Segmentation}}{\text{path:cement}}$, $\frac{\text{Digit}}{5:8}$ and $\frac{\text{Digit}}{3:8}$, our proposed framework improves the results remarkably. Compared with these three data sets, Ms-pbTL makes a less improvement on the data sets German and $\frac{\text{WQ}}{\text{level 5: 7}}$.

What's more, we need to pay attention to the results of $\frac{\text{Digit}}{5:8}$ especially. In the experiments of this data set, though Transfer SVM fails to excel the standard SVM, Ms-pbTL still outperforms standard SVM which illustrates three important points as follows. Firstly, general transfer learning can not exert its benefit all the time. Secondly, even though the whole-based transfer learning has been ineffective, the part-based transfer learning can still be effective. Thirdly, the part-based model can help avoid negative transfer.

Overall speaking, experimental results of real data sets show that the combination of the part-based model and transfer learning can promote the learning efficacy and obtain a higher accuracy with the help of multi-source learning.

3.2 Varying the Number of Source Tasks

Here we intend to study the effect of transfer learning caused by varying the number of source tasks. Our purpose here is to observe the changes of the experimental results about Ms-pbTL with the increase of the number of source tasks. We use two different ways to generate the target task and source tasks.

SCITOS-G5 is one four-class data set which records the wall-following navigation task of one mobile robot. However, because of the sparse of the data of two class, Slight-Right-Turn and Slight-Left-Turn, we just use the data from other two classes, Move-Forward and Sharp-Right-Turn to generate one binary-classes data set to run our experiments. We divide the data set SCITOS-G5 into several parts to generate the target task and source tasks by its first feature, US1, which is the ultrasound sensor at the front of the robot. Details can be seen in Table 4.

Then, for the other set of experiments, in order to acquire enough sub-data sets with different labels to generate the target task and source tasks, we reuse the ten-class data set, Handwritten Digit here. Similar to the experiments of $\frac{\text{Digit}}{5:8}$ in the last part, we come to the target task and source tasks by its class labels as shown in Table 4.

Table 4. Summary of SCITOS-G5 and $\frac{\text{Digit}}{5:8}$

	SCITOS-G5	$\frac{\text{Digit}}{5:8}$
Total number of samples	4302	5620
Dimensions	24	64
Number of classes	2	10
Rule of the target task	US1 $\geq$ 2.1	Digit 5 and 8
Training size of the target task	400	500
Testing size of the target task	349	612
Rule of the source task A	1.5 $\leq$ US1 $<$ 2.1	Digit 3 and 8
Size of the source task A	615	1126
Rule of the source task B	1.3 $\leq$ US1 $<$ 1.5	Digit 5 and 9
Size of the source task B	716	1120
Rule of the source task C	1.0 $\leq$ US1 $<$ 1.3	Digit 2 and 6
Size of the source task C	770	1115
Rule of the source task D	0.8 $\leq$ US1 $<$ 1.0	Digit 7 and 0
Size of the source task D	619	1120
Rule of the source task E	US1 $<$ 0.8	Digit 1 and 4
Size of the source task E	833	1139

Note that we only use the source task A to run the experiments of Transfer SVM and pbTL. Then with the increase of the number of source tasks in Ms-pbTL, we intend to add one more source task into our experiments every time from the source tasks B to E orderly. Furthermore, due to the fact that we divide every data set into three parts randomly by their features, we run the experiments of every data set for ten times and get the mean of them as the final scores. Certainly, standard deviation (Std) will be also calculated synchronously. Detailed results have been given in Table 5. Note that Ms-pbTL, Ms3-pbTL, Ms4-pbTL and Ms5-pbTL represent the results of the experiments about the part-based transfer learning with two, three, four and five source tasks.

Table 5. Accuracy rates of different methods (%)

	SCITOS-G5	$\frac{\text{Digit}}{5:8}$
SVM	81.95	58.99
Transfer SVM	82.52	50.98
pbTL	$84.84_{\pm1.43}$	$88.97_{\pm4.06}$
Ms-pbTL	$86.59_{\pm1.04}$	$94.59_{\pm1.37}$
Ms3-pbTL	$87.51_{\pm1.61}$	$95.53_{\pm1.10}$
Ms4-pbTL	$88.82_{\pm1.63}$	$95.52_{\pm1.44}$
Ms5-pbTL	$88.62_{\pm2.04}$	$95.31_{\pm1.47}$

According to the results of SCITOS-G5 in Table 5, from Ms-pbTL to Ms4-pbTL, we can see that, as the number of source tasks increases, the increasing degrees of experimental results come to decrease and the negative growth happens to Ms5-pbTL finally. The experiments of $\frac{\text{Digit}}{5:8}$ meet the similar condition as well. The results of this data set reach the peak in Ms3-pbTL and then begin to decrease.

In general, though both Ms-pbTL and Ms3-pbTL perform well, the increasing degrees of experimental results become progressively less obvious with increasing number of source tasks. Therefore, we can derive the following conclusion. Too many source tasks can not lead to a better outcome for transfer learning. The most important point of improving the effectiveness of transfer learning is to select the source tasks which are more similar to the target task rather than use as many source tasks as we can.

4   Conclusion and Future Work

In this paper, we propose a new learning framework, multi-source part-based transfer learning. From our experiments on real data sets, this framework is proved to be more useful and effective than traditional transfer learning. We conclude the reasons about its feasibility with the following points. Firstly, the part-based model lets us have chance to take advantage of different latent knowledge on one task. Secondly, it also decreases the influence of irrelevant and useless features. Thirdly, the multi-source principle makes us obtain more knowledge from different source tasks to learn the target task. At the same time, it also helps avoid negative transfer.

In the future, how to split one task into several interrelated parts more logically is an interesting direction to study. At the same time, experiments in our paper show that the increase of the number of source tasks does not always improve transfer learning, and therefore it may still be a challenge to study how to select the optimal combination of multiple source tasks to promote transfer learning.

Acknowledgements. This work is supported by the Scientific Research Foundation for the Returned Overseas Chinese Scholars, State Education Ministry, and Shanghai Knowledge Service Platform Project (No. ZF1213).

References

1. Caruana, R.: Multitask Learning. Machine Learning 28, 41–75 (1997)
2. Dai, W., Yang, Q., Xue, G.R., Yu, Y.: Boosting for transfer learning. In: Proceedings of the 24th International Conference on Machine Learning, pp. 193–200 (2007)
3. Evgeniou, T., Pontil, M.: Regularized multi-task learning. In: Proceedings of the 10th International Conference on Knowledge Discovery and Data Mining, pp. 109–117 (2004)
4. Xu, Z., Sun, S.: An algorithm on multi-view adaboost. In: Proceedings of the 17th International Conference on Neural Information Processing, pp. 355–362 (2010)
5. Zhang, Q., Sun, S.: Multiple-view multiple-learner active learning. Pattern Recognition 43(9), 3113–3119 (2010)
6. Thrun, S., Pratt, L.: Learning to Learn. Kluwer Academic Publishers (1997)
7. Thrun, S.: Explanation-Based Neural Network Learning: A Lifelong Learning Approach. Kluwer Academic Publishers, Boston (1996)
8. Cao, B., Liu, N.N., Yang, Q.: Transfer learning for collective link prediction in multiple heterogenous domains. In: Proceedings of the 27th International Conference on Machine Learning, pp. 159–166 (2010)
9. Rohrbach, M., Stark, M., Szarvas, G., Gurevych, I., Schiele, B.: What helps where and why? Semantic relatedness for knowledge transfer. In: Proceedings of the 23rd Computer Society Conference on Computer Vision and Pattern Recognition, pp. 910–917 (2010)
10. Wu, P., Dietterich, T.G.: Improving SVM accuracy by training on auxiliary data sources. In: Proceedings of the 21st International Conference on Machine Learning, pp. 871–878 (2010)
11. Davis, J., Domingos, P.: Deep transfer via second-order Markov logic. In: Proceedings of the 26th International Conference on Machine Learning, pp. 217–224 (2009)
12. Ross, D.A., Zemel, R.S.: Learning parts-based representations of data. Journal of Machine Learning Research 7, 2369–2397 (2006)
13. Bar-Hillel, A., Hertz, T., Weinshall, D.: Object class recognition by boosting a part-based model. In: Proceedings of the IEEE Computer Society Conference on Computer Vision and Recognition, pp. 702–709 (2005)
14. Perkins, D.N., Salomon, G.: Transfer of learning. The Journal of International Encyclopedia of Education 11, 6452–6457 (1992)
15. Crammer, K., Kearns, M., Wortman, J.: Learning from multiple sources. Journal of Machine Learning Research 9, 1757–1774 (2008)

16. Luo, P., Zhuang, F., Xiong, Y., He, Q.: Transfer learning from multiple source domains via consensus regularization. In: Proceedings of the 17th ACM Conference on Information and Knowledge Management, pp. 103–112 (2008)
17. Xu, Z., Sun, S.: Part-based transfer learning. In: Proceedings of the 8th Internaional Symposium on Neural Networks, pp. 434–441 (2011)
18. Lima, N.H.C., Neto, A.D.D., Melo, J.D.: Creating an ensemble of diverse support vector machines using Adaboost. In: Proceedings of the International Joint Conference on Neural Networks, pp. 2342–2346 (2009)
19. Shawe-Taylor, J., Sun, S.: A review of optimization methodologies in support vector machines. Neurocomputing 74(17), 3609–3618 (2011)

Dimensionality Reduction Using Stacked Kernel Discriminant Analysis for Multi-label Classification

Muhammad Atif Tahir[1], Ahmed Bouridane[2], and Josef Kittler[3]

[1] College of Computer and Information Sciences
Al-Imam Mohammad Ibn Saud Islamic University, Riyadh, KSA
`mtahir@ccis.imamu.edu.sa`

[2] Computer and Electronics Security Systems, Northumbria University, UK
`ahmed.bouridane@northumbria.ac.uk`

[3] Centre for Vision Speech and Signal Processing, University of Surrey, UK
`j.kittler@surrey.ac.uk`

Abstract. Multi-label classification in which each instance may belong to more than one class is a challenging research problem. Recently, a considerable amount of research has been concerned with the development of "good" multi-label learning methods. Despite the extensive research effort, many scientific challenges posed by e.g. curse-of-dimensionality and correlation among labels remain to be addressed. In this paper, we propose a new approach to multi-label classification which combines stacked Kernel Discriminant Analysis using Spectral Regression (SR-KDA) with state-of-the-art instance-based multi-label (ML) learning method. The proposed system is validated on two multi-label databases. The results indicate significant performance gains when compared with the state-of-the art multi-label methods for multi-label classification.

Keywords: Multi-label Classification, Dimensionality Reduction, KDA using Spectral Regression.

1   Introduction

A conventional multi-class classification system assigns each instance x a single label l from a set of disjoint labels L. However, in many modern applications such as text classification [1, 2], image/video categorisation [3] etc, each instance is to be assigned to a subset of labels $Y \subseteq L$. This problem is known as multi-label learning. Figures 1 shows some examples of multi-label images.

There is a considerable amount of research concerned with the development of "good" multi-label learning methods [4, 5, 1, 6–8]. Despite the extensive research effort devoted to the problem of multi-label learning, there still exist many scientific challenges. They include (i) the curse-of-dimensionality, as multi-label learning involves data with a large number of features, and (ii) capturing the correlation among classes. The curse of dimensionality especially can severely degrade the performance of the learning techniques especially methods based on

Z.-H. Zhou, F. Roli, and J. Kittler (Eds.): MCS 2013, LNCS 7872, pp. 283–294, 2013.
© Springer-Verlag Berlin Heidelberg 2013

a   b

Fig. 1. Examples of multi-label images. (a) Two concepts Horse and People appear in the same image. (b) Four concepts Bicycle, Cow, Motorcycle and People appear in the same image.

nearest neighbour principle [9]: keeping the number of training samples limited and increasing the number of features will eventually result in badly performing classifiers. Interestingly, most state-of-the-art multi-label methods, including those referred to as instance-based, are designed to focus mainly on the second problem and a very limited effort has been devoted to the curse-of-dimensionality problem. The aim of this paper is to focus on the first aspect, and tackle this problem by dimensionality reduction using stacked kernel discriminant analysis (Stacked-KDA).

Spectral methods have emerged as a powerful tool for dimensionality reduction and manifold learning. Spectral Regression combined with Kernel Discriminant Analysis (SR-KDA) introduced by Cai et al [10] has been successful in classification tasks such as multi-class face, text, spoken letter recognition and visual category recognition [11]. The method combines the spectral graph analysis and regression for an efficient large matrix decomposition in KDA. It has been demonstrated that it can achieve an order of magnitude speed-up over the eigendecomposition while producing smaller error rate compared to state-of-the-art classifiers.

In this paper, we propose a new approach to ML classification which combines SR-KDA with state-of-the-art instance-based ML learning method. Stacked KDA is used as a dimensionality reduction approach while instance-based method (MLkNN [4]) is used for multi-label classification. Interestingly, a great benefit of the proposed approach is that both the curse-of-dimensionality and the correlation problems can be tackled simultaneously [12]. The curse-of-dimensionality is handled by using feature extraction while the correlation problem is solved by stacked generalisation as well as the multi-label classification at later stage that inherently takes correlation among labels into account. To the best of our knowledge, this is the first study that aims to use KDA using spectral regression (with/without stacking) as a dimensionality reduction method for multi-label classification.

The proposed approach is applied to two multi-label data sets. We validate the advocated approach experimentally and demonstrate that it yields significant performance gains when compared with the state-of-the art multi-label methods. In addition, the complexity analysis of the proposed dimensionality reduction technique has indicated several orders of magnitude speed-up over the traditional KDA for multi-label classification.

The paper is organised as follows. In Section 2, we review the state-of-the-art methods for multi-label classification. Section 3 presents the proposed approach. Experiments are described in Section 4 followed by the results obtained and their discussion in Section 5. Section 6 concludes the paper.

2   Related Work

The sparse literature on multi-label classification, driven by problems in text classification, bioinformatics, music categorisation, and image/video classification, has recently been summarised by Tsoumakas et al [5]. This research can be divided into two different groups: i) *problem transformation* methods, and ii) *algorithm adaptation* methods. The problem transformation methods aim to transform a multi-label classification task into one or more single-label classification problems [13, 3], or label ranking [14] tasks. The algorithm adaptation methods extend traditional classifiers to handle multi-label concepts directly [4, 15, 7]. In this section, we briefly review the state-of-the-art work in this research area.

Multi-label classification can be reduced to the conventional binary classification problem. This approach is referred to as *binary relevance* (BR) learning in the literature [5]. In BR learning, the original data set is divided into $|Y|$ data sets where $Y = \{1, 2, ..., N\}$ is the finite set of labels. BR learns one binary classifier $h_a : X \rightarrow \{\neg a, a\}$ for each concept $a \in Y$. BR learning is theoretically simple and has a linear complexity with respect to the number of labels. Its assumption of label independence makes it attractive to situations where new examples may not be relevant to any known subset of labels or where label relationships may change over the test data [6].

As already pointed out, instance-based approaches are also quite popular in multi-label classification. In [4], a lazy learning approach (MLkNN) is proposed and is based on the principle of BR learner. This method is derived from the popular k-Nearest Neighbour (kNN) algorithm and bayesian inference. Given an instance x and its associated label set $y \subseteq Y$, it finds the k nearest neighbours of x in the training data. Let $N(x)$ be the set of k nearest neighbours of x. Then, based on the label sets of these neighbors, a membership counting vector C_x is defined as: $C_x(l) = \sum_{b \in N(x)} t_b(l), l \in Y$, where $C_x(l)$ counts the number of neighbors of x belonging to the l-th class, $t_b(l)$ is the category vector for x. Given a query instance q, MLkNN first identifies its k nearest neighbors $N(q)$ in the training set. Let H_1^l be the event that q has label l while H_0^l be the event that q does not belong to l. In addition, let $E_j^l (j \in 0, 1, ..., k))$ denote that among the k nearest neighbors of q, there are exactly j instances which have label l. Therefore, based on the membership counting vector C_q, the category vector t_q is determined using the following bayesian principle:

$$t_q(l) = argmax_{b \in 0,1} P(H_b^l) P(E_{C_q(l)}^l | H_b^l) \tag{1}$$

where $P(H_b^l)$ and $P(E_{C_q(l)}^l | H_b^l)$ are prior and conditional label observation probabilities. It has been shown in [4] that these probabilities can be directly estimated from the training set based on frequency count.

In [16], a multi-label dimensionality reduction method called MDDM (Multi-label Dimensionality reduction via Dependence Maximization) is proposed. This method finds the lower-dimensionality feature space by maximizing the dependence between the original feature description and class labels associated with the same object. A closed-form solution is then derived for MDDM which enables the multi-label dimensionality reduction process to become both effective and efficient. In [17], a multi-label dimensionality reduction method (MLSI) is introduced based on supervised latent semantic indexing. This method maps the input features into a new feature space that retains the information of the original inputs and also captures the dependency of output dimensions. In [18], a general framework is proposed to extract shared subspaces in multi-label classification. In [19], a joint learning framework is studied in which dimensionality reduction and multi-label classification are performed simultaneously.

3   Multi-label Dimensionality Reduction Using Stacked Kernel Discriminant Analysis

Dimensionality reduction is a pre-processing procedure which maps the original data space X to a low dimensional space Z. Normally, dimensionality reduction is represented by a transformation matrix $H \in \mathcal{R}^{m \times l}$ such that

$$z = H^T x \text{ for } x \in X \tag{2}$$

where l is usually much smaller than m. In this section, we discuss the proposed dimensionality reduction method using stacked KDA. We first briefly discuss KDA using Spectral Regression followed by the proposed system of dimensionality reduction that involves stacked KDA. One advantage of the dimensionality reduction is the lower cost of the learning process since the learning process will be performed in the lower dimensional space. Also the effects of noisy or redundant features can be reduced. Further, by introducing stacking during feature extraction, the instances belonging to multiple classes will capture the correlation for mutli-label classification.

3.1   KDA Using Spectral Regression (SR-KDA)

Linear Discriminant Analysis (LDA) is one of the most popular dimensionality reduction algorithm [10, 20] where projection vectors are obtained by maximizing the between class covariance and simultaneously minimizing the within class covariance. Mathematically, the objective function of LDA is defined as follows:

$$w_{opt} = argmax \frac{w^T S_b w}{w_T S_t w} \tag{3}$$

where S_b and S_t denote between and within class scatter matrix. The optimal w's are the eigenvectors corresponding to the non-zero eigenvalue of eigen-problem $S_b w = \lambda S_t w$.

To extend LDA to the nonlinear case, Let $\mathbf{x}_i \in \mathcal{R}^d, i = 1, \cdots, m$ be training vectors represented as an $m \times m$ kernel matrix K such that $K(x_i, x_j) = \langle \Phi(x_i), \Phi(x_j) \rangle \rangle$, where $\Phi(x_i)$ $\Phi(x_j)$ are the embeddings of data items x_i and x_j. If ν denotes a projective function into the kernel feature space, then the objective function for KDA is

$$\max_{\nu} D(\nu) = \frac{\nu^T C_b \nu}{\nu^T C_t \nu} \tag{4}$$

where C_b and C_t denote the between-class and total scatter matrices in the feature space respectively. Equation 4 can be solved by the eigen-problem $C_b = \lambda C_t$. It is proved in [21] that equation 4 is equivalent to

$$\max_{\alpha} D(\alpha) = \frac{\alpha^T KWK\alpha}{\alpha^T KK\alpha} \tag{5}$$

where $\alpha = [\alpha_1, \alpha_2, \ldots .\alpha_m]^T$ is the eigen-vector satisfying $KWK\alpha = \lambda KK\alpha$. $W = (W_l)_{l=1,\ldots n}$ is a $(m \times m)$ block diagonal matrix of labels arranged such that the upper block corresponds to positive examples and the lower one to negative examples of the class. Each eigenvector α gives a projection function ν into the feature space.

It is shown in [10] that instead of solving the eigen-problem in KDA, the KDA projections can be obtained by the following two linear equations

$$W\phi = \lambda \phi$$
$$(K + \delta I)\alpha = \phi \tag{6}$$

where ϕ is an eigenvector of W, I is the identity matrix and $\delta > 0$ is a regularisation parameter. $W = (W_l)_{l=1,\ldots n}$ is a $(m \times m)$ block diagonal matrix of labels arranged such that the upper block corresponds to positive examples and the lower one to negative examples of the class. Eigenvectors ϕ are obtained directly from the Gram-Schmidt method. Since $(K + \delta I)$ is positive definite, the Cholesky decomposition is used to solve the linear equations in (6). Thus, for the resolution of linear system (6), the system becomes

$$K^*\alpha = \phi \Leftrightarrow \begin{cases} R^T \beta = \phi \\ R\alpha = \beta \end{cases} \tag{7}$$

i.e., first solve the system to find vector β and then vector α. In summary, SR-KDA only needs to solve a set of regularised regression problems and there is no eigenvector computation involved. This results in great improvement of computational cost and allows to handle large kernel matrices. After obtaining α, new data items are calculated from : $f(x) = \sum_{i=1}^{m} \alpha_i K(x, x_i)$ where $K(x, x_i) = \langle \Phi(x), \Phi(x_i) \rangle \rangle$ and classification can be carried out on the projected data.

Complexity Analysis. The computation of SR-KDA involves two steps: (i) response generation which is the cost of the Gram-Schmidt method (ii) regularised regression which involves solving $(c - 1)$ linear equations using the Cholesky decomposition where c is the number of classes. As in [22], we use the term flam, a compound operation consisting of one addition and one multiplication, to measure the operation counts. The cost of the Gram-Schmidt method requires $(mc^2 - \frac{1}{3}c^3)$ flams. The Cholesky decomposition requires $\frac{1}{6}m^3$ flams and the $c-1$ linear equations can be solved with m^2c flams. Thus, the computational cost of SRKDA excluding the cost of Kernel Matrix K is $\frac{1}{6}m^3 + m^2c + mc^2 - \frac{1}{3}c^3$ which can be approximated as $\frac{1}{6}m^3 + m^2c$. Comparing to the cost of ordinary KDA $(\frac{9}{2}m^3 + m^2c)$, SR-KDA significantly reduces the dominant part and achieves an order of magnitude (27 times) speed-up.

3.2   Proposed Dimensionality Reduction Technique Using Stacked KDA

Let X denote a set of instances and let $Y = \{1, 2, ..., N\}$ be a set of labels. Given a training set $S = \{(x_1, y_1), ..., (x_m, y_m)\}$ where $x_i \in X$ is a single instance and $y_i \subseteq Y$ is the label set associated with x_i, the goal is to design a multi-label learner that predicts a set of labels from an unseen example. Motivated by the fact that there should exist some relation between the features and labels associated with the same instance, we attempt to find a lower-dimensional feature space using the concept of Stacking. Stacked generalisation or (stacking) is a type of ensemble method that uses the outputs of one classifier as inputs to another classifier [23]. In this paper, this concept is used to capture the correlation among labels during supervised dimensionality reduction.

The proposed system is demonstrated in Figure 2. It consists of a two-stage process. The objective of the first stage is to capture correlation among labels. In this stage, N projections $(\alpha_1, \alpha_2, ...\alpha_N)$ are obtained by applying SR-KDA on binary data sets since multi-label data can be divided into N binary data sets using binary relevance model. During this stage, the projections of training data are obtained using leave one out cross validation. Cross validation is necessary so that the projections of training data remain unbiased and accurately reflect the true performance in the first stage. Without cross validation, the projections obtained from training data do not reflect the true representation of test data necessary for dimensionality reduction in the second stage and to capture correlation among labels.

In the second stage, the projected scores obtained from the first stage and the original features are combined and normalised. The main objective of this stage is to extract a small number of co-related features so that the learning using nearest neighbour classifiers can be performed more efficiently. In this stage, N projections are obtained which results in N new features. Thus, the new feature vector consists of small number of features that also consider correlations among different labels. Instance-based multi-label classifier (MLkNN) is then trained to obtain the predicted label sets.

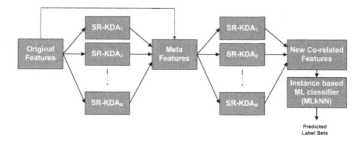

Fig. 2. Proposed system for multi-label classification using Stacked KDA

Complexity Analysis. In many practical situations the number of concepts can be very high, and the learning of independent binary classification tasks may become computationally expensive, especially if a kernel-based learning model is adopted. This problem problem can easily be solved by using SR-KDA in binary classification tasks as its time complexity scales linearly with respect to the number of concepts. The main computational part results from cholesky decomposition which is independent of the number of labels (See Figure 3). The total computational cost of SR-KDA using the Cholesky factorisation for all concepts is $\frac{1}{6}m^3 + m^2Nc$. Compared to the cost of ordinary KDA for multi-label classification, $(N \times (\frac{9}{2}m^3 + m^2c))$, SR-KDA achieves a several orders of magnitude ($27N$ times) speed-up over KDA which is massive for large scale multi-label datasets.

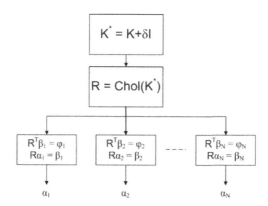

Fig. 3. Multi-label Dimensionality Reduction using SR-KDA. The most computational part i.e. Cholesky decomposition is performed only once irrespective of the number of labels.

4   Experimental Setup

4.1   Datasets

We experimented with 2 multi-label datasets. Table 1 shows certain standard statistics of these datasets. The image dataset "scene" is concerned with semantic indexing of images of still scenes [13]. The "yeast" data set contains 14 functional classes of 2417 genes of Yeast Saccharomyces Cerevisiae [15] where each gene is represented by a 103-dimensional feature vector. The publicly available feature vectors are used in this paper [1]. A short description of these datasets is presented below. All reported results are estimated from 5×2 fold cross validation and the paired t-test is then used to determine their significance under a value of 0.05.

Table 1. Standard and multi-label statistics for the data sets used in the experiments. DS = Distinct Subsets.

Datasets	Domain	Samples	Features	Labels	DS
Scene	Vision	2407	294	6	15
Yeast	Biology	2417	103	14	198

4.2   Evaluation Criterion

Multi-label classification requires different evaluation measures, compared to the traditional single-label classification. In this paper, 5 different evaluation measures are used to compare the proposed approach with the state-of-the-art multi-label classification methods. These measures include Hamming Loss, Accuracy, Classification Accuracy, and Micro/Macro F_1 and are described below.

4.3   Benchmark Methods

In order to show the effectiveness of the proposed approach, the features obtained from the proposed method (Stacked-KDA) are used to improve the performance of instance-based multi-label classifier MLkNN [4]. These classifiers are then compared with several state-of-the-art multi-label classifiers (RAkEL, CLR, ECC, BPMLL). Decision Tree (C4.5) is used as a base classifier in RAkEL [3] while Linear SVM is used as a base classifier in ECC [6] and CLR[14]. BPMLL is run with 0.05 learning rate, 100 epochs and the number of hidden units equal to 20% of the input units as recommended in [2]. For the training of these multi-label classifiers, the Mulan [24] and MEKA [2] open-source libraries in Java for multi-label classification are used. The proposed method is also compared with state-of-the-art multi-label dimensionality reduction methods MDDM (non-linear) [16] and multi-label linear discriminant analysis [25]. Since MDDM is already proved to be more effective than Principal Component Analysis (PCA), Locality Preserving Projections (LPP), and Multi-label informed latent semantic indexing MLSI [17], due to lack of space, we only show a comparison with MDDM. For KDA, KDA-stacked, and MDDM, RBF kernel is used with γ set to $1/A$ where A is the average squared euclidean distance between all elements of the kernel matrix. For KDA, regularisation parameter δ is set to default value 0.01.

---

[1] All feature vectors can be downloaded from `http://mlkd.csd.auth.fr/multilabel.html`

[2] `http://www.cs.waikato.ac.nz/~jmr30/software`

5   Results and Discussion

In this section, we first compare and discuss the results obtained using the proposed method with other dimensionality reduction techniques including LDA, multi-label LDA [25], and MDDM [16] followed by a comparison with state-of-the-art multi-label classifiers.

5.1   Comparison with State-of-the-Art Dimensionality Reduction Techniques

Tables 2-3 compare the performance of the proposed technique. It is observed from these tables that when the stacked-KDA is used as a dimensionality reduction technique, a significant improvement has been achieved in all measures and in both data sets except for hamming loss measure in Yeast when compared with MLkNN. The features obtained from the proposed method also provide better discrimination when compared with LDA, MDDM, and MLDA.

Although, it is not a general rule, the proposed method seems to have performance advantage over MLkNN (i.e. by using original features) when the number of features is high. For example, in yeast which does not suffer from the curse-of-dimensionality problem in the majority of classes, only 2.7% improvement is achieved when accuracy is used as evaluation measure. In contrast, in scene, which consists of large number of features, 16.0% increase in performance is obtained. This supported our earlier argument that the number of features can significantly degrade the performance of the nearest-neighbor rule. Tables 2 and 3 also show that impact of stacking as better discrimination is observed when compared with SR-KDA i.e. without stacking. This is due to capturing of correlation among labels during supervised dimensionality reduction.

Table 2. Comparison of proposed method with other dimensionality reduction techniques for Scene. For each evaluation criterion, ↓ indicates "the smaller the better" while ↑ indicates "the higher the better". ∗ means significantly better than all other methods except those which are marked as +.

	H-Loss ↓	Accuracy ↑	Clas-Acc ↑	Micro F$_1$ ↑	Macro F$_1$ ↑
MLkNN [4]	0.092	0.644	0.604	0.714	0.718
LDA	0.154	0.502	0.378	0.586	0.598
MLDA [25]	0.126	0.635	0.589	0.646	0.659
MDDM [16]	0.091	0.672	0.636	0.724	0.729
SR-KDA (Proposed)	0.083	0.756	0.712	0.767	0.779
Stacked SR-KDA (Proposed)	0.079∗	0.767∗	0.727∗	0.776∗	0.787∗

5.2   Comparison with other Multi-label Classifiers

Figure 5.2 shows the comparison of the proposed method with state-of-the-art multi-label classifiers. It is observed that by using the proposed technique, significant performance gains have been observed in majority of evaluation measures for both yeast and scene. Classification accuracy which is a very strict evaluation

Table 3. Comparison of proposed method with other dimensionality reduction techniques for Yeast

	H-Loss ↓	Accuracy ↑	Clas-Acc ↑	Micro F_1 ↑	Macro F_1 ↑
MLkNN [4]	**0.198***	0.499	0.165	0.633	0.352
LDA	0.276	0.393	0.031	0.544	0.393
MLDA [25]	0.263	0.440	0.157	0.565	0.405
MDDM [16]	0.199	0.503	0.177	0.635	0.356
SR-KDA (Proposed)	0.213	0.500	0.198	0.628	**0.427$^+$**
Stacked SR-KDA (Proposed)	0.202	**0.513**	**0.214***	**0.640$^+$**	0.408

measure and requires the predicted set of labels to be the exact match of the actual labels ranks first in all data sets. The accuracy which is defined by how close the actual set of labels is to the predicted set of labels is quite high using the proposed approach. Overall, there is improvement of 12.1%, in accuracy when compared with the other best multi-label classifier for scene. Figure 5.2 shows that when Hamming Loss is used as an evaluation measure which is defined as the percentage of labels that are misclassified, our proposed method and Godbole have achieved best performance.

The proposed method also compares favorably when Micro/Macro F_1 are used as evaluation measures. The two averaging procedures (Micro and Macro) bias the results differently. The micro-averaging tends to over-emphasise the performance for the largest categories, while macro-averaging over-emphasizes the performance on the smallest categories. It is observed that the performance of Macro F_1 is comparable to other ML classifiers. This can be explained by the fact that these data sets contain as few as one example for some categories, and since Macro F_1 gives an equal weight to every category, misclassification of these few samples due to nearest-neighbour rule in MLkNN can drop the performance.

5.3 Discussion

The results presented in this paper show the merit of Stacked-SRKDA as dimensionality reduction technique. The presented method avoids expensive eigen-value decomposition and thus makes it possible to use KDA in large scale experiments as a dimensionality reduction technique. For multi-label classification, the proposed method leads to an improvement in the majority of data sets and evaluation measures when compared with other multi-label classifiers that use more complex processes to model label correlations. In addition, SR-KDA also inherits the convenient property of data visualization, since it allows low dimensional views of the data vectors. This makes an intuitive analysis possible, which is helpful in many practical applications. In summary, considering both accuracy and efficiency, the proposed approach is very attractive compared to the other state-of-the-art approaches. Furthermore, the proposed method is easy-to-use and no tedious parameters tuning is required. For SR-KDA, regularisation parameter δ is set to 0.01 in all data sets and for RBF kernel, the gamma value is set to the inverse of average squared euclidean distance between all elements of the kernel matrix.

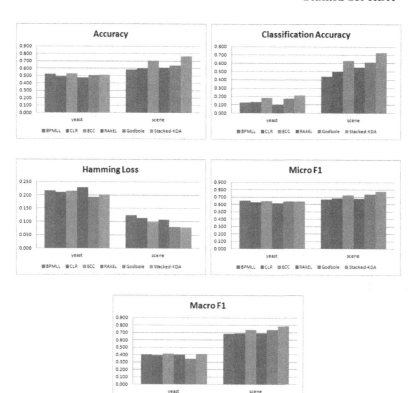

Fig. 4. Comparison of the proposed method with state-of-the-art multi-label classifiers

6   Conclusion

In this paper, we have proposed a stacked-based dimensionality reduction method to simultaneously solve the two major problems in multi-label categorisation i.e. curse-of-dimensionality and label correlation . The curse-of-dimensionality is handled by using feature extraction while the correlation problem is solved by stacked generalisation. The presented approach is then applied to scene and yeast multi-label datasets. It has been shown that the presented approach provides a very accurate and efficient solution when compared with the state-of-the-art multi-label methods.

References

1. Godbole, S., Sarawagi, S.: Discriminative methods for multi-labeled classification. In: Dai, H., Srikant, R., Zhang, C. (eds.) PAKDD 2004. LNCS (LNAI), vol. 3056, pp. 22–30. Springer, Heidelberg (2004)
2. Zhang, M.L., Zhou, Z.H.: Multilabel neural networks with applications to functional genomics and text categorization. IEEE Transactions on KDE 18(10), 1338–1351 (2006)

3. Tsoumakas, G., Katakis, I., Vlahavas, I.: Random k-labelsets for multi-label classification. IEEE Transactions on KDE 23(7), 1079–1089 (2011)
4. Zhang, M.L., Zhou, Z.H.: ML-KNN: A lazy learning approach to multi-label learning. Pattern Recognition 40(7), 2038–2048 (2007)
5. Tsoumakas, G., Katakis, I., Vlahavas, I.: Mining Multi-label Data. In: Data Mining and Knowledge Discovery Handbook, 2nd edn. Springer (2009)
6. Read, J., Pfahringer, B., Holmes, G., Frank, E.: Classifier chains for multi-label classification. In: Buntine, W., Grobelnik, M., Mladenić, D., Shawe-Taylor, J. (eds.) ECML PKDD 2009, Part II. LNCS (LNAI), vol. 5782, pp. 254–269. Springer, Heidelberg (2009)
7. Cheng, W., Hullermeier, E.: Combining instance-based learning and logistic regression for multilabel classification. Machine Learning 76(2-3), 211–225 (2009)
8. Kittler, J., Tahir, M.A., Bouridane, A.: Multilabel classification using heterogeneous ensemble of multi-label classifiers. Pattern Recognition Letters 33, 513–523 (2012)
9. Hastie, T., Tibshirani, R.: Discriminant adaptive nearest neighbor classification. PAMI 18, 607–616 (1996)
10. Cai, D., He, X., Han, J.: Efficient kernel discriminant analysis via spectral regression. In: Proc. of ICDM, Omaha, NE (October 2007)
11. Tahir, M.A., Yan, F., Barnard, M., Awais, M., Mikolajczyk, K., Kittler, J.: The university of surrey visual concept detection system at imageCLEF@ICPR: Working notes. In: Ünay, D., Çataltepe, Z., Aksoy, S. (eds.) ICPR 2010. LNCS, vol. 6388, pp. 162–170. Springer, Heidelberg (2010)
12. Pena, J.M., Zhang, M.L., Robles, V.: Feature selection for multi-label naive bayes classification. Information Sciences 179(19), 3218–3229 (2009)
13. Boutell, M.R., Luo, J., Shen, X., Brown, C.M.: Learning multi-label scene classification. Pattern Recognition 37(9), 1757–1771 (2004)
14. Furnkranz, J., Hullermeier, E., Mencía, E.L., Brinker, K.: Multilabel classification via calibrated label ranking. Machine Learning 23(2), 133–153 (2008)
15. Elisseeff, A., Weston, J.: A kernel method for multi-labelled classification. In: Advances in NIPS, vol. 14 (2002)
16. Zhang, Y., Zhi-Hua, Z.: Multi-label dimensionality reduction via dependence maximization. ACM Transactions on Knowledge Discovery from Data 4(3) (2010)
17. Kai, Y., Shipeng, Y., Volker, T.: Multi-label informed latent semantic indexing. In: Proc. of SIGIR. ACM (2005)
18. Shuiwang, J., Lei, T., Shipeng, Y., Jieping, Y.: Extracting shared subspace for multi-label classification. In: Proc. of KDD (2008)
19. Ji, S., Ye, J.: Linear dimensionality reduction for multi-label classification. In: Proc. of IJCAI (2009)
20. Fukunaga, K.: Introduction to Statistical Pattern Recognition. Academic Press (1990)
21. Baudat, G., Anouar, F.: Generalized discriminant analysis using a kernel approach. Neural Computation 2(12), 2385–2404 (2000)
22. Stewart, G.W.: Matrix Algorithms. Basic Decomposition, vol. I. SIAM (1998)
23. Wolpert, W.: Stacked generalization. Neural Networks 5(2) (1992)
24. Tsoumakas, G., Spyromitros-Xioufis, E., Vilcek, J., Vlahavas, I.: Mulan: A java library for multi-label learning. Journal of Machine Learning Research 12, 2411–2414 (2011)
25. Wang, H., Ding, C., Huang, H.: Multi-label linear discriminant analysis. In: Daniilidis, K., Maragos, P., Paragios, N. (eds.) ECCV 2010, Part VI. LNCS, vol. 6316, pp. 126–139. Springer, Heidelberg (2010)

Ensemble of Feature Chains for Anomaly Detection

Lena Tenenboim-Chekina, Lior Rokach, and Bracha Shapira

Department of Information Systems Eng. and Telekom Innovation Laboratories
at Ben-Gurion University of the Negev, Israel

Abstract. Along with recent technological advances more and more new threats and advanced cyber-attacks appear unexpectedly. Developing methods which allow for identification and defense against such unknown threats is of great importance. In this paper we propose new ensemble method (which improves over the known cross-feature analysis, CFA, technique) allowing solving anomaly detection problem in *semi-supervised* settings using well established *supervised learning algorithms*. Theoretical correctness of the proposed method is demonstrated. Empirical evaluation results on Android malware datasets demonstrate effectiveness of the proposed approach and its superiority against the original CFA detection method.

Keywords: ensemble methods, machine learning, anomaly detection, probabilistic methods, network monitoring, Android, malware.

1   Introduction

Anomaly detection refers to the problem of findings patterns in data that do not conform to expected behavior [1]. Numerous anomaly detection techniques have been developed over the years and implemented in various domains, such as fault-detection, healthcare applications, and intrusion detection systems. Different types of anomaly detection methods exist depending on application domain, problem and data types, system location etc. The prime differentiation between various techniques is according to the type of utilized detection algorithm: supervised, semi-supervised or unsupervised.

Anomaly detection methods based on *supervised* learning algorithms can be used when the training data includes instances labeled as either `normal' or `abnormal'. Then, a learning algorithm can be applied to distinguish between these types of data, and hence discover anomalies. The major shortcoming of this approach is that it requires examples of anomalous data, which often do not exist or are very scarce (leading to imbalance class distributions, a known machine learning problem). Furthermore, it requires manual labeling of instances. Moreover, it mainly allows detection of known attack patterns. In the case of *semi-supervised* problem settings, only `normal' instances are available for training, and thus, only the normal behavior can be learned and modeled. Instances that deviate from the learned 'normal' models can then be considered as anomalous. *Unsupervised* anomaly detection techniques detect anomalies in an unlabeled dataset under the assumption that the typical normal

Z.-H. Zhou, F. Roli, and J. Kittler (Eds.): MCS 2013, LNCS 7872, pp. 295–306, 2013.

instances will be much more common than abnormal ones and are looking for instances with less fit to the rest of the data.

In this paper we propose a new technique allowing solving *semi-supervised* anomaly detection problem using *supervised* learning methods for which numerous well established and quick algorithms exist. The idea of this technique was inspired by two existing methods: cross-feature analysis [2] and Classifier Chains [3]. Similarly to cross-feature analysis (CFA) the new technique estimates the probability of a feature getting a certain value, given the values of other features. The estimated probabilities of all the features are then combined into the entry vector's probability following the chaining approach, initially suggested in [3] for solving multi-label classification problems. The proposed approach is theoretically justified and hence is expected to improve the accuracy performance over the original CFA method.

We evaluate the proposed method experimentally on 15 datasets representing network behavior of five real and ten self-developed mobile malware applications and their benign versions. Specifically, we aim to detect mobile malware applications of a new recently appeared self-updating type [4]. The initial version of such malware applications which is hosted on official marketplace sites is absolutely benign and does not contain any malware by itself. Instead after the application is downloaded and installed on end user device the update procedure is initiated and the package containing actual malicious payload is downloaded from the attacker's server insensibly to the user. The update action can be scheduled for any specific or random time in the future, or even be initiated remotely by sending a command message to the devices, using, for instance, Google's push notification service. This new technique allows malware applications to stay undiscovered on the market despite recently deployed scanning service [5] designed to flag malicious applications before they can be downloaded by end users. Additionally, such a self-updating capability makes it possible for malware developers to simultaneously penetrate new threats into numerous devices. The new threats can even exploit system vulnerabilities which were unknown at the time of the development of the initial application version. Developing methods which allow for identification and defense against this new emerging malware type is of great importance. Malware activities of this type and most others regularly affect the application's network behavior and can be detected by monitoring network behavior patterns. Thus, we focus on monitoring applications network behavior and aim to detect its unexplained changes any time they occur. Evaluation results demonstrate that deviations from an application's normal behavior can be detected quickly and accurately. In addition, the proposed ensemble algorithm allows for better detection and lower false positive rates.

The rest of the paper is organized as follows. Section 2 describes the existing methods for anomaly detection. Section 3 presents our new method. Section 4 presents the conducted experiments and their results. Lastly, Section 5 concludes the paper and outlines future research.

2   Related Work

Our task relates to the family of semi-supervised anomaly detection methods which assume that the training data consists of "normal" instances only (or mainly from normal instances while abnormal instances are negligible). These types of problems can be solved, for example, with one-class support vector machines (SVMs), the local outlier factor (LOF) method or clustering based techniques [1, 6]. In the literature there are several attempts to use probabilistic methods for anomaly detection. In particular, Bayesian networks [7] and cross feature analysis [2, 6]. Generally speaking all these methods are based on the notion of likelihood. The idea is to evaluate the likelihood of getting a current behavior given the historical behavior of the application. Formally, it can be defined as follows

$$P(event\ x\ is\ normal) = P(f_1, f_2, \dots, f_L | T),$$

where $\{f_1, f_2, \dots, f_L\}$ is the features vector, L is the total number of features and T is a training set of normal events. If the estimated likelihood is relatively low then we define the current behavior as abnormal and we suspect that it might be due to malicious activity. In order to estimate the likelihood we utilize probabilistic supervised learning methods. Given a training set, these methods can induce a model that estimates the probability of a feature getting a certain value, given the values of all other features. We examine two different ways to estimate the likelihood using probabilistic supervised learning methods: original cross-feature analysis [2] (CFA) and its improved versions referred to as Feature Chains (FC) and Ensemble of Feature Chains (EFC).

2.1   Cross-Feature Analysis

The cross-feature analysis approach was initially presented by Huang *et al.* [2] and then further analyzed by Noto *et al.* [6]. Both of these works have found this approach successful and useful for anomalies detection. Differently from Huang *et al.* [2] which consider discrete features only and from Noto *et al.* [6] who mainly focus on methods for measuring and combining the contributions of each feature predictor, we developed an improved version of cross-feature analysis technique which can handle both numeric and nominal features and is suitable for running on mobile devices. As well, we precisely implemented the original CFA version for comparison purposes. In the following the general idea of cross-feature analysis and the original CFA technique are presented followed by the description of the proposed improvements.

The main assumption underling the cross-feature analysis approach is that in normal behavior patterns, strong correlations between features exist and can be used to detect deviations caused by abnormal activities. The basic idea of a cross-feature analysis method is to explore the correlation between one feature and all the other features. Formally, cross-feature analysis approach tries to solve the classification problem C_i: $\{f_1, \dots, f_{i-1}, f_{i+1}, \dots, f_L\} \rightarrow \{f_i\}$, where $\{f_1, f_2, \dots, f_L\}$ is the features vector and L is the total number of features. Such a classifier is learned for each feature i, where $i = 1, \dots L$. Thus, an ensemble of learners for each one of the features represents the model through which each features vector will be tested for "normality". The procedure utilized for online analysis of each individual instance is described below.

When a features vector representing a normal event is tested against C$_i$, there is a higher probability for the predicted value to match (for discrete features) or be very similar (for numeric features) to the observed value. However, in the case of a vector representing abnormal behavior, the probability of such a match or similarity is much lower. Thus, by applying all the features models to a tested vector and combining their results, a decision about vector normality can be derived. The more different the predictions are from the true values of the corresponding features, the more likely that the observed vector comes from a different distribution other than the training set (i.e., represents an anomaly event). A threshold distinguishing between normal and anomalous vectors can be computed by calculating the lower bound of output values from normal events.

For each predictor C_i the probability of the corresponding feature value of a vector x to come from a normal event is computed. For numeric features this probability, noted $P(f_i(x)$ is normal$)$, is calculated as the following:

$$P(f_i(x) \ is \ normal) \ = \ 1 - log_{10}(\frac{C_i(x)}{f_i(x)}) \tag{1}$$

where, $C_i(x)$ is the predicted value and $f_i(x)$ is the actual observed value. Note that if the result of the logarithm function above is greater than one, it is converted to one. Thus, the calculated probability is always in the range [0, 1]. For the nominal features the estimated probability for the true class is utilized. In [2] two options for combining predictions of all features into the final decision are examined: *Average Match Count* and *Average Probability*. The second option which computes the average of probabilities over all classifiers, as follows

$$Average \ Probability = \frac{\sum_{i=1}^{L} P(f_i | f_1, \dots, f_{i-1}, f_{i+1}, \dots, f_L)}{L} \tag{2}$$

was found by the authors as providing better performance results than *Average Match Count*. Thus, we evaluate this approach in our experiments. Events with the *Average Probability* below the threshold learned on normal data are classified as anomaly.

3   Feature Chains

In this section we describe the proposed improvement over the original CFA method which is referred to as *Feature Chains*. First we describe a single feature chain model. Then we show how to build an ensemble of feature chains.

3.1   A Single Feature Chain Model

This new method for likelihood estimation of an observed features vector $\{f_1, f_2, \dots, f_L\}$ was inspired by a successful algorithm for multi-label classification called *Classifier Chains* [3]. Classifier Chains method was recently proposed for solving multi-label classification problem using binary classifiers in a way that overcomes the label independence assumption. According to the Classifier Chains algorithm, a single binary classifier is associated with each one of the predefined labels in the dataset and all these classifiers are linked in an ordered chain. The feature space of each classifier in the chain is extended with the 0/1 label associations of all previous classifiers. Following this idea we suggest to perform the chaining on the input features (as opposed to the labels chain that was performed in the original Classifier Chains algorithm), for estimating likelihood of the observed features vector in the following way:

$$P(f_1, f_2, \dots, f_L) = \prod_{i=1}^{L} P(f_i | f_1, f_2, \dots, f_{i-1}) \tag{3}$$

Note that Equation (3) is justified by the following equivalence which can be derived by applying Bayes rule $(P(A|B) = \frac{P(A,B)}{P(B)})$ on the features conditional probabilities, calculated at the right side term of the Equation (3):

$$P(f_1) * P(f_2|f_1) * P(f_3|f_1, f_2) * \dots * P(f_L|f_1, f_2, \dots, f_{L-1}) =$$

$$= P(f_1) * \frac{P(f_2, f_1)}{P(f_1)} * \frac{P(f_3, f_2, f_1)}{P(f_2, f_1)} * \dots * \frac{P(f_L, f_{L-1}, \dots, f_1)}{P(f_{L-1}, \dots, f_1)} \tag{4}$$

By reducing the corresponding denominator and numerator in Equation (4) we will be left by only the last term $P(f_L, f_{L-1}, \dots, f_1)$ which is equivalent to the left side of Equation (3). Differently, from the Equation (2) used in the original version of Cross-Feature Analysis method the Equation (3) used by Feature Chains approach has a theoretical justification. Thus, it potentially can be used for the likelihood estimation of $P(f_1, f_2, \dots, f_L)$ and we expect for a practical improvement in the anomaly detection accuracy.

Similarly to the Cross Feature Analysis method, every conditional probability term in Equation (3) is estimated using any probabilistic supervised learning algorithms that can provide the conditional probability of the target feature given the input features. If the examined feature is nominal then classification methods (such as SVM or classification trees) should be used and if the target feature is numeric, then regression methods should be used. Certain methods such as Neural Networks and Classification and Regression tree (e.g., CART) can be used for both nominal and numeric target features.

In the case of numeric features, the distance between actual and predicted values is used as a proxy to the estimated probability for getting the actual value. Various scaling methods can be used to convert the distance to a probability. For example, one of the options is the *log distance* approach as proposed by [2] and presented in Equation (1). Another approach followed by [6] is to calculate the distance as the difference in actual and predicted values divided by the range of that feature's value (i.e., the maximum distance is 1.0). Yet, another option is to divide the difference in value by the mean of the observed values for that feature.

One concern about the proposed feature chain algorithm is the low dimensionality of the data used at the beginning of chaining process. It is true that using only a few attributes in the beginning is oversimplification. But on the other hand the probability can be estimated more accurately due to the low number of parameters to estimate.

In addition we try to overcome this oversimplification by using ensemble of Feature Chains as explained below, thus allowing averaging over numerous different chains.

3.2   Ensemble of Feature Chains

It should be noted of course that the order of features in the chain may have an effect on the model's accuracy. As with any learning algorithm, some models may overestimate the probability value and others underestimate the probability value. A convenient option for solving this issue is using an ensemble of Feature Chain models where each of the models is learned on a different chain of randomly ordered features. This approach was proved successful in the case of Classifier Chains method [3]. Additionally, it is known that ensemble methods are able to improve the prediction performance over a single classifier [8]. Note that Feature Chains method can occasionally be referred to as ensemble method because it involves multiple models. However, none of these models is capable for predicting the likelihood of the entry instance and therefore we use the term ensemble strictly in the sense of combining the final (i.e. instance-related) predictions of multiple models. Below the EFC learning process is described.

An Ensemble of Feature Chains trains m Feature Chain models $C_1, C_2, ..., C_m$. Each of the C_k is trained with a random features ordering in the chain. Hence predictions of each C_k model depend on underlying features order and are likely to be diverse in border-line cases. For combining the predictions of all the models several approaches exist. In this paper we examine a simple and popular majority voting approach, according to which binary decisions (0 – for normal and 1 – for anomalous instance) of all distinct FC models are summarized and divided by total number of models m, so that the output, referred to as *anomaly votes score* is normalized into the range of [0, 1]. A threshold is used to derive the final decision such that an instance is marked as anomaly if its *anomaly votes score* is above the defined threshold t.

4   Experimental Studies

This section presents the evaluation of the proposed detection methods. First, the data aggregated from several real and self-developed malware applications utilized in this experiment is described. Then, the system evaluation processes is described and the observed results are presented.

4.1   Evaluated Malware

For the evaluation of the proposed methods we experimented with five real and ten self-written Trojan malware. Each malware has two versions: the original benign application requesting network access permission for various purposes (such as displaying advertisements, high scores update, information or data sharing, etc.) and the repackaged version of the original application with injected malware code utilizing network communication for malicious purposes.

For the experiments with the real malware, five infected applications and their benign versions were used. The infected applications and the corresponding versions of the benign application were obtained from a repository collected by crawling the official and various alternative Android markets for over a year and a half. We used two applications injected with PJApps Trojan - Fling and CrazyFish; two applications injected with Geinimi Trojan - Squibble Lite and ShotGun; and one sample of Droid-KungFu-B malware found within the OpenSudoku game. The PJApps Trojan sends sensitive information containing the IMEI, Device ID, Line Number, Subscriber ID, and SIM serial number to a web server, and retrieves commands from a remote command and control server. Similarly, the Geinimi Trojan transmits information from the device to the server and may be instructed to perform certain actions. The Droid-KungFu-B malware targets rooted phones and requests for the root privilege; then, with or without the root privilege, it collects and steals the sensitive phone information, such as IMEI, phone model, etc. All the infected applications are mobile games which exploit network communication for certain purposes, such as online advertisements or score updates.

Malware applications with the advanced self-updating capabilities have just started to appear and there are not yet enough known real malware samples of this type. Thus, for the purposes of this paper, we have created the malware packages using two different types of self-updating behavior (type 1, entry application update and type 2, injection of compiled malicious component) and infected several open-source applications with these packages.

The utilized open-source applications are: APG, K-9 Mail, Open WordSearch, Rattlesnake Free and Ringdroid. The APG, the Android Privacy Guard application, provides OpenPGP functionalities, such as encryption and signing of emails. It uses network connections for public and secret keys management. K-9 Mail is an open source email client for Android. Open WordSearch is a game application that uses network connections to synchronize global high scores. Rattlesnake Free is also a game application utilizing network connections for online advertisements. Ringdroid is an application for recording and editing sounds and creating ringtones directly on the Android phone. It uses network connections to share ringtones and other sounds created by users. Each one of these applications was infected and evaluated using the created malware of both types. To simulate malicious behavior within the created malware, we choose to implement some simple malicious behavior patterns of known malware, such as stealing a user's contacts list, recent calls details, and user's GPS location which are sent out to a remote server.

An application infected with the malware component of type 1 will present an "update is available" notification to the user when the corresponding command is received by the device. When a user agrees to install the update, it will download a malicious version of the same application from a remote server and replace the benign version with the malicious one. At this stage the user is presented with a list of permissions to be granted to the new application version, which actually could differ from those granted to the original application. Once installed, the new malicious version will wait for an external command. When the command is received, it steals the user's contacts list and sends it to a remote server.

An application infected with malware component of type 2 will silently download a precompiled malicious payload when the corresponding command is received by the device, and then continue to load and execute malicious code without any notification to the user. The malicious payload will first steal the user's contacts list and send it to a remote server and then continue to report the user's location and recent call details to the server every specified time period (set to two minutes for our experiments).

For data aggregation all the malware applications and their benign counterparts were executed on the specially designated devices and their network behavior features were collected. The list of the utilized features is presented at Table 1. Initially, a benign version of each evaluated application was installed and executed on a device for two days. Then, it was injected\replaced by the malicious version, which was executed for at least one hour.

Table 1. The list of utilized features

No.	Feature	Brief Description
1	avg_sent_bytes	Represent the average amount of data sent or received by an
2	avg_rcvd_bytes	application at the observed time interval (of 1 min.)
3	avg_sent_pct	Represent the average portion of sent and received amount of
4	avg_rcvd_pct	data at the observed time interval (of 1 min.)
5	pct_avg_rcvd_bytes	Represents the portion of average received amount of data at the observed time interval (of 1 min.)
6	inner_ sent	Average time intervals between send\receive events occurring
7	inner_ rcvd	within the time interval of less than 30 seconds.
8	outer_ sent	Average time intervals between send\receive events occurring
9	outer_ rcvd	within the time interval above or equal to 30 seconds.

4.2   Experimental Setup

We implemented all the evaluated methods in Java using Weka [9] open source library. The Decision/Regression tree (REPTree Weka's implementation) algorithm was used as base learning algorithm for CFA and FC methods, as it can handle both nominal and numeric target features. The decision threshold values were learned on a separate set of labeled data examples during the calibration experiments. The values allowing preserving an acceptably low level of false positive alarms (below 20%) were determined as follows: 0.7 for CFA method and 0.001 for FC method.

The ensemble methods are known for their capability to improve the prediction performance over a single classifier in exchange for more computational resources and longer execution times. Thus, for the sake of a fair comparison, we compare the EFC, utilizing the REPTree method as the base learner of each single chain model, with CFA and FC methods set to use Rotation Forest [10] ensemble as their base learner. The versions of CFA and FC methods utilizing the ensemble algorithm as their base learner are denoted CFA-IE and FC-IE correspondingly (IE stands for Internal Ensemble). Rotation Forest is a recently proposed but already well-known successful method for building classifier ensemble using independently trained decision trees. The Rotation Forest was set to use the REPTree algorithm as its base learner.

The majority voting threshold of the EFC method was set to a commonly used intuitive value of 0.5. Influence of the ensemble models number on the performance accuracy was analyzed on the calibration datasets and $m=50$ was selected as always providing stable optimal results. Respectively, the number of iterations for the Rotation Forest was set to 50, also.

For learning the "normal" patterns first 30 records (not counting a few bootstrapping records) of each benign application were used. The rest of the normal data and observed traces of malicious versions were used for testing the methods detection performance. To evaluate the detection capabilities of the proposed methods the following standard measures were employed: True Positive Rate (TPR) measure (also known as detection rate), which determines the proportion of correctly detected instances relating to application's malicious behavior and the False Positive Rate (FPR) measure (also known as false alarm rate), which determines the proportion of mistakenly detected anomalies in an actually normal application behavior. Note that sometime significant deviations in normal application's behavior can be caused by changes in user's behavior. Thus a certain level of false alarms might be acceptable especially for applications with diverse network functionality.

4.3   Results

Initially we compare the new methods, FC and EFC, to the original CFA method. Results of these algorithms for all the evaluated benign\malware application pairs are presented in Table 2. The best result for each evaluation measures on a particular application dataset is marked in bold separately for the FC vs. CFA and EFC vs. CFA pair-wised comparisons.

Table 2. Malware Detection Results – New Methods vs. Original CFA

	Application name	TPR (%)			FPR (%)		
		CFA	FC	EFC	CFA	FC	EFC
Real malware	Fling	66.8	**69.0**	**67.9**	**0**	4.2	3.5
	OpenSudoku	100	100	100	0	0.0	0
	ShotGun	100	100	100	**0**	4.8	4.8
	Squibble	77.5	**95.0**	**97.5**	15.8	15.8	15.8
	Crazy Fish	90.6	**100**	**100**	**0**	7.7	7.7
Self-update 1	APG	**100**	92.3	92.3	0	0.0	0
	K-9 Mail	91.7	**100**	**100**	**0**	2.0	0
	WordSearch	100	100	100	6.3	6.3	6.3
	Rattlesnake	92.3	92.3	92.3	**8.1**	12.2	**6.5**
	Ringdroid	100	100	100	0	0.0	0
Self-update 2	APG	100	100	92.9	**0**	4.3	0
	K-9 Mail	66.7	**83.3**	**91.7**	**0**	2.9	0
	WordSearch	100	100	100	8.3	8.3	8.3
	Rattlesnake	83.3	**100**	**100**	**8**	16.0	8.0
	Ringdroid	92.3	**100**	**100**	0	0.0	0

Additionally, we perform an experiment comparing EFC with CFA and FC methods utilizing ensemble algorithm as their base learner. Results of these algorithms are presented in Table 3. The best result for each evaluation measures on a particular dataset is marked in bold separately for the FC-IE vs. CFA-IE and EFC vs. CFA-IE pair-wised comparisons.

Table 3. Malware Detection Results – Ensemble Methods

	Application	TPR (%)			FPR (%)		
		CFA-IE	FC-IE	EFC	CFA-IE	FC-IE	EFC
Real malware	Fling	65.8	**68.5**	**67.9**	**0.7**	2.8	3.5
	OpenSudoku	100	100	100	0	0	0
	ShotGun	99.3	**100**	**100**	0	7.1	4.8
	Squibble	82.5	**92.5**	**97.5**	0	15.8	15.8
	Crazy Fish	94.9	**100**	**100**	0	**0**	7.7
Self-update 1	APG	**100**	**100**	92.3	0	0	0
	K-9 Mail	100	100	100	38.8	**0**	**0**
	WordSearch	100	100	100	6.3	6.3	6.3
	Rattlesnake	92.3	**100**	92.3	36.6	**9.8**	**6.5**
	Ringdroid	80.0	**100**	**100**	16.7	**0**	**0**
Self-update 2	APG	92.9	**100**	92.9	0	0	0
	K-9 Mail	25	**83.3**	**91.7**	40	**0**	**0**
	WordSearch	100	100	100	8.3	8.3	8.3
	Rattlesnake	100	100	100	32	**9.3**	**8.0**
	Ringdroid	100	100	100	16.7	**0**	**0**

Lastly, we evaluate detection performance of EFC method with respect to the number of ensemble models. The TPR and FPR results on two of the evaluated applications are presented in Fig. 1. For all other evaluated applications similar results were observed.

Statistical significance of the difference between algorithms' results was determined by Wilcoxon signed-ranks test [11]. The exact confidence level is mentioned specifically for each comparison at the results discussion.

Generally, it can be seen that for almost all malicious applications, the high level of deviation (80-100% of anomalous instances) from the normal network behavior was detected by all the evaluated methods. Additionally, it can be seen that the FPR of all the detection algorithms is below 10% in most cases.

Comparing the performance of the proposed Feature Chain and the original CFA approaches, it can be seen that FC significantly outperforms the CFA in terms of TPR (the difference is statistically significant at 0.05 confidence level). It provides higher detection rate on 7 datasets and lower detection rate on 1 dataset only. However, at the same time it suffers from much higher false alarms rate than CFA method (the difference is also statistically significant at 0.05 confidence level). Yet, the EFC approach successfully overcomes this drawback: aggregation of numerous FC models into a composite ensemble model allows reduction of FPR to statistically indistinguishable difference comparing to CFA method, while preserving the very high detection rate on all the datasets.

We continue by comparing the performance of the EFC, FC-IE and CFA-IE methods. As can be seen (in Table 3) both EFC and FC-IE methods provide the highest detection rate on most datasets. There are slight differences between TPR results of these two methods on a few datasets, however this difference is statistically insignificant. On the other hand, it can be seen that CFA-IE method archives lower TPR on 8 and 6 datasets comparing with FC-IE and EFC methods correspondingly. In some of these cases the difference in the achieved detection rates is very meaningful and could lead to much later identification of the malware. At the same time, the CFA-IE method outperforms (in terms of TPR) the FC-IE and EFC methods in 0 and 1 cases, only, correspondingly. The difference between CFA-IE and FC-IE detection rate is statistically significant at 0.01 confidence level. Additionally, considering the FPR of the ensemble algorithms, it can be seen that CFA-IE method has unexpectedly high level (above 20%) of false alarms on several datasets, while both EFC and FC-IE preserve relatively low FPR values.

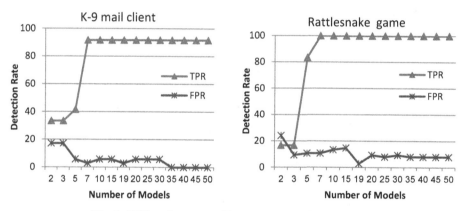

Fig. 1. EFC performance with respect to number of models

Considering, the EFC performance with respect to the number of ensemble models (as depicted in Fig. 1) it can be seen that high and stable level of True Positive Rate is achieved at relatively low number of models, $m \geq 7$. It can be seen also that larger number of models leads to lower False Positive Rate. However, for achieving a stable low FPR level, a larger number of models, regularly $m \geq 30$, is needed.

Summarizing the above comparison we conclude that the proposed Feature Chains technique allows for significant improvement of the detection performance over the original CFA method. However, it suffers from a higher False Positives Rate. At the same time, the two evaluated ensemble versions of the new Feature Chains methods, EFC and FC-IE, allow for significant reduction in the false alarms rate (the difference is statistically significant at 0.05 confidence level), while preserving the high True Positive Rate. Hence, the results justify using the proposed ensemble methods, FC-IE or EFC, for anomaly detection.

5   Summary and Conclusions

This paper presented a novel probabilistic method for solving semi-supervised anomaly detection problems and its ensemble version. The new method is based on the known cross-feature analysis and classifier chaining methods. It can handle numeric and nominal features and is suitable for running on mobile devices. The presented method can be used for solving various semi-supervised anomaly detection problems. Theoretical correctness of the proposed method was demonstrated.

Empirical evaluation of the proposed methods on the variety of datasets demonstrated effectiveness of the proposed approach for the defined problem: a high TPR along with low FPR could be achieved. The proposed Ensemble of Feature Chains and Feature Chains using internal ensemble proved superior to the original CFA method and its ensemble version.

Among our future research directions are evaluation of the present methods on more datasets from different domains and comparison with other anomaly detection methods.

References

1. Chandola, V., Banerjee, A., Kumar, V.: Anomaly detection: A survey. ACM Comput. Surv. 41(3), 1–58 (2009)
2. Huang, Y.A., Fan, W., Lee, W., Yu, P.S.: Cross-feature analysis for detecting ad-hoc routing anomalies. In: IEEE 23rd Int. Conf. on Distributed Computing Systems, pp. 478–487 (2003)
3. Read, J., Pfahringer, B., Holmes, G., Frank, E.: Classifier Chains for Multi-label Classification. In: Buntine, W., Grobelnik, M., Mladenić, D., Shawe-Taylor, J. (eds.) ECML PKDD 2009, Part II. LNCS, vol. 5782, pp. 254–269. Springer, Heidelberg (2009)
4. Symantec blog: http://www.symantec.com/connect/blogs/androiddropdialer-identified-google-play
5. Google mobile blog, android and security: http://googlemobile.blogspot.co.il/2012/02/android-and-security.html
6. Noto, K., Brodley, C., Slonim, D.: Anomaly detection using an ensemble of feature models. In: Proc. of the 10th IEEE International Conf. on Data Mining, pp. 953–958 (2010)
7. Ye, N., Xu, M., Emran, S.M.: Probabilistic networks with undirected links for anomaly detection. In: Proceedings of the IEEE Systems, Man, and Cybernetics Information Assurance and Security Workshop, West Point, NY, pp. 175–179 (2000)
8. Rokach, L., Maimon, O.: Ensemble Methods for Classifiers. In: Data Mining and Knowledge Discovery Handbook. Springer US (2005)
9. Weka 3: Data Mining Software in Java, http://www.cs.waikato.ac.nz/ml/weka/
10. Rodriguez, J.J., Kuncheva, L.I., Alonso, C.J.: Rotation Forest: A New Classifier Ensemble Method. IEEE Transactions on Pattern Analysis and Machine Intelligence 28(10), 1619–1630 (2006)
11. Demsar, J.: Statistical comparisons of classifiers over multiple data sets. Journal of Machine Learning Research 7, 1–30 (2006)

Soft-Voting Clustering Ensemble

Haishen Wang, Yan Yang*, Hongjun Wang, and Dahai Chen

School of Information Science & Technology
Southwest Jiaotong University
Chengdu, 610031, P.R. China
{hshwang,dahaichen}@my.swjtu.edu.cn,
{yyang,wanghongjun}@swjtu.edu.cn

Abstract. Clustering ensemble is a framework for combining multiple based clustering results of a set of objects without accessing the original feature of the objects. The majority voting method is widely used in clustering ensemble because of its simplicity, robustness and stability. In general, the existing voting methods only accept hard clustering results as input. In this paper we propose a new algorithm, Soft-Voting Clustering Ensemble (SVCE), which has better flexibility and generalization. The theory of SVCE is illustrated and the algorithm of SVCE is stated in detail firstly. Then 15 UCI datasets are used for the experiment and the results show that the proposed method has a better performance than state of the art ensemble methods in most cases, such as Majority Voting, Weighted Majority Voting, CSPA, MCLA, HGPA.

Keywords: Clustering ensemble, Majority voting, Soft-Voting Clustering Ensemble.

1 Introduction

Clustering ensemble [1] is widely used in data mining, information retrieval, knowledge-reuse [2], multiviews clustering [3], distributed computing [4] and other fields. Clustering ensemble is a framework for combining multiple based clustering results of a set of objects without accessing the original feature of the objects. In general, clustering ensemble can be considered as two step processes: generating based clusterings and consensus function [5]. In the first step, all objects are partitioned into several separate clusters, known as based clustering results, which are generated either by different algorithms or by the same algorithm with different initialization. In the second step, all based clustering results are combined by a consensus function to get the final result.

Depending on the way that objects are assigned to clusters, clustering methods are divided into two kinds: hard clustering and soft clustering. In hard clustering, the degree of membership between an object and a cluster is 0 or 1, showing that every object only belongs to one cluster. In soft clustering, every object belongs to any cluster with different degrees of membership. Soft clusterings output a

---

* Corresponding author.

Z.-H. Zhou, F. Roli, and J. Kittler (Eds.): MCS 2013, LNCS 7872, pp. 307–318, 2013.
© Springer-Verlag Berlin Heidelberg 2013

matrix of membership degrees instead of a label vector for all objects; often these degrees of a object sum up to one.

Clustering ensemble can go beyond what is typically achieved by a single clustering algorithm in several respects [6]. First, ensemble model improves robustness and has better average performance across the domains and datasets. Second, ensemble model has novelty that finding a combined solution unattainable by any single clustering algorithm. Third, ensemble model provides a higher stability and confidence estimation that clustering solutions with lower sensitivity to noise, outliers or sampling variations and clustering uncertainty can be assessed from ensemble distributions. Forth, ensemble model has a better parallelization and scalability, because of parallel clustering of data subsets with subsequent combination of results and the ability to integrate solutions from multiple distributed sources of data or attributes.

There are five popular consensus functions used in clustering ensemble [7], for example co-association method, majority voting method, hypergraph methods, mutual information method and mixture model method. In co-association method, co-association matrix is used as similarity matrix, and one can use numerous similarity-based clustering algorithms by applying them to the co-association matrix. In major voting method, the first step is to solve a label correspondence problem and the second step is to use a simple voting to assign objects in clusters. In hypergraph method, all clustering results are represented as hyperedges on a graph with N vertices, each hyperedge describes a set of objects belonging to the same cluster and a consensus function is formulated as a solution to hypergraph partitioning problem. In mutual information approach, the objective function for a clustering ensemble can be formulated as the mutual information between the empirical probability distribution of labels in the consensus partition. In mixture model method, the consensus clustering is derived from a solution of the maximum likelihood problem for a finite mixture model of the ensemble of partitions.

In terms of soft clusterings ensemble, Punera et al. proposed several consensus algorithms that work on soft clusterings and experimented with many real-life datasets to empirically show that using soft clusterings as input does offer significant advantages [8]. Yang et al. stated a method to combine soft clusterings based on fuzzy similarity measure and showed promising results compared to general clustering ensemble methods based on crisp clusterings [9]. Zhai et al. presented a dual boosting for fuzzy clustering ensemble, which is efficient in stability and accuracy [10].

Strehl and Ghosh first proposed clustering ensemble in 2002 and they proposed three graph-theoretic approaches for finding the consensus clustering [1]. Fred et al. explored evidence accumulation for combining the results of multiple clusterings [11] and showed that the evidence accumulation clustering performs better compared to other combination methods [12]. Zhang et al. stated a new algorithm called spectral clustering ensemble, which provides necessary diversity and high quality of component learners, and overcomes the shortcomings of spectral clustering [13]. Fern et al. defined the clustering ensemble selection problem, and achieved a better performance by selecting a subset of based clustering results to form a small ensemble [14]. Domeniconi et al. addressed the problem of combining multiple weighted clusters, making use of the weight vectors associated with the clusters [15]. Ayad et al. presented a more general formulation of the voting problem as a regression problem with multiple-response and multiple-input variables [16]. Wang et al. proposed Bayesian cluster ensembles, which is mixed-membership model for learning clustering ensemble and is applicable to all the primary variants of the problem [17]. Yang et al. illustrated the novel semi-supervised consensus clustering ensemble algorithm based on multi-ant colonies, which incorporates pairwise constraints as well [18].

Most of popular ensemble approaches only accept hard clustering results as input. In order to form an ensemble of soft clustering using the methods mentioned above, we have to "harden" the results. This leads to the loss of the valuable information produced by soft clustering. In this paper we propose SVCE algorithm that accept the results of soft clustering and use the information adequately.

Compared with state of the art ensemble methods, there are two advantages of SVCE.

(1) SVCE has better flexibility and generalization. It not only can do an ensemble in its own way, but also can accept hard clustering results. In addition, membership degree of SVCE is [0,1] instead of {0,1}.

(2) SVCE has higher robustness. By making full use of the valuable information produced by soft clustering, used as input, SVCE obtains a better performance.

In the rest of this paper, related work on majority voting is described in Section 2. The theory and algorithm of SVCE are introduced formally in Section 3. The experimental setup and results comparison with various ensemble methods are showed in Section 4. Finally, this paper is concluded in Section 5.

2   Related Work

2.1   Fuzzy C-Means

Fuzzy c-means (FCM) is a popular soft clustering method, which is proposed by Dunn [19] and expanded by Bezdek [20]. FCM shows how to group objects into a special number of different clusters. The purpose of this method is to find the best c-partitions at the same time every object can belong to several clusters with different membership degrees.

The objective function of FCM Algorithm is

$$J(U, C_1, ..., C_c) = \sum_{i=1}^{c} J_i = \sum_{i=1}^{c} \sum_{j}^{n} u_{ij}^{q} d_{ij}^{2} \tag{1}$$

where u_{ij} is the membership degree of object j and cluster i, and it is between 0 and 1, and $\sum_{i=1}^{c} u_{ij} = 1$; C_i is the i-th cluster center, c is the number of all clusters; $d_{ij} = ||C_i - X_j||$ is the Euclidean distance between object j and cluster

i; The q is weight exponent, which can be any real number greater than 1. The number q is larger, the fuzziness is greater. So q can indicate the grade of fuzzy.

However this objective function is too complex to use for computation. So two necessary conditions are used to replace it, which can make the objective function get smallest value. They are as follows,

$$C_i = \frac{\sum_{j=1}^n u_{ij}^q x_j}{\sum_{j=1}^n u_{ij}^q}, \tag{2}$$

$$u_{ij} = \frac{1}{\sum_{k=1}^c (\frac{d_{ij}}{d_{kj}})^{2/(q-1)}}. \tag{3}$$

Then we do a iteration by using this two equation and update the membership degree u_{ij} and the cluster center C_i. This iteration can be ended when $\|C_{t+1} - C_t\| < \varepsilon$, where ε is a termination criterion, which is between 0 and 1, and it is very small, and t is the iteration steps.

2.2   Majority Voting

Majority voting [21] is a simple and intuitive ensemble technique. Essentially, the ensemble chooses the cluster for object which is chosen by the majority of based clustering results.

Let us define the decision of the m-th clustering H_m as $h_{m,i} \in \{0,1\}$ ($m = 1, 2, 3, ..., M, i = 1, 2, 3, ..., C$), where M is the number of based clustering results and C is the number of clusters. If m-th clustering result chooses cluster i, then $h_{m,i} = 1$, and 0 otherwise. The majority voting result in an ensemble decision for cluster k if

$$\sum_{m=1}^M h_{m,k} = \max_i \sum_{m=1}^M h_{m,i} . \tag{4}$$

There are many works on majority voting for its simpleness and stability. Breiman developed the bagging predictors, which is a method to generate multiple versions of a predictor and use these to get an aggregated predictor. The aggregation averages over the versions when predicting a numerical outcome and does a majority voting when predicting a cluster [22]. Dietterich proposed the Bayesian voting as follows,

$$P(H_m|S) \propto P(S|H_m) \times P(H_m) \tag{5}$$

and the Bayesian voting primarily addresses the statistical component of ensembles [23]. Stepenosky et al. used majority voting and decision templates to form an ensemble and explored the feasibility of a diagnostic tool for early diagnosis of Alzheimers disease [24]. Zhou et al. developed four methods of voting, weighted-voting, selective voting and selective weighted-voting to do the ensemble and improved the clustering performance [25]. Fu et al. proposed a fuzzy majority voting scheme and offered a decision model based on fuzzy set theory for fuzzy clustering ensemble [26]. Tumer et al. developed the voting active clusters for combining multiple based clusterings into a single unified "ensemble" clustering which is robust against missing data and does not need to collect all objects in one central location [27]. Rokach et al. made a summary that all existing ensemble techniques, and voting has been mentioned as an important ensemble method [28]. Toman et al. proposed a method using a generalization of weighted majority voting scheme to locate the optic disc in retinal images automatically and achieved better performance [29].

3   Soft-Voting Clustering Ensemble

In this section we first introduce Soft-Voting from input, output and other aspects. Then we describe the function of SVCE and make a comparison between SVCE and Majority Voting Clustering Ensemble (MVCE) in detail.

3.1   Soft-Voting

An object is partitioned into the cluster supported by most partitions in majority voting. Similar to majority voting, Soft-Voting assigns an object to the cluster with the highest membership degree.

As mentioned above, the input of Soft-Voting is different from majority voting. Now it is a matrix of membership degrees instead of a group of labels. If an object has a group of membership degrees just like (0.7, 0.2, 0.1), it means that the possibility that this object belongs to the first cluster is 0.7, the second is 0.2, and the third is 0.1. So this object most possibly belongs to the first cluster. If a voting is needed, this object is completely assigned to the first cluster, and the group of membership degrees becomes (1, 0, 0). If there are many groups of membership degrees for one object, all groups have to be combined before voting.

Suppose that all the based clustering results of one dataset are independent. If the probability that object j belongs to cluster c in partition m_1 is $P(c-m_1)$ and the probability that object j belongs to cluster c in partition m_2 is $P(c-m_2)$, the probability that object j belongs to cluster c in both partition m_1 and partition m_2 at the same time is $P(c-m_1m_2)$, given by

$$P(c - m_1m_2) = P(c - m_1) \times P(c - m_2). \tag{6}$$

In this processing, we use the product of two entities in same position as a new membership degree. For example, there are two partitions of 3 objects,

$$m1 = \begin{vmatrix} 0.8\ 0.1\ 0.1 \\ 0.2\ 0.7\ 0.1 \\ 0.3\ 0.2\ 0.5 \end{vmatrix}, m2 = \begin{vmatrix} 0.7\ 0.1\ 0.2 \\ 0.1\ 0.8\ 0.1 \\ 0.2\ 0.2\ 0.6 \end{vmatrix}.$$

As for the object 1, three values (0.8×0.7, 0.1×0.1, 0.1×0.2) are used as new membership degrees, because they show the probability that the two partitions support this object assigned into the same cluster.

The result of the two partitions combined by Soft-Voting is

$$
m1 \times m2 = \begin{vmatrix} 0.8 \times 0.7 \ 0.1 \times 0.1 \ 0.1 \times 0.2 \\ 0.2 \times 0.1 \ 0.7 \times 0.8 \ 0.1 \times 0.1 \\ 0.3 \times 0.2 \ 0.2 \times 0.2 \ 0.5 \times 0.6 \end{vmatrix} = \begin{vmatrix} 0.56 \ 0.01 \ 0.02 \\ 0.02 \ 0.56 \ 0.01 \\ 0.06 \ 0.04 \ 0.30 \end{vmatrix}.
$$

The result needs to be normalized. As for the object 1, the sum of probability (0.56, 0.01, 0.02) is not equal to 1.0, the solution is that we use $0.56/(0.56+0.01+0.02)$ to replace 0.56, $0.01/(0.56+0.01+0.02)$ to replace 0.01, $0.02/(0.56+0.01+0.02)$ to replace 0.02. Then the sum of membership degrees of the first object is 1.0. After having the normalized result, we choose cluster for every object by Soft-Voting.

3.2 Clustering Ensemble Based on Soft-Voting

For SVCE, we repeatedly use formula (6) to do an ensemble among all based clustering results, and use Soft-Voting to choose cluster for every object.

Let us assume that there are M partitions of a given dataset $X = \{x_1, x_2, ..., x_N\}$ into C clusters. Each of these M partitions is represented by an $N \times C$ membership matrix U^m $(m = 1, ..., M)$. u_{ij}^m is the element of U_{ij}^m and means the degree of membership of x_j to the i-th cluster of the m-th partition.

The purpose is to choose the cluster for every object with the highest membership degree. Having M partitions, first to all, an ensemble is formed among the M partitions as formula (6). In the process of ensemble, the u_{ij} is normalized whenever two partitions are combined.

The main technique of SVCE is showed by formula(7),

$$
\hat{x}_j = arg \ \max_i \prod_{m=1}^{M} u_{ij}^m \qquad (i = 1, 2, ..., C, j = 1, 2, 3, ..., N) \tag{7}
$$

where $\hat{x}_j$ means the cluster label of object x_j.

In this function, we note that u_{ij} belongs to $[0,1]$, which is produced by soft clustering. The hard clustering result is a special case of soft clustering result. SVCE use soft clustering result as input, while the input of MVCE is only hard clustering, then SVCE is more flexible than MVCE. In addition, SVCE also accepts the input of MVCE and gets a similar ensemble result.

Because majority voting can not use the results of soft clustering as input, we do a improvement on majority voting, which makes it accept soft clustering results and use the matrix of membership degrees as input.

The following example illustrates that SVCE can get a better result as compared to modified MVCE in general. Let us assume that for object x_j there are two partitions: $p1 = (u_1, u_2, u_3)$, $p2 = (v_1, v_2, v_3)$, where $u_1 > u_2 > u_3$ and $v_1 > v_2 > v_3$. The group of membership degrees produced by modified MVCE is $MV = (m_1, m_2, m_3)$, and the group of membership degrees produced by SVCE is $SV = (s_1, s_2, s_3)$. So this object most probably belongs to the first cluster. Obviously, u_1 and v_1 are more close to 1, the result is more accurate.

Let us define that a_1, a_2, a_3, a_4 respectively represents the accurate ratio of partition $p1$, partition $p2$, modified MVCE and SVCE. And accurate ratios of these four partitions are calculated as follows,

$$a_1 = \frac{u_1}{u_1+u_2+u_3} = \frac{u_1}{1} = u_1,$$

$$a_2 = \frac{v_1}{v_1+v_2+v_3} = \frac{v_1}{1} = v_1,$$

$$a_3 = \frac{m_1}{m_1+m_2+m_3} = \frac{u_1+v_1}{(u_1+v_1)+(u_2+v_2)+(u_3+v_3)} = \frac{u_1+v_1}{1+1} = \frac{u_1+v_1}{2},$$

$$a_4 = \frac{u_1 \times v_1}{(u_1 \times v_1)+(u_2 \times v_2)+(u_3 \times v_3)} = \frac{u_1}{(u_1 \times \frac{v_1}{v_1})+(u_2 \times \frac{v_2}{v_1})+(u_3 \times \frac{v_3}{v_1})} > \frac{u_1}{u_1+u_2+u_3} = a_1,$$

$$a_4 = \frac{u_1 \times v_1}{(u_1 \times v_1)+(u_2 \times v_2)+(u_3 \times v_3)} = \frac{v_1}{(v_1 \times \frac{u_1}{u_1})+(v_2 \times \frac{u_2}{u_1})+(v_3 \times \frac{u_3}{u_1})} > \frac{v_1}{v_1+v_2+v_3} = a_2.$$

a_3 is between a_1 and a_2, because a_3 is the average of a_1 and a_2. a_4 is greater than both a_1 and a_2, so a_4 is greater than a_3, that is to say, SVCE has a higher accurate ratio than modified MVCE when $u_1 > u_2 > u_3$ and $v_1 > v_2 > v_3$. The results of the based clustering are considered to meet this precondition. Though the results are produced by different methods, they have the similarity, which can be illustrated by experiment. So SVCE has a higher robustness and better average performance than modified MVCE.

The steps of SVCE is illustrated in Algorithm 1.

Algorithm 1: SVCE
Input: X: a dataset has N objects
       C: the number of clusters in one FCM
       M: the number of based clustering results
Output: the cluster labels of N objects
Steps:
    (1) Use FCM algorithm M times to obtain based clustering
        results:
        a.Get dataset and find the number of clusters;
        b.Obtain M based clustering results by using FCM.
    (2) Choose a clustering result as standard, relabeling
        others. Then combine all results and find the cluster
        label for every object.
        a.Randomly choose a clustering result as the standard,
          then process other (M-1) results and make all
          clustering results have the identical labels ;
        b.From 1 to M, do the multiplication as formula(6),
          normalizing and saving based clustering results;
        c.Refering to the final result, assign every object
          into the cluster which this object belongs to with
          the highest membership degree.

4   Empirical Study

All ensemble methods, including CSPA, MCLA, HGPA, SVCE, MVCE, WMVCE(Weighted Majority Voting Clustering Ensemble [25]), are applied to 15 datasets and F-measure is used to evaluate the results. In this section, the experiments and results are showed in detail.

4.1   Experiments

15 datasets are used for this experiment, which are summarized in Table 1. 14 datasets are from UCI Data Repository and the remaining one is an artificial dataset.

Table 1. The instances, classes, features and source of each dataset

Dataset	Instances	Classes	Features	Source
2d4c	200	4	2	Artificial
Balance-scale	625	3	4	UCI
Contraceptive-method-choice	1473	3	9	UCI
Diabetes	768	2	8	UCI
Glass	214	6	9	UCI
Heart-statlog	270	2	13	UCI
Ionosphere	351	2	34	UCI
Iris	150	3	4	UCI
Liver-disorders	345	2	7	UCI
Pima-Indians-diabetes	768	2	8	UCI
Segment	210	7	19	UCI
Sonar	208	2	60	UCI
SPECTF-heart	267	2	44	UCI
Vehicle	846	4	18	UCI
Wine	178	3	14	UCI

In this experiment, FCM algorithm is used to produce based clustering results. For FCM, the weight exponent q is set as 2, membership degrees u_{ij} are initialized by random number between 0 and 1, termination criterion ε is set as 0.0001. Different based clusterings are generated by changing the initialization of membership degrees u_{ij}, which are used as input of SVCE. Then a final clustering result is obtained by SVCE. Because other methods do not accept soft clustering results, all these based clustering results are "hardened". The hardened results are sent to other ensemble methods, and the final results are obtained respectively. Finally, a comparison is made among all results.

In the process of voting-based cluster ensemble methods, unifying clusters label is crucial. In order to achieve the most consistent labeling of clusters in a partition, we must solve an assignment problem equivalent to maximum weight bipartite matching problem. Equivalent matching problem is constructed from a contingency table between two partitions. A contingency matrix contains a number of cluster label co-occurrences counted for two partitions of the same set of objects [30]. We use the method mentioned in [30] to achieve relabeling.

F-measure is used to evaluate the results, and the best method is found by comparing the values of F-measure. The F-measure is an external evaluation method, which combines the ideas of precision and recall. The precision and recall of cluster t respect to class s are defined as

$$precision(s,t) = \frac{n_{st}}{n_s}, \tag{8}$$

$$recall(s,t) = \frac{n_{st}}{n_t}, \qquad (9)$$

where N_{st} is the number of class s in cluster t, N_t is the number of members of cluster t and N_s is the number of members of class s.

The f-measure of class s is then given as

$$F(s) = \frac{2 \times precision(s,t) \times recall(s,t)}{precision(s,t) + recall(s,t)}. \qquad (10)$$

The F-measure value is larger, the result is better.

4.2 Results

In order to get more exact results, the experiment is conducted 10 times repeatedly with same conditions. The comparison among the 6 ensemble methods is made from two aspects: the average value of F-measure and the variance value of F-measure. The average value reflects the performance in general and the variance value reflects the degree of stability.

The comparison among 6 ensemble methods is made from the aspect of the average F-measure value. Then we compare 6 ensemble methods in the aspect of variance F-measure value. Table 2 shows all the results.

Table 2. The performance of 6 methods on 15 datasets (boldface is the highest)

Dataset	SVCE	MVCE	WMVCE	CSPA	HGPA	MCLA
2d4c	**0.976**±0.0004	0.967±0.0006	0.971±0.0003	0.957±0.0091	0.302±0.0426	0.966±0.0006
Balance-scale	0.589±0.1026	**0.680**±0.1630	**0.680**±0.0940	0.570±0.7735	0.414±0.0156	0.564±0.8226
Contraceptive-method-choice	**0.402**±0.0000	0.372±0.0003	0.385±0.0001	0.392±0.0021	0.349±0.0413	0.395±0.0000
Diabetes	**0.633**±0.0001	0.606±0.0002	0.620±0.0002	0.557±0.0269	0.514±0.0000	0.616±0.0001
Glass	**0.423**±0.0002	0.391±0.0007	0.420±0.0013	0.417±0.0341	0.357±0.0250	0.382±0.0096
Heart-statlog	**0.590**±0.0004	0.541±0.0003	0.569±0.0005	0.566±0.0000	0.516±0.0000	0.575±0.0002
Ionosphere	**0.713**±0.0001	0.658±0.0002	0.653±0.0001	0.668±0.0000	0.590±0.0000	0.706±0.0002
Iris	**0.884**±0.0019	0.748±0.0051	0.877±0.0029	0.828±0.0259	0.643±0.0125	0.874±0.0014
Liver-disorders	0.357±0.0038	0.345±0.0072	0.420±0.0056	0.513±0.0008	**0.520**±0.0121	0.349±0.0002
Pima-Indians-diabetes	**0.635**±0.0015	0.617±0.0007	0.626±0.0011	0.565±0.0160	0.508±0.0000	0.606±0.0001
Segment	0.565±0.0089	0.562±0.0075	0.565±0.0062	0.540±0.0564	0.490±0.0136	**0.571**±0.0788
Sonar	0.549±0.0015	0.516±0.0073	0.523±0.0047	0.535±0.0000	0.504±0.0000	**0.552**±0.0067
SPECTF-heart	0.661±0.0012	0.639±0.0047	**0.671**±0.0039	0.575±0.0084	0.617±0.0538	0.641±0.0168
Vehicle	**0.431**±0.0037	0.407±0.0001	0.421±0.0001	0.380±0.0019	0.278±0.0007	0.379±0.0076
Wine	**0.699**±0.0027	0.681±0.0098	0.690±0.0043	0.671±0.0113	0.533±0.0101	0.674±0.0039
Best	10	1	2	0	1	2
Worst	0	2	0	1	12	0

Table 2 shows that SVCE has 10 highest F-measure values in the 15 datasets and no worst value. Thus SVCE has a better performance in general among 6 methods in the aspect of average value. From Table 2, we also note that the variance value of SVCE is smaller, which indicates stability of SVCE is higher. So the performance of SVCE is the best in 6 methods mentioned above.

From Table 2, we obtain the Fig. 1, which more clearly shows the average performance of 6 ensemble methods.

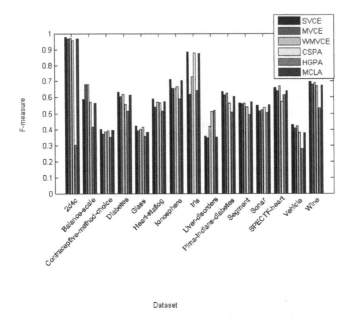

Fig. 1. The average F-measure values of 6 methods on 15 datasets

All the results show that SVCE is a better ensemble method in general, because SVCE uses the soft clustering results as input. By making full use of the valuable information, it obtains a better ensemble result.

5   Conclusion

In this paper, we propose a new ensemble method - SVCE, which accept the results of soft clustering. SVCE is more generalized that the membership degree SVCE processed is [0, 1], while the membership degree of other ensemble methods is just a binary strategy {0, 1}. SVCE is flexible, because it also can use the results of hard clustering as input. The experiment shows that SVCE get a better ensemble result than other ensemble methods.

In this experiment, it also has been proved soft clustering method is more suitable to represent an object. In future work, we will try to study more on soft clustering algorithms and use them to produce based clustering results with diversity. We also will try to extend this mode to a semi-supervised model, by adding a little labeled data in the process of ensemble.

Acknowledgements. This work is supported by the National Science Foundation of China (Nos. 61170111, 61003142, 611734002 and 61262058), the Fundamental Research Funds for the Central Universities(Nos. SWJTU11ZT08 and SWJTU12CX092) and the Research Fund of Traction Power State Key Laboratory, Southwest Jiaotong University (No. 2012TPL_T15).

References

1. Strel, A., Ghosh, J.: Cluster Ensembles - A Knowledge Reuse Framework for Combining Multiple Partitions. Journal of Machine Learning Research 3(3), 583–617 (2002)
2. Ghosh, J., Strel, A., Merugu, S.: A Consensus Framework of Integrating Distributed Clustering under Limited Knowledge Sharing. In: NSF Workshop on Next Generation Data Mining, pp. 99–108 (2002)
3. Kreiger, A.M., Green, P.: A Generalized Rand-Index Method for Consensus Clustering of Separate Partitions of the Same Data Base. Journal of Classification 16(1), 63–89 (1999)
4. Merugu, S., Ghosh, J.: Privacy-Preserving Distributed Clustering Using Generative Models. In: 3rd IEEE International Conference on Data Mining, pp. 211–218 (2003)
5. Vega-pons, S., Ruiz, J.: A Survey of Clustering Ensemble Algorithms. Journal of Pattern Recognition and Artificial Intelligence 25(3), 337–372 (2011)
6. Ghosh, J., Acharya, A.: Cluster Ensembles. WIREs Data Mining and Knowledge Discovery 1(4), 305–315 (2011)
7. Topchy, A., Jain, A.K., Punch, W.: Clustering Ensembles: Models of Consensus and Weak Partitions. IEEE Transactions on Pattern Analysis and Machine Intelligence 27(12), 1866–1881 (2005)
8. Punera, K., Ghosh, J.: Consensus Based Ensembles of Soft Clusterings. Journal of Applied Artificial Intelligence 22(7-8), 780–810 (2008)
9. Yang, L., Lv, H., Wang, W.: Soft Cluster Ensemble Based on Fuzzy Similarity Measure. In: IMACS Multiconference on Computational Engineering in Systems Applications, vol. 2, pp. 1994–1997 (2006)
10. Zhai, S., Luo, B., Guo, Y.: A Fuzzy Clustering Ensemble Based on Dual Boosting. In: International Conference on Fuzzy Systems and Knowledge Discovery, pp. 237–241 (2007)
11. Fred, A.L.N., Jain, A.K.: Data Clustering Using Evideence Accumulation. In: 16th International Conference on Pattern Recognition, vol. 4, pp. 276–280 (2002)
12. Fred, A.L.N., Jain, A.K.: Combining Multiple Clusterings Using Evidence Accumulation. IEEE Transactions on Pattern Analysis and Machine Intelligence 27(6), 835–850 (2005)
13. Zhang, X., Jiao, L., Liu, F., Bo, L., Gong, M.: Spectral Clustering Ensemble Applied to SAR Image Segmentation. IEEE Transactions on Geoscience and Remote Sensing 46(7), 2126–2136 (2008)
14. Fern, X.Z., Lin, W.: Cluster Ensemble Selection. In: 8th SIAM International Conference on Data Mining. Applied Mathematics 130, vol. 2, pp. 787–797 (2008)
15. Domeniconi, C., Al-Razgan, M.: Weighted Cluster Ensembles: Methods and Analysis. ACM Transactions on Knowledge Discovery from Data 2(4), 1–40 (2009)
16. Ayad, H.G., Kamel, M.S.: On Voting-Based Consensus of Cluster Ensembles. Pattern Recognition 43(5), 1943–1953 (2010)
17. Wang, H., Shan, H., Banerjee, A.: Bayesian Cluster Ensembles. Statistical Analysis and Data Mining 4(1), 54–70 (2011)
18. Yang, Y., Wang, H., Lin, C., Zhang, J.: Semi-supervised Clustering Ensemble Based on Multi-ant Colonies Algorithm. In: Li, T., Nguyen, H.S., Wang, G., Grzymala-Busse, J., Janicki, R., Hassanien, A.E., Yu, H. (eds.) RSKT 2012. LNCS (LNAI), vol. 7414, pp. 302–309. Springer, Heidelberg (2012)
19. Dunn, J.C.: A fuzzy relative of the isodata process and its use in detecting compact well-separated clusters. Journal of Cybernetics 3(3), 32–57 (1973)

20. Bezdek, J.C.: Pattern Recognition with Fuzzy Objective Function Algorithm. Plenum Press, New York (1981)
21. Dimitriadou, E., Weingessel, A., Hornik, K.: Voting-Merging: An Ensemble Method for Clustering. In: Dorffner, G., Bischof, H., Hornik, K. (eds.) ICANN 2001. LNCS, vol. 2130, pp. 217–224. Springer, Heidelberg (2001)
22. Breiman, L.: Bagging Predictors. Machine Learning 24(2), 123–140 (1996)
23. Dietterich, T.G.: Ensemble Methods in Machine Learning. In: Kittler, J., Roli, F. (eds.) MCS 2000. LNCS, vol. 1857, pp. 1–15. Springer, Heidelberg (2000)
24. Stepenosky, N., Green, D., Kounios, J., Clark, C.M., Polikar, R.: Majority Vote and Decision Template Based Ensemble Classifiers Trained on Event Related Potentials for Early Diagnosis of Alzheimer's Disease. In: IEEE International Conference on Acoustics, Speech and Signal Processing, pp. 901–904 (2006)
25. Zhou, Z., Tang, W.: Clusterer Ensemble. Knowledge-Based System 19(1), 77–83 (2006)
26. Fu, Y., Yang, Y., Liu, Y.: A Decision Model for Fuzzy Clustering Ensemble. In: International Conference on Intelligent Systems and Knowledge Engineering, pp. 753–758 (2007)
27. Tumer, K., Agogino, A.K.: Ensemble Clustering with Voting Active Clusters. Pattern Recognition Letters 29(14), 1947–1953 (2008)
28. Rokach, L.: Ensemble-Based Classifiers. Artificial Intelligence Review 33(1-2), 1–39 (2010)
29. Toman, H., Kovacs, L., Jonas, A., Hajdu, L., Hajdu, A.: Generalized Weighted Majority Voting with An Application to Algorithms Having Spatial Output. In: Corchado, E., Snášel, V., Abraham, A., Woźniak, M., Graña, M., Cho, S.-B. (eds.) HAIS 2012, Part II. LNCS, vol. 7209, pp. 56–67. Springer, Heidelberg (2012)
30. Topchy, A.P., Law, H.C., Jain, K., Fred, L.: Analysis of consensus partition in cluster ensemble. In: 4th IEEE International Conference on Data Mining (ICDM 2004), pp. 225–232 (2004)

Randomized Bayesian Network Classifiers

Qing Wang[1,2] and Ping Li[2]

[1] School of Computer Science, Fudan University, Shanghai, China
wangqing@fudan.edu.cn
[2] School of Management Science and Engineering, Anhui University of Technology,
Ma'anshan, China
pingli@ahut.edu.cn

Abstract. In this paper, we propose Randomized Bayesian Network Classifiers (RBNC). It borrows the idea of ensemble learning by constructing a collection of semi-naive Bayesian network classifiers and then combines their predictions as the final output. Specifically, the structure learning of each component Bayesian network classifier is performed by just randomly choosing the parent of each attribute in addition to class attribute, and parameter learning is performed by using maximum likelihood method. RBNC retains many of naive Bayes' desirable property, such as scaling linearly with respect to both the number of instances and attributes, needing a single pass through the training data and robust to noise, etc. On the 60 widely used benchmark UCI datasets, RBNC outperforms state-of-the-art Bayesian classifiers.

1 Introduction

A Bayesian network [1] encodes the joint probability distribution of a set of variables as a directed acyclic graph (DAG) and a set of conditional probability tables (CPTs). Its modularity and intuitive graphical representation make it an attractive model for real world problems, and their use for classification has received considerable attentions [2,3]. Assume that $X_1, X_2, ..., X_a$ are a attributes (corresponding to attribute nodes in a Bayesian network). An instance I is represented by a vector $(x_1, x_2, ..., x_a)$, where x_i is the value of X_i. Let C represent the class variable (corresponding to the class node in a Bayesian network). We use c to represent the value that C takes and $c(I)$ to denote the class label of I. A Bayesian network classifier predicts the class label of instance I using Equation 1.

$$c(I) = \arg \max_{c \in C} P(c)P(x_1, x_2, ..., x_a|c) \tag{1}$$

Assume that all attributes are independent given the class, that is,

$$c(I) = \arg \max_{c \in C} P(c) \prod_{i=1}^{a} P(x_i|c) \tag{2}$$

This assumption is called conditional independence assumption and the resulting classifier is called a naive Bayesian classifier, or simply naive Bayes.

Naive Bayes is the simplest form of Bayesian network classifier and has been widely applied to many real world applications [4,5,6,7,8]. Despite the fact that the conditional

Z.-H. Zhou, F. Roli, and J. Kittler (Eds.): MCS 2013, LNCS 7872, pp. 319–330, 2013.

independence assumptions are often inaccurate, the naive Bayes classifier has several properties that make it surprisingly useful in practice. In particular, both the time and space complexity grow linearly with respect to both the number of instances and attributes, the learning can be done with a single pass through the training data and the performance is robust to noise, etc.

It is obvious that the conditional independence assumption in naive Bayes is rarely true. To relax this assumption, many techniques have been proposed. Extending its structure is a direct way to overcome the limitation of naive Bayes, since attribute dependencies can be explicitly represented by adding arcs. Learning Bayesian networks has become an active research in the past decade [3,9,10]. The goal of learning a Bayesian network is to determine both the structure of the network (structure learning) and the set of CPTs (parameter learning). Since the number of possible structures is extremely huge, structure learning often has high computational complexity. Thus, heuristic and approximate learning algorithms are the realistic solution. A variety of learning algorithms have been proposed, such as TAN [2], BNC[9], HNB [11], $\hat{f}$CLL[3], AnDE[6], etc. Most of these algorithms achieve improved accuracy over naive Bayes. However, this is achieved at the cost of **increasing the order of computational complexity** which severely limits its applicability in practice, especially for large-scale and high-dimensional data.

In fact, a model that could relax conditional independence assumption and also retain many of naive Bayes' desirable computational and theoretical properties, is more desirable. In this paper, we present a new model Randomized Bayesian Network Classifiers (RBNC). It borrows the idea from ensemble learning paradigms by constructing a collection of semi-naive Bayesian network classifiers and then combining their predictions as the final output. Specifically, the structure learning of each component Bayesian network classifier is performed by just randomly choosing the parent of each attribute in addition to class attribute, and parameter learning is performed by using maximum likelihood method (i.e. frequency counting). Our experimental results show that RBNC demonstrates remarkable accuracy compared to other state-of-the-art algorithms.

The rest of the paper is organized as follows. We first introduce the related work. Then we present our new model RBNC, followed by the description of our experimental setup and results in detail. Finally, the paper is concluded in section 5.

2   Related Work

Numerous techniques have been proposed to improve or extend naive Bayes, mainly in two approaches: selecting or forming new attribute subsets in which attributes are conditionally independent, and extending the structure of naive Bayes to represent attribute dependencies.

The idea of selecting a subset of attributes or forming new attributes is to convert the data to a new form that satisfies the conditional independence assumption. Of the proposed techniques, selective naive Bayes (SBC) by [12] demonstrates a remarkable improvement over naive Bayes. SBC uses forward selection to find a good subset of attributes, and then uses this subset to construct a naive Bayes.

Learning Bayesian networks has become an active research in the past decade. The goal of learning consists of determining both the structure of the network and the set of CPTs. Since the number of possible structures is extremely huge, structure learning often has high computational complexity. Moreover, learning unrestricted Bayesian network seems to not necessarily lead to a classifier with good performance. Thus, heuristic and approximate learning algorithms are the realistic solution. For example, [2] proposed Tree Augmented Naive Bayes (TAN), a structure learning algorithm that learns a maximum spanning tree from the attributes, but retains naive Bayes model as a part of its structure to bias towards the estimation of conditional distribution. BNC-2P [9], on the other hand, is a heuristic discriminative structure learning method with conditional log likelihood as scoring function. Although the structures in TAN and BNC-2P are selected discriminatively, the parameters are trained via maximum likelihood training for computational efficiency.

Factorized conditional log-likelihood ($\hat{f}$CLL) [3] is the most recently proposed score function for learning Bayesian network classifiers. It is an approximation of the conditional log-likelihood criterion, and is devised in order to guarantee decomposability over the network structure as well as efficient estimation of the optimal parameters. This discriminative criteria achieves the same time and space complexity as the log-likelihood scoring function. The experimental results show that $\hat{f}$CLL trained TAN achieves improved accuracy over other discriminatively trained Bayesian network classifiers.

Hidden Naive Bayes (HNB) [13,11] using a predefined network structure to take the influences from all attributes into account. In HNB, each attribute X_i has a hidden parent X_{hp_i} which combines the influences from all other attributes. The classifier corresponding to an HNB on an instance $I = (x_1, ..., x_a)$ is defined as follows:

$$c(I) = \arg\max_{c \in C} P(c) \prod_{i=1}^{a} P(x_i | X_{hp_i}, c) \qquad (3)$$

where

$$P(x_i | X_{hp_i}, c) = \sum_{j=1, j \neq i}^{a} w_{ij} P(x_i | x_j, c) \qquad (4)$$

The weight w_{ij} is defined by the conditional mutual information between two attributes X_i and X_j. The hidden parent X_{hp_i} for X_i is essentially a mixture of the weighted influences from all other attributes. Since there is no structure learning, learning an HNB is mainly about estimating the parameters from the training data. To create the hidden parent of an attribute, HNB needs to compute the conditional mutual information for each pair of attributes.

The most recent work on improving naive Bayes is AnDE (averaged n-dependence estimators) [6] which is an generalization of the well-known AODE (averaged one-dependence estimators) [5] algorithm. In AnDE, an ensemble of n-dependence classifiers are learned and the prediction is produced by aggregating the predictions of all qualified classifiers. An x-dependence estimator means that the probability of an attribute is conditioned by the class variable and at most x other attributes. In AnDE, a n-dependence classifier is built for every combination of n attributes, in which the given

n attributes are set to be the parent of all other attributes. AnDE predicts the class label of instance I using Equation 5.

$$c(I) = \arg\max_{c \in C} \sum_{s \in S^n} P(c, \mathbf{s}) \prod_{j=1, j \notin s}^{a} P(x_j | c, \mathbf{s}) \tag{5}$$

where S^n indicates the set of all size-n subsets of $\{x_1, ..., x_a\}$. The experimental results show that the bias-variance trade-off for A2DE results in strong predictive accuracy over a wide range of data sets. Another reason for the authors presenting primarily results for A2DE is because the computational complexity (both space and time) of AnDE($n \geq 3$) is very high and defeats their Weka implementation on most data sets [6]. The ensemble size is a (the number of attributes) for both AODE and A2DE.

Table 1 shows the training time and space complexity of some algorithms discussed.

Table 1. Computational complexity of algorithms

Algorithm	Training Complexity		Testing Complexity	
	Time	Space	Time	Space
NB	$O(ta)$	$O(kav)$	$O(ka)$	$O(kav)$
TAN	$O(ta^2 + k(av)^2 + a^2 \log a)$	$O(k(av)^2)$	$O(ka)$	$O(kav^2)$
HNB	$O(ta^2 + k(av)^2)$	$O(k(av)^2)$	$O(ka^2)$	$O(k(av)^2)$
AODE	$O(ta^2)$	$O(k(av)^2)$	$O(ka^2)$	$O(k(av)^2)$
AnDE	$O(t\binom{a}{n+1})$	$O(k\binom{a}{n+1}v^{n+1})$	$O(kn\binom{a}{n})$	$O(k\binom{a}{n+1}v^{n+1})$
RBNC-n	$O(Ntan)$	$O(Nkav^{n+1})$	$O(Nkan)$	$O(Nkav^{n+1})$

k is the number of classes.
a is the number of attributes.
v is the average number of values for an attribute.
t is the number of training examples.
n is the number of parent nodes except class.
N is the number of component models of RBNC.

3   The RBNC Algorithm

In this section, we introduce the RBNC family of algorithms and analyze its computational complexity.

3.1   Algorithm Definition

Instead of searching for a single Bayesian network classifier model by optimizing some (discriminative or generative) score on data, RBNC randomly constructs multiple Bayesian network classifier models and then simply average their probability predictions as the final output.

We focus on *augmented naive Bayes classifiers*, that is, Bayesian network classifiers where the class attribute has no parents and all attributes have at least the class attribute as parent. In addition, we introduce a parameter n to control the maximum number of

Algorithm 1. RBNC algorithm

Input: Training data D, where $< X_1, ..., X_a >$ and C represent a input attributes and class attribute, respectively. Maximum number of parents (except class) per node n and number of component models N.
Output: A set of Bayesian network classifier models E.

Initialize $E = \{\}$.
$/*$ structure learning $*/$
for $i = 1$ **to** N **do**
    Generate a random permutation $< A_1, .., A_a >$ of the given a input attributes.
    Initialize an empty Bayesian network model M_i with $a + 1$ node.
    For M_i, set class attribute C as parent for all other attributes.
    for $j = 2$ **to** a **do**
        if $j \leq n$ **then**
            For M_i, set all attributes in $\{A_1, ..., A_{j-1}\}$ as parent of attribute A_j.
        else
            For M_i, randomly select n attributes in $\{A_1, ..., A_{j-1}\}$ as parent of attribute A_j.
        end if
    end for
    $E = E \bigcup M_i$.
end for
$/*$ parameter learning $*/$
Compute the CPTs for all $M_i \in E$ on data D using maximum likelihood.
return E

parents per node in the network. The structure of each component Bayesian network classifier in RBNC is constructed by just randomly choosing n other attributes as the parents for each attribute in addition to class attribute. To ensure the generated structure is DAG, first, all the attributes are ordered, then each attribute can only select those ahead of it as parents. The parameters in each component network are set to their maximum likelihood values, i.e. observed frequency counting over the data. The detailed learning process of RBNC is depicted in Algorithm 1.

RBNC predicts the class label of instance I using:

$$c(I) = \arg\max_{c \in C} \sum_{q=1}^{N} P_q(c|I) \tag{6}$$

where $P_q(c|I)$ is the posterior probability estimation of the q-th component model in RBNC, and is defined as:

$$P_q(c|I) = P(c) \prod_{i=1}^{a} P(x_i|\pi_i, c) \tag{7}$$

where π_i is the set of parents values of attribute X_i.

It should be noted that RBNC-0 is just naive Bayes and in RBNC-n ($n \geq 1$), each component models define a weaker conditional independence assumption than naive Bayes, as it is necessarily true if the naive Bayes' assumption is true and may also be true when the naive Bayes' assumption is not. As this is a weaker assumption than Equation 2, the bias of the model should be lower than that of naive Bayes. However, it is derived from higher-dimensional probability estimates and hence its variance should be higher.

Similar to AnDE, RBNC utilizes parameter n that transforms the approach between a low-variance high-bias learner (naive Bayes) and a high-variance low-bias learner with Bayes optimal asymptotic error. So, RBNC actually defines a family of algorithms. Successive members of the family will be best suited to differing quantities of data, starting with low variance for small data set, with successively lower bias but higher variance suiting to increasing data quantities.

3.2   Computational Complexity of RBNC

Each component Bayesian network model in RBNC forms an $(n+2)$-dimensional probability table containing the observed frequency for the given combination of $n + 1$ attribute values and the class labels. The space complexity of the table is $O(kav^{n+1})$ and the time complexity of compiling it is $O(tan)$, as we need to update each entry for the combination of the $n + 1$ attribute-values for every instance. The time complexity for classifying a single instance is $O(kan)$ as we need to consider each attribute for the combination of n parent attributes within each class.

Assume the number of component models in RBNC is N, then for RBNC, the space complexity is $O(Nkav^{n+1})$, time complexity of compiling it is $O(Ntan)$ and classifying a single instance is $O(Nkan)$.

4   Experiments and Results

4.1   Experiment Setup

We conduct our experiments under the framework of Weka [14] on a PC with Intel Core 2 Duo P8600 2.4G CPU and 4G RAM. In our experiments, we use the 60 well-recognized datasets from the UCI repositories[15], which include all the datasets recommended by Weka and the benchmark datasets used by related works [2,9,10,3]. A brief description of the data sets is in Table 2. Numeric variables are discretized using supervised discretization method implemented in Weka. Missing values are also processed using the mechanism in Weka, which replaces all missing values with the modes and means from the training data. In addition, all the preprocessing is done with the default parameters in Weka implementation.

We compared RBNC-n (n=1,2,3, and ensemble sizes N are all set to 20) with the following algorithms:

1. The naive Bayes classifier (**NB**).
2. The discriminatively trained tree-augmented naive Bayes (**TAN-$\hat{f}$CLL**) algorithm using factorized conditional log-likelihood [3].
3. The Hidden naive Bayes classifier (**HNB**) [11].

Table 2. Description of the data sets used for experiments

Datasets	Size	Attribute	Classes	Datasets	Size	Attribute	Classes
adult	48842	15	2	ionosphere	351	35	2
albalone	4177	9	28	iris	150	5	3
anneal	898	39	6	kr-vs-kp	3196	37	2
anneal.ORIG	898	39	6	labor	57	17	2
audiology	226	70	24	letter	20000	17	26
australian	690	15	2	lymph	148	19	4
autos	205	26	7	mofn	1324	11	2
badges	294	11	2	mushroom	8124	23	2
balance-scale	625	5	3	nursery	12960	9	5
breast-cancer	286	10	2	optical	5620	65	10
breast-w	699	10	2	ozone	2536	73	2
car	1728	7	4	page-blocks	5473	11	5
chess	28056	7	18	pendigital	10992	17	10
cleve	296	14	2	pima	768	9	2
cmc	1473	10	3	primary-tumor	339	18	21
colic	368	23	2	segment	2310	20	7
colic.ORIG	368	28	2	shuttle	5800	10	7
corral	128	7	2	sick	3772	30	2
credit-a	690	16	2	sonar	208	61	2
credit-g	1000	21	2	soybean	683	36	19
dermatology	366	35	6	spambase	4601	58	2
diabetes	768	9	2	splice	3190	62	3
ecoli	336	8	8	tic-tac-toe	958	10	2
flare	1066	11	2	vehicle	846	19	4
glass	214	10	7	vote	435	17	2
heart-c	303	14	5	vowel	990	14	11
heart-h	294	14	5	waveform-5000	5000	41	3
heart-statlog	270	14	2	wine	178	14	3
hepatitis	155	20	2	yeast	1484	10	10
hypothyroid	3772	30	4	zoo	101	18	7

4. The Averaged one-dependence estimators (**AODE**) and Averaged two-dependence estimators (**A2DE**). We do not present the results of AnDE ($n \geq 3$) since even the computational requirements of A3DE defeat the Weka implementation except in cases of low dimensional data, and this is the same issue encountered by [6].

5. The Random Forests classifier with both the default setting of 10 trees (**RF-10**) and with 100 trees (**RF-100**).

The naive Bayes, HNB, AODE, A2DE and RF are already implemented in Weka, and the source code of TAN-$\hat{f}$CLL algorithm is available at the author's homepage http://kdbio.inesc-id.pt/~asmc/software/fCLL.html. So, we only implemented RBNC within the Weka framework and uploaded the **source codes** of RBNC at[1]. We used the **laplace estimation** to avoid the zero-frequency problem for all compared methods. In our experiment, the performance of an algorithm on each data set has been calculated via 10 runs of 10-fold stratified cross validation.

4.2   The Effect of Varying n within RBNC

To investigate how increasing n within the RBNC framework affects performance as the quantity of data increases, we form learning curves for NB, RBNC-1, RBNC-2, RBNC-3 and RBNC-4 on the Adult and Nursery dataset, respectively.

---

[1] http://homepage.fudan.edu.cn/wangqing/files/2011/10/rbnc1.zip

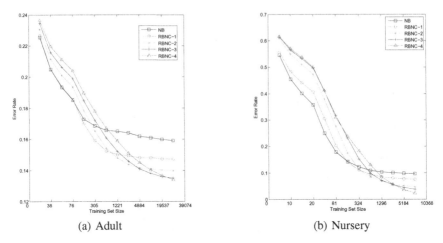

(a) Adult   (b) Nursery

Fig. 1. Learning curves as function of training set size

First, 10% instances are selected at random as a test set and training sets were sampled from the remaining instances. The training sets size consist of $\frac{1}{2^{11}}$, $\frac{1}{2^{10}}$,..., 1 fraction of the remaining instances, respectively. This process is repeated 100 times and each algorithm is evaluated on the resulting training-test set pairs. The learning curves of error rate for NB, RBNC-1, RBNC-2, RBNC-3 and RBNC-4 are presented in Figure 1.

Figure 1 clearly show the predicted trade-off for increasing n. At the smallest data size, where low variance is more important than low bias, error rate is minimized by n = 0 (NB) and increases as n increases. At the largest data size, where low bias is most important, this dimensionality is reversed.

4.3 Experimental Results

Table 3 shows the comparison results of two-tailed t-test with a 95% confidence level between each pair of algorithms on data, in which each entry $w/t/l$ means that the

Table 3. Summary of experimental results under pairwise two-tailed t-test with 95% confidence level. Each cell contains the number of wins, ties and losses between the algorithm in that row and the algorithm in that column.

$w/t/l$	RBNC-3	RBNC-2	RBNC-1
RBNC-2	1/44/15	–	–
RBNC-1	4/33/23	1/38/21	–
NB	4/24/32	2/25/33	1/31/28
TAN-$\hat{f}$CLL	6/28/26	8/30/22	14/31/15
HNB	3/40/17	8/39/13	13/42/5
AODE	5/33/22	6/39/15	9/45/6
A2DE	6/37/17	9/40/11	17/41/2
RF-10	6/29/25	8/27/25	19/20/21
RF-100	11/33/16	19/23/18	25/26/9

algorithm at the corresponding row wins in w data sets, ties in t data sets, and loses in l data sets, compared to the algorithm at the corresponding column. Table 4 and 5 show the detailed accuracies of the algorithms on each data set. The mean accuracy and standard deviation, together with the overall rank on all data sets are summarized at the bottom of the table.

From the experimental results, we can see that RBNC-n algorithms can achieve substantial improvement over naive Bayes (32 wins and 4 losses, 33 wins and 2 losses, 28 wins and 1 losses, respectively). This results show that many data sets in our experiments contain strong dependencies, and conditional independence assumption failed to capture these dependencies. In addition, RBNC-1 and RBNC-2 are comparable to AODE (6 wins and 9 losses) and A2DE (11 wins and 9 losses), respectively. For Random Forests algorithms, RBNC-n (n=1, 2, 3) all outperform RF-10 and RBNC-3 significantly outperforms RF-100 (16 wins and 11 losses). Overall, the performance of RBNC-3 is the best among all the algorithms compared. Considering that RBNC-n scales linearly with respect to both the number of instances and attributes of the training data, RBNC-n are overall more efficient.

To study the robustness of our algorithm, we test it on these 60 UCI data sets under artificial noise in the class labels. Following the method in [16], the noisy version of each training data set is generated by choosing 10% instances and changing their class labels to other incorrect labels randomly. Due to space limited, we do not list the detailed results of the accuracy and standard deviation on each data set here. The experimental results show that the RBNC algorithms are all robust to noise and also achieve substantial improvement over naive Bayes. And RBNC-3 still to be the best among the algorithms compared.

To further understand the working mechanism of RBNC-n and the difference compared with Random Forests, we use the bias-variance decomposition to analysis them. The results again demonstrate that with n increasing, RBNC-n evolves from low variance coupled with high bias through to high variance coupled with low bias. The bias terms for RBNC-n (n=1, 2, 3) and RF (10 and 100) are 0.0963, 0.0760, 0.0696, 0.0663 and 0.0669, respectively. So RBNC-3 could achieve the same level of bias compared with Random Forests. This is of interest because it demonstrates that it is possible to create low-bias high-variance generative learners without discriminative learning.

5 Conclusion

In this paper, we propose the RBNC family of algorithms which utilize a single parameter n to control over a bias-variance trade-off, such that higher values of n are appropriate for greater numbers of training cases. RBNC retains many of naive Bayes' desirable property, such as the time and space complexity are linear with respect to both the number of training instances and attributes, the learning can be done by a single pass through the training data and the performance is robust to noise. Our experimental results show that RBNC has a better overall performance compared to the state-of-the-art Bayesian network classifier algorithms. Considering the simplicity and efficiency, RBNC is a promising model that could be used in many applications.

Table 4. Accuracy and standard deviation of the ten algorithms on the 60 UCI data sets

Data sets	RBNC-3	RBNC-2	RBNC-1	NB	TAN-$\widehat{f}$CLL	HNB	AODE	A2DE	RF-10	RF-100
adult	86.48±0.39	86.00±0.39	85.27±0.39	84.07±0.42	86.21±0.37	84.76±0.36	85.20±0.41	85.95±0.39	85.27±0.42	85.66±0.37
albalone	25.15±1.83	25.53±1.80	26.16±1.82	26.18±1.97	25.23±1.89	24.79±1.92	26.49±1.77	26.44±1.84	23.6±1.66	23.42±1.72
anneal	99.10±0.90	98.95±1.04	98.63±1.18	96.13±2.16	98.98±1.10	98.29±1.26	98.05±1.37	98.31±1.24	99.2±0.94	99.48±0.75
anneal.ORIG	93.97±2.47	94.03±2.45	93.39±2.58	92.66±2.72	94.47±2.32	94.82±2.23	93.35±2.53	93.31±2.55	94.72±2.13	94.9±1.99
audiology	76.61±6.22	76.61±6.44	76.04±6.53	71.40±6.37	69.48±6.62	73.15±6.00	71.66±6.42	71.88±6.47	75.95±8.23	79.97±6.85
australian	86.42±3.57	86.74±3.34	86.23±3.44	85.38±3.49	85.58±3.81	85.04±4.04	86.09±3.54	86.19±3.63	84.12±3.97	84.9±4.12
autos	88.48±6.95	85.56±7.32	81.65±8.65	72.3±10.31	80.55±8.68	85.51±7.62	81.14±8.50	82.56±7.90	86.28±7.26	87.11±6.89
badges	100.0±0.0	100.0±0.0	100.0±0.0	100.0±0.0	100.0±0.0	100.0±0.0	100.0±0.0	100.0±0.0	99.97±0.33	100.0±0.0
balance-scale	68.7±3.72	68.75±3.74	69.47±3.81	71.08±4.29	70.91±4.06	69.02±3.76	69.34±3.82	68.75±3.70	69.92±3.98	68.99±3.76
breast-cancer	74.59±6.97	73.58±6.36	71.93±6.96	72.94±7.71	67.29±7.35	70.23±6.49	72.73±7.01	73.97±6.19	69.17±6.8	70.09±7.36
breast-w	96.45±2.21	96.98±1.87	97.24±1.69	97.25±1.79	97.21±1.82	96.27±2.12	96.97±1.87	96.55±2.12	95.19±2.64	95.67±2.57
car	95.27±1.41	93.92±1.77	91.74±2.06	85.46±2.56	94.44±1.79	93.01±2.02	91.41±2.06	93.56±1.70	93.18±1.86	94.7±1.66
chess	75.57±0.81	67.31±0.85	55.37±0.85	36.11±0.81	55.85±0.76	58.89±0.86	52.72±0.91	65.88±0.84	65.03±1.11	69.57±0.93
cleve	82.12±6.52	82.19±6.48	83.17±6.58	83.47±6.80	81.21±6.53	82.13±6.56	83.34±6.35	82.80±6.51	80.13±7.52	81.33±7.08
cmc	52.70±3.84	53.29±4.01	53.37±3.58	52.17±3.80	54.48±3.55	54.66±3.42	51.80±3.54	52.38±3.75	50.12±4.29	50.84±4.16
colic	82.39±6.38	82.39±5.87	82.91±5.51	81.39±5.74	79.59±6.25	81.60±6.04	82.37±5.73	82.28±5.90	82.2±6.52	82.99±5.72
colic.ORIG	76.28±6.39	77.15±5.99	76.42±6.07	74.46±6.81	71.31±5.23	68.49±6.39	75.66±6.65	75.79±6.09	71.93±5.92	73.73±5.35
corral	99.23±2.99	99.69±1.89	95.74±5.76	86.60±8.87	99.85±1.08	99.85±1.08	88.72±9.01	96.05±5.96	98.68±3.69	99.77±1.71
credit-a	86.96±3.76	86.88±3.59	86.00±3.87	86.25±4.01	85.36±3.83	85.38±4.01	86.55±3.82	86.83±3.59	83.77±4.38	84.46±3.94
credit-g	75.56±3.81	76.21±3.87	75.63±3.88	75.43±3.84	73.68±4.31	76.04±3.64	76.45±3.89	76.31±3.61	72.92±3.67	74.31±3.58
dermatology	98.11±2.16	98.44±1.91	98.20±2.03	97.93±2.12	96.69±2.81	98.61±1.85	97.82±2.29	97.85±2.27	95.9±3.06	97.07±2.62
diabetes	77.98±4.38	78.20±4.32	78.37±4.48	77.85±4.67	78.54±4.15	76.73±4.05	78.07±4.56	78.58±4.34	76.11±4.33	76.61±4.69
ecoli	83.91±4.93	83.67±4.72	84.27±5.14	85.57±4.95	84.86±4.95	82.54±4.99	85.16±5.05	84.02±4.87	82.99±5.59	82.93±5.38
flare	82.35±2.99	82.31±3.01	81.43±3.24	80.19±3.41	82.40±2.68	82.70±2.13	81.83±3.12	82.33±2.80	80.89±2.71	80.94±2.52
glass	79.09±8.90	78.86±8.90	76.44±8.51	74.39±7.95	77.64±8.76	78.27±8.97	76.49±7.71	76.59±8.80	76.25±8.79	77.32±8.21
heart-c	81.78±6.10	82.02±6.49	83.07±6.22	83.60±6.42	81.28±6.54	82.38±6.79	83.26±6.19	82.97±6.37	80.96±6.4	81.13±5.56
heart-h	84.67±5.90	84.39±6.17	84.53±6.01	84.46±5.92	84.84±5.94	85.31±5.92	84.43±5.92	84.60±6.09	82.25±6.66	81.4±6.66
heart-statlog	82.19±6.33	82.56±6.54	83.74±6.29	83.74±6.25	82.70±6.41	82.37±6.54	83.33±6.61	82.52±6.60	83.52±6.93	83.59±7.01
hepatitis	85.51±8.93	85.45±9.40	84.86±9.56	84.22±9.41	84.79±9.31	87.90±8.09	85.43±8.96	85.58±8.55	83.92±9.05	85.02±8.76
hypothyroid	99.04±0.50	99.00±0.51	98.82±0.54	98.48±0.59	99.28±0.39	99.13±0.41	98.74±0.55	98.79±0.49	99.11±0.46	99.29±0.4
ionosphere	93.68±3.88	93.19±4.01	92.91±3.92	90.77±4.76	92.85±4.23	92.62±4.12	92.97±4.32	92.99±4.16	92.45±3.83	93.5±3.92

Table 5. (*Continued*) Accuracy and standard deviation of the ten algorithms on the 60 UCI data sets. Bottom rows of the table present the mean values and overall ranks, respectively.

Data sets	RBNC-3	RBNC-2	RBNC-1	NB	TAN-fCLL	HNB	AODE	A2DE	RF-10	RF-100
iris	93.07±6.06	93.60±5.60	93.33±5.61	94.47±5.61	93.60±5.91	93.33±5.92	93.20±5.76	93.33±5.76	94.07±5.76	94.4±5.74
kr-vs-kp	93.10±1.65	91.35±1.65	89.73±1.74	87.79±1.91	94.21±1.37	92.35±1.32	91.03±1.66	93.08±1.41	98.87±0.61	99.27±0.44
labor	95.47±8.74	94.93±9.00	95.27±9.61	93.13±10.56	95.57±9.36	95.60±9.47	95.43±8.80	94.67±10.04	92.47±11.5	92.8±10.98
letter	93.65±0.53	92.35±0.56	86.59±0.72	74.00±0.88	85.94±0.67	89.77±0.63	88.76±0.70	90.82±0.64	90.21±0.75	93.5±0.54
lymph	88.40±8.18	87.79±8.15	86.52±8.39	84.97±8.30	83.60±9.73	86.78±8.47	87.53±8.13	87.61±8.67	79.65±9.2	82.89±8.49
mofn	95.67±1.92	93.88±2.16	89.31±2.06	85.28±1.95	92.01±1.93	95.19±2.18	88.60±2.10	92.54±2.10	99.69±0.6	99.99±0.08
mushroom	100.0±0.00	99.96±0.07	99.77±0.16	95.52±0.78	99.77±0.15	99.96±0.06	99.95±0.07	99.99±0.03	100.0±0.0	100.0±0.0
nursery	96.19±0.44	95.25±0.47	92.51±0.59	90.30±0.72	93.41±0.55	94.25±0.50	92.73±0.62	94.73±0.52	98.36±0.38	99.17±0.32
optical	96.44±0.71	95.85±0.81	94.81±0.86	92.42±1.07	95.92±0.81	96.35±0.79	96.99±0.71	97.12±0.68	91.67±1.14	96.18±0.77
ozone	89.56±1.93	86.53±2.29	83.10±2.56	80.04±2.71	94.16±1.37	88.96±1.92	88.59±2.05	93.91±1.49	97.05±0.6	97.08±0.61
page-blocks	97.41±0.64	97.18±0.69	95.73±0.79	93.53±0.98	96.30±0.79	96.91±0.69	97.27±0.67	97.35±0.65	97.14±0.67	97.29±0.64
pendigital	98.45±0.38	98.43±0.37	97.14±0.49	87.89±0.99	96.71±0.63	97.83±0.43	97.94±0.45	98.23±0.40	96.36±0.63	97.64±0.46
pima	78.22±4.30	78.16±4.35	78.50±4.11	77.97±4.27	78.84±4.07	77.03±4.09	78.31±4.16	79.06±4.05	76.59±4.54	77.28±4.44
primary-tumor	49.00±5.49	47.88±5.53	48.67±5.99	47.20±6.02	45.76±6.35	47.85±6.06	47.87±6.37	48.20±6.11	39.68±5.96	41.3±6.05
segment	96.66±1.14	96.60±1.14	95.84±1.32	91.71±1.68	93.15±1.98	96.86±1.08	95.77±1.24	95.90±1.20	96.25±1.11	96.96±1.06
shuttle	99.82±0.13	99.86±0.12	99.88±0.11	99.35±0.30	99.83±0.14	99.84±0.14	99.84±0.13	99.82±0.13	99.74±0.17	99.76±0.16
sick	97.53±0.80	97.30±0.79	97.27±0.80	97.10±0.84	97.41±0.75	97.55±0.76	97.39±0.79	97.47±0.76	97.64±0.75	97.77±0.73
sonar	86.12±7.30	86.31±7.23	85.16±7.43	85.16±7.52	85.16±7.62	84.63±7.64	86.60±6.91	86.84±6.84	79.72±9.02	82.83±7.8
soybean	94.41±2.38	94.25±2.40	93.66±2.64	92.20±3.23	94.33±2.45	94.67±2.25	93.31±2.85	93.43±2.69	92.8±2.45	93.81±2.72
spambase	92.72±1.09	91.79±1.17	91.10±1.17	90.24±1.23	92.73±1.11	92.42±1.13	93.36±1.12	94.19±1.03	94.01±1.15	94.88±1.02
splice	87.40±1.77	94.07±1.28	95.93±1.12	95.42±1.14	94.83±1.05	96.13±0.99	96.12±1.00	96.41±0.85	90.09±1.78	95.88±1.14
tic-tac-toe	92.79±2.67	86.37±3.55	73.63±4.02	69.64±4.40	74.76±3.71	77.46±3.53	73.86±3.93	90.79±3.17	92.11±2.59	97.06±1.76
vehicle	73.76±3.53	73.35±3.73	71.89±3.15	62.52±3.81	70.98±3.98	73.31±3.38	72.31±3.62	73.06±3.71	72.99±4.4	73.19±3.5
vote	95.77±2.92	94.78±3.20	92.94±3.54	90.21±3.95	90.76±3.73	94.36±3.20	94.52±3.19	94.75±3.29	95.95±2.83	96.18±2.85
vowel	89.34±2.98	87.75±3.35	83.21±3.81	65.23±4.53	89.16±3.39	89.11±2.98	80.88±3.81	83.99±3.41	88.98±3.01	90.13±2.73
waveform-5000	85.21±1.42	84.26±1.36	82.47±1.35	80.72±1.50	81.26±1.45	86.26±1.45	86.03±1.56	86.37±1.48	80.58±1.73	84.99±1.47
wine	98.88±2.66	98.32±2.94	98.88±2.26	98.82±2.30	98.08±3.13	98.14±2.89	98.37±2.92	98.32±2.94	97.97±3.24	98.65±2.55
yeast	60.13±3.34	60.26±3.60	59.79±4.01	59.16±3.80	59.88±3.68	59.64±3.82	59.72±3.86	59.63±4.01	59.1±3.56	59.59±3.41
zoo	97.03±5.19	96.05±5.60	84.67±6.39	93.21±7.35	93.20±6.83	97.11±4.97	94.66±6.38	94.66±6.38	91.75±7.78	93.55±6.83
Mean	86.37±3.44	85.91±3.47	84.67±3.63	82.12±3.92	84.65±3.60	85.34±3.44	84.74±3.67	85.81±3.56	84.95±3.71	86.08±3.43
Overall Rank	3.72	4.48	5.73	7.59	5.91	5.23	5.75	4.63	7.01	4.95

Acknowledgements. We would like to thank the anonymous reviewers for their help-ful comments. This work was supported by the Natural Science Foundation of Anhui Provincial Education Department under Grant No.KJ2013A053.

References

1. Pearl, J.: Probabilistic reasoning in intelligent systems: Networks of plausible inference. Morgan Kaufmann, San Francisco (1988)
2. Friedman, N., Geiger, D., Goldszmidt, M.: Bayesian network classifiers. Machine Learning 29, 131–163 (1997)
3. Carvalho, A.M., Roos, T., Oliveira, A., Myllymaki, P.: Discriminative learning of bayesian networks via factorized conditional log-likelihood. Journal of Machine Learning Research 12, 2181–2210 (2011)
4. Domingos, P., Pazzani, M.J.: On the optimality of the simple bayesian classifier under zero-one loss. Machine Learning 29(2), 103–130 (1997)
5. Webb, G.I., Boughton, J.R., Wang, Z.: Not so naive bayes: Aggregating one-dependence estimators. Machine Learning 58(1), 5–24 (2005)
6. Webb, G.I., Boughton, J.R., Zheng, F., Ting, K., Salem, H.: Learning by extrapolation from marginal to full-multivariate probability distributions: Decreasingly naive bayesian classification. Machine Learning 86(2), 233–272 (2012)
7. Salem, H., Suraweera, P., Webb, G.I., Boughton, J.R.: Techniques for efficient learning without search. In: Tan, P.-N., Chawla, S., Ho, C.K., Bailey, J. (eds.) PAKDD 2012, Part I. LNCS, vol. 7301, pp. 50–61. Springer, Heidelberg (2012)
8. Wu, X.D., Kumar, V., Quinlan, J.R., Ghosh, J., Yang, Q., Motoda, H., McLachlan, G.J., Ng, A., Liu, B., Yu, P.S., Zhou, Z.H., Steinbach, M., Hand, D.J., Steinberg, D.: Top 10 algorithms in data mining. Knowledge and Information Systems 14(1) (2008)
9. Grossman, D., Domingos, P.: Learning bayesian network classifiers by maximizing conditional likelihood. In: Proceedings of the 21st International Conference on Machine Learning, pp. 46–53 (2004)
10. Jing, Y.S., Pavlovi, V., Rehg, J.M.: Efficient discriminative learning of bayesian network classifier via boosted augmented naive bayes. In: Proceedings of the 22nd International Conference on Machine Learning, pp. 369–376 (2005)
11. Jiang, L., Zhang, H., Cai, Z.: A novel bayes model: hidden naive bayes. IEEE Transations on Knowledge and Data Engineering 21(10), 1361–1371 (2009)
12. Langley, P., Sage, S.: Induction of selective bayesian classifiers. In: Proceedings of the Uncertainty in Artificial Intelligence, pp. 399–406 (1994)
13. Zhang, H., Jiang, L.X., Su, J.: Hidden naive bayes. In: The Twentieth National Conference on Artificial Intelligence (AAAI 2005), pp. 919–924 (2005)
14. Witten, I.H., Frank, E.: Data Mining: Practical Machine Learning Tools and Techniques with Java Implementations. Morgan Kaufmann, San Francisco (2000)
15. Blake, C., Merz, C.J.: UCI repository of machine learning databases. Department of ICS, University of California, Irvine,
   http://www.ics.uci.edu/~mlearn/MLRepository.html
16. Breiman, L.: Random forests. Machine Learning 45, 5–32 (2001)

A Novel Pattern Rejection Criterion Based on Multiple Classifiers

Wei-Na Wang[1], Xu-Yao Zhang[2], and Ching Y. Suen[1]

[1] CENPARMI, Concordia University, Montreal, Canada
{wein_wa,suen}@encs.concordia.ca
[2] Institute of Automation, Chinese Academy of Sciences, Beijing, China
xyz@nlpr.ia.ac.cn

Abstract. Aiming at improving the reliability of a recognition system, this paper presents a novel SVM-based rejection measurement (SVMM) and voting based combination methods of multiple classifier system (MCS) for pattern rejection. Compared with the previous heuristic designed criteria, SVMM is more straight-forward and can make use of much more information from the training data. The voting based combination methods for rejection is a preliminary attempt to adopt MCS for rejection. Comparison of SVMM with other well-known rejection criteria proves that it achieves the highest performance. Two different methods (structural modification and dataset re-sampling) are used to build MCSs. The basic classifier is the convolution neural network (CNN) which has achieved promising performances in numerous applications. Rejection based on MCS is then evaluated on MNIST and CENPARMI digit databases. Specifically, different rejection criteria (FRM, FTRM and SVMM) are individually combined with MCS for pattern rejection. Experimental results indicate that these combinations improve the rejection performance consistently and MCS built by dataset re-sampling works better than that with structural modification in rejection.

Keywords: Rejection criterion, SVMM, MCS, CNN, soft voting, handwritten digit recognition.

1 Introduction

In pattern recognition, the recognition rate is always an important factor in evaluating the performance of a classifier and plenty of classifiers or multiple classifier systems have achieved high recognition rates based on different datasets like MNIST, CENPARMI and so forth in the past decades. However, although the recognition accuracy of some models has reached error rates of less than 1% on the benchmark MNIST dataset [1, 2, 3, 4] and CENPARMI numeral dataset [5], it is still impossible to reach a 100% recognition accuracy. And a low percentage of errors in recognition could still cause a huge loss in real-life systems, like check-reading in the banks; hence the reliability of a classifier is as important as recognition accuracy, as defined below:

Z.-H. Zhou, F. Roli, and J. Kittler (Eds.): MCS 2013, LNCS 7872, pp. 331–342, 2013.
© Springer-Verlag Berlin Heidelberg 2013

$$Recognition\ rate = \frac{Number\ of\ correct\ samples}{Total\ number\ of\ testing\ samples}$$

$$Rejection\ rate = \frac{Number\ of\ rejected\ samples}{Total\ number\ of\ testing\ samples}$$

$$Reliability = \frac{Number\ of\ correct\ samples\ among\ \text{nonrejected ones}}{Total\ number\ of\ testing\ samples - \text{number of rejected samples}}$$

In order to improve the reliability of a classifier, some confusing patterns must be rejected before entering the testing loop in order to prevent errors. That is why some useful rejection criteria are produced to determine and filter out the confusing samples. To evaluate the effectiveness of rejection, we can draw a curve in the coordinate system whose x-axis is the number of rejected samples and the y-axis is reliability. A good rejection criterion can achieve a higher reliability with fewer samples rejected. So in this case, we expect the curve to be as close to the top left corner as possible.

In this paper, our main goal is to improve the reliability of recognition systems by detecting the confusing samples that may easily cause error. To accomplish this goal, we have designed a novel rejection criterion, called SVM-based Measurement (SVMM), which learns the optimal rejection boundary from the training data. Brief descriptions of this criterion as well as several other well-known rejection criteria are presented in Sections 2. After that, we first attempt to use Multiple Classifier System (MCS) for the purpose of pattern rejection. It is implemented by using voting methods to combine decisions from different classifiers. Both hard voting and soft voting are considered and details are followed in Section 3. Section 4 reports all the experimental results and analyses. Specifically, the newly proposed rejection criterion verified and compared with other rejection criteria on MNIST numeral dataset. MCS based rejections with both hard voting and soft voting are evaluated on the same dataset and also CENPARMI numeral dataset with MCSs differing in structural modification and dataset re-sampling. At last, we provide our concluding remarks in Section 5.

2   Rejection Criteria

Pattern rejection can be viewed as a two-class recognition problem, which takes the output values of a classifier as features to recognize a pattern as a confusing one to reject or a clear one to accept. Generally, for a regular classifier, the output is always a vector consisting of confidence values or probabilities of possible classes. Given a pattern x, suppose the output vector of the classification is (c is the number of possible classes)

$$\{f_1, f_2, \ldots, f_c\}, \ f_i \geq 0, \ i = 1, 2, \ldots, c \tag{1}$$

After that, this pattern is classified according to $x \in class\ arg \max_{1 \leq i \leq c} f_i$. In case that the outputs are negative, normalization can be used to guarantee that all the values are positive (e.g. $f_i = f_i - f_{min}$, $f_{min} = \min_{1 \leq i \leq c} f_i$).

2.1   Existing Rejection Criteria

In the research field of rejection, some traditional rejection criteria have been studied before and have reached high recognition rates as well as high reliability. In this section, some useful criteria are presented.

The first rank confidence value (FR) and the second rank confidence value (SR) can be described as

$$FR = \max_{1 \le i \le c} f_i, \quad SR = \max_{1 \le i \le c, f_i \neq FR} f_i \tag{2}$$

FR is expected to be much larger than all the other output values for a clear sample. Besides, the gap between FR and SR is also viewed as a useful index, to reflect the quality of a sample. That is why First Rank Measurement (FRM) and First Two Rank Measurement (FTRM) have been proposed for rejection [6].

FRM is one of the most useful criteria, which takes into account only FR of the output vector. It rejects samples by setting a threshold T_1 to FR and accepts those satisfying $FR \ge T_1$.

FTRM is another important index for rejection. Unlike FRM, it emphasizes the gap between FR and SR. It sets a threshold T_2 to the gap and accepts only the samples satisfying $FR - SR \ge T_2$.

Besides these two well-known rejection criteria, He et al propose a novel LDA measurement (LDAM) in [6, 7], which relies on the principle of Fisher Linear Discriminant Analysis. They apply the principle of LDA on outputs for the rejection option as a one dimensional application which shifts the Fisher criterion to

$$J(w) = \frac{S_B}{S_W} = \frac{(\mu_1 - \mu_2)}{\Sigma_{12}} \tag{3}$$

where μ_1 and μ_2 are the centers of two classes and Σ_{12} is within-class scatter.

Then they define two classes for rejecting and accepting samples: $G^{(1)} = \{\hat{f}_1\}$ and $G^{(2)} = \{\hat{f}_2, ..., \hat{f}_c\}$, in order to maximize the separation between FR and all the other confidence values. (Here $\hat{f}_i$ are confidence values in a descending order). Thus, in LDA, $J(w)$ can be defined by:

$$J(w) = \frac{\left\{ \sum_{i=2}^{c} (\hat{f}_1 - \hat{f}_i) \right\}^2}{(c-1)^2 \Sigma_{12}} \tag{4}$$

where $\mu_1 = \hat{f}_1, \mu_2 = \frac{1}{c-1} \sum_{i=2}^{c} \hat{f}_i$, $\Sigma_1 = 0$, $\Sigma_2 = \frac{1}{c-1} \sum_{i=2}^{c} (\hat{f}_i - \mu_2)^2$ and $\Sigma_{12} = \frac{1}{2} \Sigma_2$.

Then a threshold T_3 is set and samples are accepted if they satisfy $J(w) \ge T_3$. The criterion has been proved to produce a better performance than FRM and FTRM based on eight-direction gradient feature with SVM classifier for handwritten character recognition [6, 7].

2.2   SVM-Based Rejection Measurement (SVMM)

The previous rejection criteria have been designed based on some heuristic ideas. In this section, we propose a new SVM-based rejection measurement (SVMM) to extend the rejection process into a learning based method. Specifically, rejection can be viewed as a two-class recognition problem, one stands for rejected samples and the other for accepted ones. For a classifier, the output of a sample is a vector of confidence values $\{f_1, f_2, \dots, f_c\}$, $f_i \geq 0$, $i = 1, 2, \dots, c$, as mentioned before. Then these values are extracted as features and sorted into a descending order:

$$\{\hat{f}_1, \hat{f}_2, \dots, \hat{f}_c\}, \quad \hat{f}_1 \geq \hat{f}_2 \geq \cdots \geq \hat{f}_c \tag{5}$$

The correctly and wrongly classified samples are labeled differently (correctly classified samples with label "1" while incorrectly classified ones with label "-1") and used to train an SVM classifier. Linear SVM is selected for training to locate the rejection boundary. So the decision boundary is a linear function combining all the components of the output vector, represented in Eq. (6). ($\{w_i\}_{i=0}^c$ are the coefficients of SVM)

$$T = \sum_{i=1}^c w_i \hat{f}_i + w_0 \tag{6}$$

The reason for choosing a linear kernel for SVM rather than a nonlinear one, like RBF kernel, is based on the following points:

1. A linear kernel works very fast in training and testing and an optimal linear separating boundary is a good way to avoid over-fitting.
2. A linear boundary is more meaningful physically and function (6) includes some special cases in it. For instance, FRM can be viewed as a linear boundary with $w_1 = 1$ and $w_2 = w_3 = \cdots = w_c = w_0 = 0$; while FTRM can be viewed as: $w_1 = 1$, $w_2 = -1$ and $w_3 = w_4 = \cdots = w_c = w_0 = 0$.

Note that in the training process of SVMM, the number of samples in class "1" is always much larger than that of class "-1", because the baseline accuracy of the classifier is high. In this case, the problem is an unbalanced classification problem. To solve this problem, we use different weighting functions for different classes in the "libsvm" software [8]. In the testing process, the same features are extracted and sorted into descending order, and a sample is rejected if T in Eq. (6) is smaller than a pre-defined threshold.

With this new criterion, the linear rejection boundary is located by training an SVM with training data. The main difference between SVMM and other criteria, like FRM, FTRM and LDAM, is that SVMM extends the rejection process from heuristic design to learning based procedure. Using learning based method on the training set to predict the rejection on testing samples is more straight-forward and can make use of much more information from the data.

3   Rejection with Multiple Classifier System

3.1   Construction of Multiple Classifier System (MCS)

Since convolution neural network (CNN), especially MCS based on CNN, works effectively in handwritten character recognition as shown in [4, 9, 10], it is selected as the core classifier and MCS is built on it in our strategy. The CNN classifier is based on the principle of deep learning. It processes the raw images of samples and extracts useful trainable features to classify samples into different categories [1].

Re-sampling the dataset (with Bagging [11], Boosting [12] and so forth) and changing the classifier (in structure or type [13]) are two main ways to produce committees. Many researchers have used these methods to produce a group of classifiers and applied certain combination methods for recognition. Some of them have achieved extremely high recognition rate in handwritten numeral recognition with CNN model on MNIST dataset [9, 10].

For the construction of the MCS, we select the CNN model in [4] as the basis model "M0". It has three convolution layers with 25, 50 and 100 feature maps sequentially, and one output layer which is fully connected to the last convolution layer. Two modifications have been explored: one is changing the number of feature maps in each of three convolution layers in both increasing and decreasing ways to build new models. The other is using "Bagging" method (i.e. dataset re-sampling) to randomly select samples for the training sets to train the same CNN model numerous times. The structures of the modified classifiers are listed in Tables 1 and 2, while the information of re-sampling datasets is listed in Table 3.

3.2   Rejection Based on MCS

MCS for Recognition VS Rejection. MCS with different combination methods are often used in pattern recognition to enhance the recognition rate. In handwritten numeral recognition, some researchers have yielded state-of-the-art performance in recognition based on differently designed MCSs. On the MNIST numeral dataset, a recognition rate of 99.73% is achieved with an MCS consisting of 35 classifiers [9]; Wu et al obtained an even better recognition rate of 99.77% based on a MCS with 5 CNNs based on different training sets as well as different operations of spatial pooling [10].

Although MCS has contributed a lot to recognition, it is seldom used for pattern rejection. As it is so effective in recognition, it is assumed to be useful in rejection as well. Therefore, we attempt to adopt MCS to the rejection problem. In [14, 15], the authors apply MCS for rejection based on the cascading methods and achieve high performances. In this paper, a committee approach for MCS rejection is used.

Voting Based Combination Method for MCS Rejection. For the purpose of combining multiple classifiers, voting is always a good choice for the reason that it is simple and effective. Hard voting is the simplest voting method which assigns equal weight to all votes. Another frequently used method is soft voting, which assigns a weight to each classifier according to its performance [16, 17]. For the weights part, all the rejection criteria mentioned in Section 2 can be selected for the reason that they reflect the rejection performance of a single classifier. A certain type of rejection criterion is assigned to each model in the voting procedure, and the class label with the highest voting value provides the final decision for each sample.

Suppose there are N different classifiers in the MCS, denoted as $g_1, g_2, \ldots, g_N$, for a random pattern, each classifier $g_i(i = 1,2, \ldots, N)$ would provide a prediction of the label y_i as well as an output vector $\{f_1^i, f_2^i, \ldots, f_c^i\}$. Then for each classifier, the selected rejection criterion (FRM, FTRM or SVMM) can be calculated based on the output vector $\{f_1^i, f_2^i, \ldots, f_c^i\}$, denoted as $t_i(i = 1,2, \ldots, N)$. (For the reason that LDAM does not work as effectively as the other criteria, it is not considered for combination.) The above-mentioned method is the *soft voting*. We also consider the *hard voting* method by simply setting $t_i = 1$. After that, a voting value $V_j(j = 1,2, \ldots, c)$ is calculated for each class denoted as:

$$V_j = \sum_{i=1}^N t_i I(y_i, j), \quad I(y_i, j) = \begin{cases} 1 & if \ y_i = j \\ 0 & otherwise \end{cases} \qquad (7)$$

Among V_j, a maximum voting value $V_{max} = \max_{1 \le j \le c} V_j$ can be found and a threshold T_{com} is searched and determined. A pattern is rejected if V_{max} is smaller than a threshold. As the voting values are sums of all models, the thresholds T_{com} can be any real numbers between 0 and N. But for the hard voting method, the threshold can only be an integer which cannot yield a reliability-rejection curve. The whole procedure of MCS based pattern rejection is shown in Fig. 1.

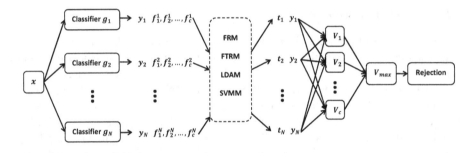

Fig. 1. Flow chart of voting based combination of MCS for pattern rejection

4   Experiments

4.1   Multiple Classifier System

Two well-known datasets are selected for these experiments including CENPARMI [18] and MNIST [19] handwritten numeral datasets. The former contains 4000 training samples and 2000 testing samples with no-fixed size while the latter contains 60000 training samples and 10000 testing samples with identical size of 28 by 28 pixels.

Firstly, structural modification (SM) method [20] is conducted to build committees. For the CENPARMI data, we increase the numbers of feature maps in each convolution layer (C1, C3 and C5) of the basic model and train all the models to 150th epoch as shown in Table 1. For the MNIST data, these numbers are slightly changed in both increasing and decreasing directions as listed in Table 2 below. Secondly, dataset re-sampling (DR) method is used on CENPARMI data. In this phase, model structure is fixed as the basic one. Different training sets are formed by randomly selecting 2000 training samples and distorting them with elastic algorithm [3]. The process is repeated four times to obtain 4 different training sets (G1-G4) with 4000 samples each, as listed in Table 3. The numbers in the first 10 columns represent the numbers of samples selected in different categories for different training sets.

Table 1. Information about modified structures in MCS with CENPARMI dataset

	M0 (basis)	M1	M2	M3
C1	25	50	50	70
C3	50	75	90	75
C5	100	120	100	100
Training Error Rate (%)	0.5	0.38	0.38	0.43
Testing Error Rate (%)	2.45	2.45	2.25	2.45

Table 2. Information about modified structures in MCS with MNIST dataset

	M0	M1	M2	M3	M4	M5	M6
C1	25	25	25	25	25	10	40
C3	50	50	50	30	80	50	50
C5	100	80	120	100	100	100	100
Training Error Rate (%)	0.36	0.34	0.31	0.34	0.26	0.34	0.29
Testing Error Rate (%)	0.62	0.63	0.61	0.6	0.58	0.63	0.61

Table 3. Information about re-sampling training sets with CENPARMI data

	0	1	2	3	4	5	6	7	8	9	Training Error Rate (%)	Testing Error Rate (%)
G1	474	462	416	350	332	394	380	370	400	422	1.65	2.80
G2	450	408	358	404	394	382	424	424	396	360	1.52	3.65
G3	458	482	408	340	372	410	392	426	386	326	1.27	3.50
G4	402	440	380	390	430	426	370	412	350	400	1.77	3.45

4.2   Comparison of Different Rejection Criteria

In the selected CNN model, the output of each sample is a 10-dimention vector consisting of confidence values for possible classes. FRM, FTRM and LDAM are used respectively as rejection criteria with the basic model. Thresholds are searched incrementally. As in CNN model, the outputs are confidence values instead of probabilities, the most appropriate starting point, step and ending point for thresholds searching vary according to different rejection criteria. For the newly proposed SVMM, "libsvm" tools are applied and the same CNN model is used as a feature extractor. Totally, there are 216 out of 60000 samples labeled "-1" while the rest are labeled "1" for the training process. Since the training set is so unbalanced with the number of samples in class "1" almost 300 times that of class "-1", the weight parameter is set to "400" for class "-1". A linear kernel is selected in order to find a linear decision boundary in the feature space. Normalization is conducted on the decision value with SVM of each sample on purpose of making the threshold-setting procedure more convenient. Then different thresholds are set for rejection. All the results are shown by the curves presenting the relationship between the number of rejected samples and reliability in Fig. 2.

Results show that, although LDAM is proved to have a better performance than FRM and FTRM in [7] based on eight-direction gradient feature with an SVM classifier, it is the least useful one in our experiment with the CNN model. The performances of FRM and FTRM which are far different in [7] are insignificantly different in CNN model "M0". So it can be concluded that these pre-defined criteria vary in performance with different classifier models or types of features.

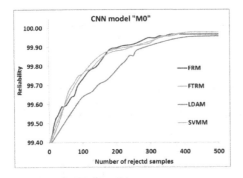

Fig. 2. Relationship between number of rejected sample and reliability in "M0"

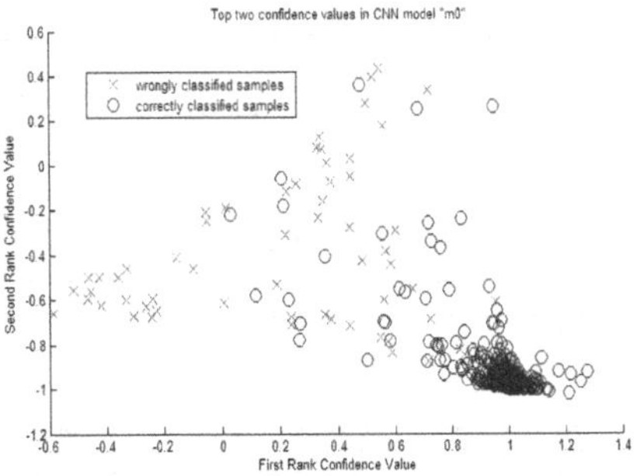

Fig. 3. Samples in FR-SR feature space

From Fig. 3, FR and SR of correctly classified samples are extremely close to 1 and -1 respectively. As a result, a line with slope "1" standing for FTRM is an optimal boundary to separate wrongly and correctly classified samples. That is why FTRM is an effective criterion for rejection. Another effective criterion FRM can also be viewed as a problem of finding a boundary parallel to the y-axis in Fig. 3, which, by observation, is less effective than FTRM. However, it is noticed that although these two criteria can be useful, many correctly classified samples will also be rejected by them no matter where the boundary is.

It is also shown in Fig. 2 that SVMM works as effective as FTRM in rejection and the two are always the relatively best ones among all of the criteria. Similar results appear when we applied these criteria to all the modified CNN models, as displayed in Fig. 4. Besides, it is noticed that the performances of FTRM and SVMM are too close to determine which one is better. The reason for this can be traced back to the training process of CNN model when the expected values in the decision layer are set to be "1" for the true class and "-1" for the other classes. Hence, FTRM is already a distinctively effective criterion to determine the quality of a sample as analyzed with Fig. 3. When we use the SVMM, which uses all the values of the output vector, FR and SR contribute much more than the others since the others are slightly different from SR. Therefore, the rejection boundary of SVMM is very close to that of FTRM. This explains the similar performances of these two criteria.

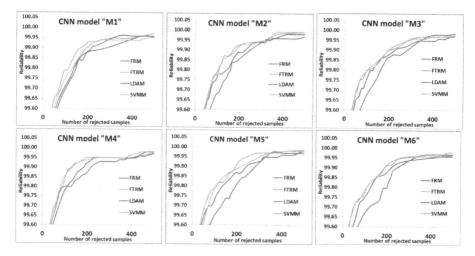

Fig. 4. Relationship between number of rejected sample and reliability in modified models

4.3   Pattern Rejection with MCS

Voting Based Combination Method. In this experiment, hard voting and soft voting rejection methods are both conducted based on the MNIST dataset with MCS built by SM.

In hard voting, a range limitation problem makes the rejection process inflexible for the reason that the thresholds can only be set to several integers. Once the maximum value (number of classifiers in the MCS) is reached, the reliability cannot be improved anymore. The highest reliability is 99.86% with 118 samples rejected when the threshold is set to "7".

In soft voting, the proposed combination method has been applied with FRM, FTRM and SVMM respectively. Since these criteria have different value ranges, different starting points, search steps and ending points are chosen. For FRM and SVMM, the starting and ending points are 0 and 1 respectively; while for FTRM, the starting and ending points are 0 and 2. The search steps for all of them are 0.1 at regular places and 0.01 at the sections where the number of rejected samples changes sharply based on different criteria. The results are shown in Fig. 5. We can find that with the combination of seven CNN models, the rejection performances are consistently improved for all rejection criteria (FTM, FTRM and SVMM).

Structural Modification (SM) and Data Re-sampling (DR). In this section, we adopt the soft voting combination rejection method with MCS on the CENPARMI handwritten numeral dataset. The MCS is constructed in two different ways including SM and DR, as presented in Section 4.1. FTRM is chosen as weight for soft voting combination and thresholds are searched from 0 with an incremental step of 0.05 until suitable reliability values are reached. The results are shown as curves displaying the relationship between number of rejected samples and reliability, presented in Fig. 6.

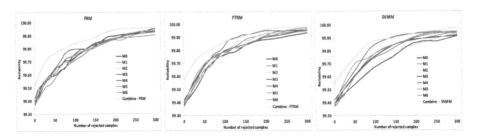

Fig. 5. Relationship between number of rejected sample and reliability with MCS and single models based on different rejection criteria

From these two figures, it is proved again that soft-voting combination method with MCS could improve the rejection performance of the system no matter which method is adopted to construct the MCS. Furthermore, it is shown in Fig. 6 that with our combination method, although MCS does not necessarily improve the recognition rate (without rejection), it can still improve the rejection performance of the whole system.

Table 4 below lists some important information about the performance of different rejection methods based on the CENPARMI dataset. In [7], it is claimed that using LDAM, a reliability of 99.67% is achieved with 175 samples rejected. With our combination methods, the MCS with SM (Com-SM) obtains a reliability of 99.78% with only 164 samples rejected and 99.89% with 180 rejected. The other MCS with DR (Com-DR) achieves the same reliability as LDAM with 6 less samples rejected and 99.73% with 179 samples rejected. Both of these two construction methods with MCS obtain better rejection results than state-of-the-art rejection method based on the same dataset.

Comparing two different construction methods of MCS (SM and DR), it is clear that the system with DR performs better than that with SM. As shown in Table 4, to reach a reliability of 99.94%, DR should reject 257 samples while SM should reject 393 samples, even if the original recognition rate (without rejection) of DR is smaller than that of SM (see Table 1 and 3). This indicates that building MCS with DR makes errors between different classifiers in the system much more diverse.

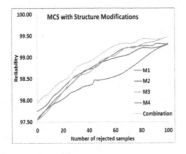
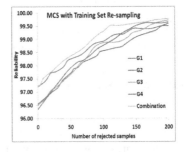

Fig. 6. Relationship between number of rejected sample and reliability with MCS built by different methods

Table 4. Rejection performances of different rejection methods based on CENPARMI dataset

Number of rejected samples	Reliability	Method
175	99.67%	[7]
164	99.78%	Com-SM
180	99.89%	Com-SM
169	99.67%	Com-DR
179	99.73%	Com-DR
393	99.94%	Com-SM
257	99.94%	Com-DR

5   Conclusion

In this paper, a novel SVM-based rejection measurement and voting based combination methods with multiple classifier system (MCS) for rejection are proposed. The main difference between SVMM and other criteria (FRM, FTRM, LDAM and so forth) is that SVMM finds the rejection boundary based on the training data rather than experiences as in those pre-defined criteria. The voting based combination method of MCS is a new attempt to adopt MCS for the purpose of rejection. In the soft voting method, different rejection criteria (FRM, FTRM and SVMM) are used as weights for different models since they reflect their rejection effectiveness. Experiments are conducted on well-known MNIST and CENPARMI digit datasets. Different MCSs are constructed with two different building methods, structural modification and dataset re-sampling. The results show that no matter what building method is chosen or what criterion is selected as weight in soft voting, rejection based on MCS can improve the rejection performance of the system consistently. It is also indicated that MCS built by dataset re-sampling works better than that by structural modification in rejection.

References

1. Lauer, F., Suen, C.Y., Bloch, G.: A trainable feature extractor for handwritten digit recognition. Pattern Recognition 40(6), 1816–1824 (2007)
2. Zhang, P., Bui, T.D., Suen, C.Y.: A novel cascade ensemble classifier system with a high recognition performance on handwritten digits. Pattern Recognition 40(12), 3415–3429 (2007)
3. Simard, P.Y., Steinkraus, D., Platt, J.C.: Best practice for convolutional neural networks applied to visual document analysis. In: Int'l Conf. on Document Analysis and Recognition, pp. 958–962 (2003)
4. Niu, X.X., Suen, C.Y.: A novel hybrid CNN-SVM classifier for recognition handwritten digits. Pattern Recognition 45(4), 1318–1325 (2012)
5. Liu, C.L., Nakashima, K., Sako, H., Fujisawa, H.: Handwritten digit recognition: investigation of normalization and feature extractor techniques. Pattern Recognition 37(2), 265–279 (2004)

6. He, C.L., Lam, L., Suen, C.Y.: A novel rejection measurement in handwritten numeral recognition on linear discriminant analysis. In: Int'l Conf. on Document Analysis and Recognition, pp. 451–455 (2009)

7. He, C.L., Lam, L., Suen, C.Y.: Rejection measurement based on linear discriminant analysis for document recognition. Int'l J. of Document Analysis and Recognition 14(3), 263–272 (2011)

8. Chang, C.C., Lin, C.J.: LIBSVM: A library for support vector machine. Journal ACM Transactions on Intelligent Systems and Technology 2(3), 27:1–27:27 (2011)

9. Ciresan, D.C., Meier, U., Gambardella, L.M., Schmidhuber, J.: Convolutional neural network committees for handwritten character classification. In: Int'l Conf. on Document Analysis and Recognition, pp. 1135–1139 (2011)

10. Wu, C., Fan, W., He, Y., Sun, J., Naoi, S.: Cascaded heterogeneous convolution neural networks for handwritten digit recognition. In: Int'l Conf. on Pattern Recognition, pp. 657–660 (2012)

11. Breiman, L.: Bagging predictors. Machine Learning 24(2), 123–140 (1996)

12. Freund, Y., Schapire, R.E.: Experiments with a new boosting algorithm. In: Int'l Conf. Machine Learning, pp. 148–156 (1996)

13. Dietterich, T.G.: Ensemble methods in machine learning. In: Kittler, J., Roli, F. (eds.) MCS 2000. LNCS, vol. 1857, pp. 1–15. Springer, Heidelberg (2000)

14. Zhang, P., Bui, T.D., Suen, C.Y.: A novel cascade ensemble classifier system with a high recognition performance on handwritten digits. Pattern Recognition 40(12), 3415–3429 (2007)

15. He, C.L., Suen, C.Y.: A hybrid multiple classifier system of unconstrained handwritten numeral recognition. Pattern Recognition and Image Analysis 17(4), 608–611 (2007)

16. Suen, C.Y., Lam, L.: Multiple classifier combination methodologies for different output levels. In: Kittler, J., Roli, F. (eds.) MCS 2000. LNCS, vol. 1857, pp. 52–66. Springer, Heidelberg (2000)

17. Lam, L., Suen, C.Y.: Optimal combinations of pattern classifiers. Pattern Recognition Letters 16(9), 945–954 (1995)

18. Suen, C.Y., et al.: Computer recognition of unconstrained handwritten numerals. Proc. IEEE 80(7), 1162–1180 (1992)

19. LeCun, Y.: The MNIST database of handwritten digits,
http://yann.lecun.com/exdb/mnist

20. Yuan, A., Bai, G., Jiao, L., Liu, Y.: Offline handwritten English character recognition based on convolutional neural network. In: Int'l Workshop Document Analysis Systems, pp. 125–129 (2012)

MRF-Based Multiple Classifier System for Hyperspectral Remote Sensing Image Classification

Junshi Xia[1], Peijun Du[2,*], and Xiyan He[1]

[1] GIPSA-lab, Grenoble Institute of Technology, 38400 Grenoble, France
[2] Jiangsu Provincial Key Laboratory of Geographic Information Science and Technology, Nanjing University, 210093 Nanjing, China
dupjrs@gmail.com

Abstract. Hyperspectral remote sensing image (HRSI) classification is a challenging problem because of its large amounts of spectral channels. Meanwhile, labeled samples for supervised classifier is very limited. The above two reasons often lead to unstable classification result and poor generalization capacity. Recent research has demonstrated the potential of multiple classifier system (MCS) for producing more accurate classification result. In addition, another vital aspect of HRSI classification is spatial contents. Markov random field (MRF), which takes the spatial dependence among neighborhood pixels based on the intensity field from observed data into consideration, is always adopted as an effective way to integrate the spatial information. In this paper, we proposed an effective framework for classifying HRSI image, called MRF-based MCS, which are based on the aforementioned two powerful algorithms. The proposed model is validated by multinomial logistic regression (MLR) classifier. Experimental results with hyperspectral images collected by the NASA Jet Propulsion Laboratory's Airborne Visible Infra-Red Imaging Spectrometer (AVIRIS) demonstrate that MRF-based MCS is a promising strategy in the context of hyperspectral image classification.

1 Introduction

Hyperspectral remote sensing image classification is a challenging problem because of its high dimensionality (hundred of bands) and limited availability of training samples. Therefore, we need some advanced approaches to generate the high performance classification result. Over the course of the past 10 years, the advent of a novel machine learning scheme named as multiple classifier system had a significant impact on improving HRSI classification accuracy [1–4]. MCS, just like its name, combines the individual classifier's output according to a certain algorithm (such as majority vote) or based on an iterative error minimization [5, 6]. The output of individual classifier can be constructed by several strategies, such as: different classifiers using the same/different training set, same

---

* Corresponding author.

Z.-H. Zhou, F. Roli, and J. Kittler (Eds.): MCS 2013, LNCS 7872, pp. 343–351, 2013.
© Springer-Verlag Berlin Heidelberg 2013

classifier using different training samples etc. Many researchers have investigated the performance of HRSI classification using different MCS approaches. For instance, Foody *et al* [7] adopted majority voting rule to combine multiple binary classifiers for mapping a specific class. Doan and Foody [8] explored the combination of soft classification methods and found these methods could improve the accuracy. A wealth of information for remote sensing image classification using MCS technologies can be found in [4].

If the spatial information of HRSI image (especially for the high spatial resolution) is not considered, the thematic map, which includes salt and pepper classification noise, looks very noisy. Accordingly, it is essential to take into account spatial information. This topic is named as *spectral-spatial* classification. Many studies have been carried out on this topic. To the best of our knowledge, *spectral-spatial* analysis for HRSI classification can be divided into the following groups [9, 10]:

- Mathematical morphology. Results of morphological operators over features from original images or calculated by feature extraction/selection methods are treated as the input of classifiers.
- Segmentation and majority vote. Firstly, the neighbored regions using a spatial or spectral segmentation are designed. Secondly, the most frequently class derived from the supervised classifier in a region is adopted as the final class.
- Composite kernels. This refers to kernel-based classifier, in which multiples kernels are used to combine spatial and spectral features.
- Markov random field.
- Other approaches consists of tensor modeling, context-based classification etc.

In many cases, the exploitation of spatial information in classification is obtained through MRF, a probabilistic model that is commonly used to integrate spatial information into image classification [11, 12]. In the MRF framework, maximum a posteriori (MAP) probability is one of the most popular statistical criteria for optimality the energy function. Tarabalka *et al* [13] used MRF method as a post-processing scheme to a probability-SVM classification map. The used MRF framework is called metropolis algorithm, based on stochastic relaxation and annealing. Li *et al* [14] integrated the spectral and spatial information into a Bayesian framework, and then used a multinomial logistic regression (MLR) classifier to learn the posterior probability for the spectral information. Finally, spatial information is considered using a multilevel logistic MRF prior.

In this paper, we present a novel framework for enhancing the performance of HRSI classification by combining MCS and MRF. In particular, we propose to integrate the multiple classification results obtained by MLR classifier with different training samples using MCS approaches. The MLR classifier is used to generate a spectral-based classification map, whereas MCS is chosen to provide an ensemble of classification maps. To generate the final classification result, we propose to aggregate further ensemble pixel-based classification maps through different MRF methods. More details can be seen in Section 2.

The structure of the remainder paper is as follows. Section 2 is a fine description of MRF-based MCS framework. In Section 3, we describe the preliminary results currently in progress to verify the performance of the proposed framework. Finally, Section 4 gives the conclusions and future directions of research.

2   MRF-Based MCS

2.1   Framework Overview

MRF-based MCS can be summarized as the flowchart in Fig.1. Firstly, we randomly select training samples (repeat M times, M is the ensemble size) to train a supervised classifier. Secondly, we adopt some rules of MCS to combine the individual classification outputs. Finally, we use MRF regularization to obtain the final result. Assuming that hyperspectral image has n pixel vectors $\mathbf{X} = \{\mathbf{x}_i \in \mathcal{R}^D, i = 1, 2, ..., n\}$, $\mathbf{Y}$ is the class label of image $\mathbf{Y} = \{y_i, i = 1, 2, ..., n\}$. $\Omega = \{\omega_1, \omega_2, ..., \omega_k\}$, k is the number of class of interest.

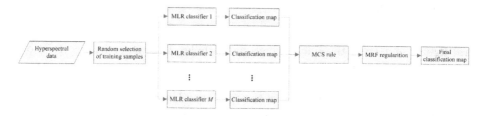

Fig. 1. The flowchart of MRF-based MCS framework

2.2   Multiple Classifier System

MCS combines class labels or probability from multiple classifiers. The final output mainly depends on the supervised classifier and the diversity among the classification results. In this paper, MLR is chosen due to its capability of offering excellent HRSI classification accuracy with short computation time [14]. This classifier can generate the class labels and probabilities, respectively. Furthermore, majority voting (MV) and Bayesian average (BA) are selected as the MCS rules, respectively. For the purpose of MRF in the next step, we need both the class label and probabilities.

MV is the most popular MCS method by which each individual classifier votes for the specific class, and the class that collects the majority votes is predicted as the final output [6]. Class probabilities, which is equal to the number of times that the class is predicted of class divided by ensemble size, is defined as follows:

$$P(\mathbf{x}_i | y_i \in \Omega_j) = \frac{\sum_{l=1}^{M} \delta(y_{l,i}, \Omega_j)}{M} \tag{1}$$

where, $y_{l,i}$ represents the class label of pixel i in $\{l^{th}, l = 1, 2, ..., M\}$ classifier, $\delta(\cdot, \cdot)$ is a Kronecker delta function, when $\alpha = \beta$, $\delta(\alpha, \beta) = 1$ and otherwise $\delta(\alpha, \beta) = 0$.

BA is used to linearly average the probabilistic derived from multiple classifiers [6, 15]. The class label is decided by which has the largest probabilities. It can be defined as follows:

$$P(\mathbf{x}_i|y_i \in \Omega) = \frac{1}{M} \sum_{j=1}^{M} P_j(\mathbf{x}_i|y_i \in \Omega), i = 1, 2, \ldots, n \qquad (2)$$

Then, the class label is decided:

$$y_i = \arg\max P(\mathbf{x}_i|y_i \in \Omega) \qquad (3)$$

where $P_j(\mathbf{x}_i|y_i \in \Omega)$ represents the probability of pixel i in $\{j^{th}, j = 1, 2, ..., M\}$ classifier.

2.3   MRF Regularization

The pixel-wise classifications only consider the image pixels as the discrete spectral signals, not treat the image pixels as a whole. And the spatial correlation between the images is ignored at all. Thus, the boundaries of the objects in image is hard to distinguish, especially for the low spatial resolution hyperspectral image. Therefore, the intergration of the spatial and spectral information is necessary for the context of hyperspectral remote sensing image classification. MRF is a powerful mathematical framework to incorporate spatial information by exploring the relationships between neighborhood image pixels. The basic principle of MRF is that the label of a pixel depends only on itself and its nearest ones among all the neighbors [16]. In some previous studies, a energy cost function with both spectral information and spatial relation is established to construct MRF model [13, 16].

Generally speaking, MRF models tend to solve the following optimization problem of a energy cost function:

$$\arg\min \sum_{i=1}^{n} E = \arg\min \sum_{i=1}^{n} E_{spectral}(\mathbf{x}_i) + E_{spatial}(\mathbf{x}_i) \qquad (4)$$

where, $E_{spectral}(\mathbf{x}_i)$ and $E_{spatial}(\mathbf{x}_i)$ are the spectral and spatial energy function, respectively. The spectral energy function is defined as:

$$E_{spectral}(\mathbf{x}_i) = -\ln P\{\mathbf{x}_i|y_i\} \qquad (5)$$

where, $P\{\mathbf{x}_i|y_i\}$ can be achieved from MCS rules. And the spatial energy function is described as:

$$E_{spatial}(\mathbf{x}_i) = -\beta \sum_{j \in C} \delta(y_i, y_j) \qquad (6)$$

where, C is the neighborhood of pixel i, β controls the importance of spatial information.

The above optimization problem, which involves pairwise and unary interaction terms, is very difficult to compute. In order to tackle this optimization problem, we exploit two different MRF models: one is based on stochastic relaxation and annealing (MRF-SA) [13], and the other is based on Graph Cut (MRF-GC) [17].

MRF-SA model firstly randomly selects a pixel in the image and computes the local energy E. Then, a new class label is assigned to this pixel and the new local energy E^{new} is calculated. Finally, we compare the new local energy with old local energy, if the new local energy is larger than the old local energy, the new class label is assigned to this pixel. Otherwise, the new class assignment is accepted with the probability $P = \exp((E^{new} - E)/T)$, T is the control parameter. The above procedures repeat N times (N is defined by the user).

Graph Cut(GC) is the development of fast algorithms for labeling MRF. GC constructs MRF model on a graph with nodes and edges and solve the minimization of an energy cost function as a maximum flow problem [16]. In this paper, α-expansion algorithm is used to solve the optimization problem of a cost function [18], because the Kronecker delta function $\delta(\cdot, \cdot)$ in spatial energy is *metric* on the space of labels. A metric means that $\delta(\cdot, \cdot)$ satisfies three conditions:

$$\delta\left(y_i, y_j\right) = 0 \Leftrightarrow y_i = y_j \tag{7}$$

$$\delta\left(y_i, y_j\right) = \delta\left(y_j, y_i\right) \geq 0 \tag{8}$$

$$\delta\left(y_i, y_j\right) \leq \delta\left(y_i, y_k\right) + \delta\left(y_k, y_j\right) \tag{9}$$

More details about α-expansion algorithm can be seen in [17]. And this algorithm yields very well approximations and is quite efficient from a computational point of view, with the computational complexity of $O(n)$ [14].

3   Experimental Results

3.1   Datasets and Experiment Design

AVIRIS dataset collected over a vegetation area of *Indian Pines*, Indiana, USA, is used to validate the performance of MRF-based MCS framework. The image size is 145 rows by 145 columns. The image includes 200 spectral bands after removing twenty water absorption bands (104-108, 150-163 and 220). The ground truth, that is used as training and test samples, consists of 16 classes with 10366 pixels, which is detailed in **Table.**1 [19]. The spatial resolution is 20 m/pixel.

Experiments achieved by MRF-based MCS are conducted to illustrate the influence of different number of training samples and the impact of ensemble size. From **Table.**1, the minor class named as *Hay-windrowed* has only 20 data points. Thus, we randomly select training samples comprised of 80, 160, 240 and

Table 1. Classes for AVIRIS image and the number of ground truth

Index	Classes	Number	Index	Classes	Number
1	Alfalfa	54	9	Hay-windrowed	20
2	Corn-no till	1434	10	Soybeans-no till	968
3	Corn-min till	834	11	Soybeans-min till	2468
4	Corn	234	12	Woods	614
5	Grass/pasture	497	13	Wheat	212
6	Grass/tree	747	14	Soybeans-clean till	1294
7	Grass/pasture-mowed	26	15	Bldg-Grass-Tree-Drives	380
8	Oats	489	16	Bldg	95

320 (5, 10, 15 and 20 per class, respectively) from the ground truth. We also investigate the impact of different ensemble size(10, 40, 70, 100) to classification accuracy. For both two MRF models, β is set to 4. For MRF-SA, T is initial set to 2. After every 10^6 iterations, T for the next $\eta + 1$ iteration is computed as $T^{\eta+1} = 0.98T^{\eta}$. In this paper, we just focus on the MCS and MRF-based MCS results, so the default parameters of individual classifier are adopt. Overall accuracy (OA), Average accuracy (AA) and kappa coefficient (κ) are treated as the quantitative indices. In order to increase the statistical significance of the results, each value of OA reported in this experiment is obtained from 10 Monte Carlo (MC) runs.

3.2   Results

Fig.2 and 3 show the average overall accuracies (OA%)(after 10 MC runs) obtained by the best of single classifiers, Single-MRF-SA, Single-MRF-GC, MV, BA, MV-MRF-SA, BA-MRF-SA, MV-MRF-GC and BA-MRF-GC using different number of training samples and ensemble size. The MV and BA are quite

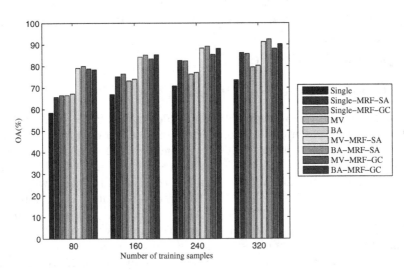

Fig. 2. Average OA of using different number of training samples, the ensemble size is fixed to 10

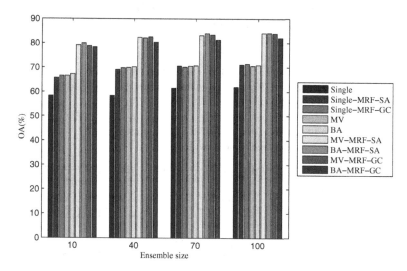

Fig. 3. Average OA of using different ensemble size, the number of training samples is fixed to 80

effective methods among the MCS algorithms. Compared with the single classifier, both MV and BA obtain more accurate results and achieve higher overall accuracies. In this dataset, BA is superior to MV. MRF regularizations on single classifier significantly improve the classification result. They indicate the effectiveness of MRF regularization. Compared to the results derived from MCS rules and single classification using MRF regularization, the four MRF-based MCS models significantly increase the classification accuracies. And MRF-SA methods outperform MRF-GC approaches. For example, when the number of training samples is 80 (5 samples per class) with the ensemble size $M = 10$, the OAs of Single, Single-MRF-SA, Single-MRF-GC, MV and BA are 58.38%, 65.73%, 66.68%, 66.56% and 67.28%, MV-MRF-SA, BA-MRF-SA, MV-MRF-GC and BA-MRF-GC gain the OAs of 79.13%, 79.99%, 78.8% and 78.36%, respectively. The standard deviations of MCS and MRF-based approaches are very small, indicated that all MCS and MRF-based methods are the stable classifiers. For different training samples (Fig.2), the general trend is that OA greatly improve when the number of training samples rise. The general trend of different ensemble size (Fig.3) is similar to the one of different training samples. OA increased slightly when ensemble size increase. Both more training samples and ensemble size can improve the classification result, but the computation time is also increased.

In order to compare the class-specific accuracies, we list all accuracies of each class using the above algorithms with the training samples of 160 and the ensemble size of 10 (only one run). The results are summarized in Table.2. It can be observed that MRF-based MCS models not only enhance the overall accuracy, but also improve all the class-specific accuracies. For instance, compared to *Single* classifier, BA-MRF-SA achieves the OA of 86.47%, with the improvement of 20.56 percentage points, *Soybeans-clean till* of BA-MRF-GC improves the accuracy from 61.75% to 99.17%. Different MRF-based MCS approaches lead to the best classification results for different class. MV-MRF-SA gains the best accuracies of class *corn,woods*, MV-MRF-GC presents the best results for class *Grass/pasture*, *Soybeans-no till*, BA-MRF-SA gives the best accuracies for class *Corn-min till,Soybeans-min till* and the other classes get the best performance using BA-MRF-GC method.

Table 2. Classification accuracies for the Indian Pines image using 160 training samples (10 samples per class) with the ensemble size $M = 10$

Class	Single	Single-MRF-SA	Single-MRF-GC	MV	BA	MV-MRF-SA	BA-MRF-SA	MV-MRF-GC	BA-MRF-GC
1	84.09	97.73	88.64	90.91	90.91	**100**	**100**	**100**	**100**
2	58.85	68.89	55.55	67.77	65.80	75.98	76.47	76.69	**80.55**
3	59.83	66.02	60.92	65.05	63.23	87.86	**88.35**	80.34	79.61
4	76.34	99.11	67.86	91.07	91.52	98.66	**100**	99.55	**100**
5	72.90	70.84	82.75	86.24	85.83	85.01	83.78	**94.25**	83.37
6	79.51	97.29	98.24	94.98	94.84	98.37	**99.19**	97.96	**99.19**
7	100	100	100	100	100	100	100	100	100
8	91.23	97.7	96.03	96.24	97.29	**99.79**	99.37	99.37	99.37
9	100	100	100	100	100	100	100	100	100
10	45.82	82.36	61.84	67.01	67.12	74.22	75.78	**76.2**	75.16
11	61.55	52.12	72.34	52.56	56.18	76.44	**82.34**	66.8	72.21
12	61.75	85.43	96.19	81.95	84.77	72.52	97.02	97.02	**99.17**
13	99.5	99.5	99.5	99.5	99.5	99.5	99.5	99.5	99.5
14	76.95	86.99	98.13	90.5	89.72	**93.85**	90.42	91.43	83.41
15	50.81	68.34	77.84	65.14	66.49	78.65	85.41	89.73	**99.46**
16	84.71	81.18	92.94	85.88	89.41	**100**	**100**	96.47	98.82
OA	65.81	75.49	77.15	73.07	73.71	83.52	**86.47**	82.8	83.56
AA	75.24	84.93	83.15	83.43	83.91	90.05	**92.35**	91.58	91.86
κ	61.59	72.48	74.07	69.87	70.55	81.41	**84.71**	85.38	81.48

4 Conclusion

In this paper, we developed a novel MRF-based MCS framework aiming at addressing hyperspectral *spectral-spatial* classification. The proposed algorithms model the class labels and probability based on multiple classifier system using MLR classifier. The final result is efficiently computed by stochastic relaxation and annealing, and α-expansion Graph Cut methods. Our experiments indicate that the proposed framework not only greatly improves the overall accuracy, but also enhance the class-specific accuracies, even for the very small training samples (5 samples per class). Future works will be directed toward testing more supervised classifiers, MCS rules and MRF-based approaches.

Acknowledgments. This work is supported by the Natural Science Foundation of China under Grant No. 41171323, the Priority Academic Program Development of Jiangsu Higher Education Institutions (PAPD), Jiangsu Provincial Natural Science Foundation under Grant No. BK2012018 and Key Laboratory of Geo-Informatics of National Administration of Surveying, Mapping and Geoinformation of China, Grant No. 201109.

References

1. Benediktsson, J.A., Chanussot, J., Fauvel, M.: Multiple classifier systems in remote sensing: from basics to recent developments. In: Haindl, M., Kittler, J., Roli, F. (eds.) MCS 2007. LNCS, vol. 4472, pp. 501–512. Springer, Heidelberg (2007)

2. Chan, J.C., Paelinckx, D.: Evaluation of Random Forest and Adaboost tree-based ensemble classification and spectral band selection for ecotope mapping using airborne hyperspectral imagery. Remote Sens. Environ. 112, 2999–3011 (2008)
3. Rodriguez-Galiano, V.F., Ghimire, B., Rogan, J., Chica-Olmo, M., Rigol-Sanchez, J.P.: An assessment of the effectiveness of a random forest classifier for land-cover classification. ISPRS J. Photogramm. 67, 93–104 (2012)
4. Du, P., Xia, J., Zhang, W., Tan, K., Liu, Y., Liu, S.: Multiple Classifier System for Remote Sensing Image Classification: A Review. Sensors 12, 4764–4792 (2012)
5. Yoav, F., Robert, E.S.: Experiments with a New Boosting Algorithm. In: Proceedings of the Thirteenth International Conference Machine Learning, Bari, pp. 148–156 (1996)
6. Kuncheva, L.I.: Combining Pattern Classifiers: Methods and Algorithms. Wiley Interscience (2004)
7. Foody, G.M., Boyd, D.S., Sanchez-Hernandez, C.: Mapping a specific class with an ensemble of classifiers. Int. J. Remote Sens. 28, 1733–1746 (2007)
8. Doan, H.T.X., Foody, G.M.: Increasing soft classification accuracy through the use of an ensemble of classifiers. Int. J. Remote Sens. 28, 4609–4623 (2007)
9. Velasco-Forero, S., Angulo, J.: Classification of hyperspectral images by tensor modeling and additive morphological decomposition. Pattern Recogn. 46, 566–577 (2012)
10. Fauvel, M., Tarabalka, Y., Benediktsson, J.A., Chanussot, J., Tilton, J.C.: Advances in Spectral-Spatial classification of hyperspectral images. Proc. IEEE (2012) (in press)
11. Farag, A., Mohamed, R., El-Baz, A.: A unified framework for map estimation in remote sensing image segmentation. IEEE Trans. Geosci. Remote Sens. 43, 1617–1634 (2005)
12. Liu, D., Kelly, M., Gong, P.: A spatial-temporal approach to monitoring forest disease spread using multi-temporal high spatial resolution imagery. Remote Sens. Environ. 101, 167–180 (2006)
13. Tarabalka, Y., Fauvel, M., Chanussot, J., Benediktsson, J.A.: SVM and MRF-Based Method for Accurate Classification of Hyperspectral Images. IEEE Geosci. Remote Sens. Lett. 7, 736–740 (2010)
14. Li, J., Bioucas-Dias, J.M., Plaza, A.: Hyperspectral Image Segmentation Using a New Bayesian Approach with Active Learning. IEEE Trans. Geosci. Remote Sens. 49, 3947–3960 (2011)
15. Giacinto, G., Roli, F.: Design of effective neural network ensembles for image classification. Image Vis. Comput. J. 19, 697–705 (2001)
16. Bai, J., Xiang, S., Pan, C.: A Graph-Based Classification Method for Hyperspectral Images. IEEE Trans. Geosci. Remote Sens. 803-816, 2113–2118 (2013)
17. Boykov, Y., Veksler, O., Zabih, R.: Efficient approximate energy minimization via graph cuts. IEEE Trans. Pattern Anal. Mach. Intel. 20, 1222–1239 (2001)
18. Bagon, S.: Matlab Wrapper for Graph Cut (December 2006), http://www.wisdom.weizmann.ac.il/~bagon
19. Tadjudin, S., Landgrebe, D.A.: Covariance estimation with limited training samples. IEEE Trans. Geosci. Remote Sens. 37, 2113–2118 (1999)

A Directed Inference Approach
towards Multi-class Multi-model Fusion

Tianbao Yang, Lei Wu, and Piero P. Bonissone

GE Global Research Center
{tyang,wul,bonissone}@ge.com

Abstract. In this paper, we propose a directed inference approach for multi-class multi-model fusion. Different from traditional approaches that learn a model in training stage and apply the model to new data points in testing stage, directed inference approach constructs (one) general direction of inference in training stage, and constructs an individual (ad-hoc) rule for each given test point in testing stage. In the present work, we propose a framework for applying the directed inference approach to multiple model fusion problems that consists of three components: (i) learning of individual models on the training samples, (ii) nearest neighbour search for constructing individual rules of bias correction, and (iii) learning of an optimal combination weights of individual models for model fusion. For inference on a test sample, the prediction scores of individual models are first corrected with bias estimated from the nearest training data points, and then the corrected scores are combined using the learned optimal weights. We conduct extensive experiments and demonstrate the effectiveness of the proposed approach towards multi-class multiple model fusion.

1 Introduction

Big data has posed great challenges in applying machine learning technologies. First, the scale of the data is too big to feed into most single-node and batch-mode machine learning algorithms. Second, the model trained on a small subset of data usually subjects to high bias and high variance.

To meet the big data challenge, a common approach is to adopt a distributed learning framework, where data and learning are distributed to different nodes in a cloud based computational network. These computational nodes are usually categorized into two types: one master node and a set of slave nodes. Each slave node will train an independent model on a subset of training data with single-node solvable scale, and make temporary decisions based on each independent model. The master node will take charge of distributing data, collecting information from slave nodes, and making the final decision, also called *model fusion*.

There are two steps involved in distributed learning framework. The first step is the distributing of multiple models in different slave nodes. The simplest way is to train each model, e.g., support vector machine (SVM)[8], neural network

Z.-H. Zhou, F. Roli, and J. Kittler (Eds.): MCS 2013, LNCS 7872, pp. 352–363, 2013.
© Springer-Verlag Berlin Heidelberg 2013

(NN) [13], decision tree [11], logistic regression (LR) [15], etc., independently on each node. The second step is to fuse multiple models and make the final decision, which has become a bottleneck problem in the distributed learning framework. There are several ways for model fusion. The simplest approach that combines the scores of multiple models with equal weights suffers from severe problems. First, each model may have substantially biased prediction and as a result adding them together may blow up the prediction bias on a test sample[1]. Second, each individual model may perform very differently since different models are learned based on different assumptions and objective functions, as a result the simple average would be very vulnerable to poorly performed models.

Although some other methods have been considering different weighting schemes to fuse multiple models, e.g., bagging, boosting, maximum margin of ensembles [10], and etc, they are studied in the traditional system on a single machine over all training samples and therefore they may not fit into the modern distributed system.

In this work, we seek an approach to directly combine multiple models with each trained on the same set (or different subset) of training samples. The proposed directed inference approach consists of three key components: (i) learning of individual models, which is same as traditional approach; (ii) nearest neighbour search for estimating the prediction bias on a test sample to correct the prediction scores of individual models; and (iii) learning of an optimal combination weights for model fusion. To make an inference on a test sample, the raw prediction scores are first computed for each model and then are corrected with estimated bias from the nearest neighbours retrieved using a distance metric and finally are added together using the learned optimal combination weights. The proposed approach can be also understood from the viewpoint of bias-variance trade-off. Combination of multiple models has shown to be effective in reducing the variance of prediction, however it could have adversary effect by increasing the bias. Therefore, the bias correction step in the proposed method helps to reduce the bias in individual models and the optimal weighting scheme further alleviate the impact of models with large bias.

We organize the remaining part of the paper as follows. In section 2, we review some related work from three angles, directed inference, bias and variance trade-off, and model fusion. In section 3 we present the proposed approach with three key components: learning of individual models, learning of a distance metric and learning of an optimal combination weights. In section 4, we present the experimental results and finally we conclude in section 5.

2   Related Work

2.1   Directed Inference

Directed (ad-hoc) inference (DAHI) approach is a new machine learning technique proposed by Vladimir Vapnik [18]. The key difference between DAHI and

---

[1] Throughout the paper, we use the terms of sample, example, instance and data point interchangeability.

traditional inductive/deductive or transductive learning is that in the testing stage, DAHI constructs a specific individual rule for each test example based on a principle concept learned in the training stage. The present work fits into the framework of DAHI by first learning multiple individual models in a single machine or in a distributed learning framework, a distance metric for retrieving a nearest neighbour and an optimal combination weights, then for each test sample by computing a bias corrected score for each individual model and then combing the multiple scores using the learned weights.

2.2   Bias and Variance Trade-Off

Bias and variance take-off is a common problem in model selection and model assessment. It has been shown that the mean square error of an estimator can be decomposed into a sum of the variance and the bias square of the estimator. Given multiple unbiased estimators, by simply averaging their prediction scores, one can obtain an estimator with dramatically reduced variance. However, if the individual models are biased, the trade-off between bias and variance may kick in, i.e. the variance of combined models may be reduced, while the bias may be blown up. One of the key motivations of the proposed approach is to reduce the bias of individual models. Given the bias and variance trade-off, it is however generally a difficult and even impossible task to construct a fixed estimator with both small bias and variance. Therefore, we resort to DAHI to construct individual rules with small bias and combine them to obtain a small variance.

Bias correction has been introduced to construct individual rules with small bias [4,2] and has shown to be effective in regression [4] and binary classification [2]. Bias correction works by subtracting an estimated bias value from the prediction score on any test example. The bias on a test sample is estimated by taking average of the bias values on training data points in the nearest neighbourhood. The underlying assumption is that in the small neighbourhood of a test example, the bias value is a constant. Previous works have used Euclidean distance or rectangle distance to retrieve a number of nearest neighbours. However, the Euclidean nearest neighbour may not share similar bias as the models may learned in a different space (e.g., kernel SVM is learned in a mapped high dimensional or infinite dimensional space).

2.3   Model Fusion

Model fusion is part of the ensemble learning process, by which multiple intelligent models are trained and combined for making a decision. Fusion is a major scheme for improving the performance by generating a more robust decision boundary based on multiple decision models. It can also be considered as a generalized model selection process, where instead of selecting the best model, fusion selects the best combination of models. The commonly used fusion methods include simple fusion, majority voting, Borda count, threshold voting, and heuristic decision rules [14,25,12], weighted average [27], fuzzy integral, fuzzy templates, and Dempster-Shafer theory[20], dynamic model selection [22], neural network (NN) based NN combination [21], local fusion [26], fuzzy combination [3], bagging [6], boosting [16], and etc.

3   A Directed Inference Approach towards Multi-class Multi-model Fusion

In this section, we present a directed inference approach towards multi-class multi-model fusion. The proposed approach consists of three key components: (i) learning of individual models, (ii) nearest neighbour search for bias correction, and (iii) learning optimal combination weights.

3.1   Learning of Individual Models

Our goal is to classify a data point into one of the K classes, denoted by $\{C_1, \ldots, C_K\}$. A common approach for multi-class classification is to cast the problem into several binary classification problems, e.g., one vs all or one vs one. In what follows, we briefly describe several methods for multi-class classification. Throughout the paper, we let $\mathbf{x}_i \in \mathbb{R}^d, i = 1, \ldots, n$ denote the feature vectors and $y_i \in \{1, \ldots, K\}, i = 1, \ldots, n$ denote their class labels. Without incurring confusion, we also use $y_i \in \{0, 1\}^K$ to denote a K-dimensional vector with only one entry equal to 1 indicating the class label.

Support Vector Machine (SVM) [8] constructs a hyperplane in the linear form $f(\mathbf{x}) = \mathbf{w}^\top \mathbf{x} + c$ by maximizing the margin from the hyperplane to the nearest training data point. It categorizes any data point into one of the two classes by checking the sign of the prediction score $\mathbf{w}^\top \mathbf{x} + c$. In addition to linear classification, SVM can also perform non-linear classification by using the kernel trick, which is equivalent to mapping data points into high dimensional or infinite dimensional feature spaces. In the experiments, we choose LibSVM [7] to run kernel SVM with RBF kernel. To perform multi-class classification, it follows one vs one scheme by constructing $K(K-1)/2$ binary classifiers and finally outputs a vector of scores that sum up to one, with each element indicating the confidence of assigning the data point into the corresponding class.

Neural Network (NN) [13] models the relationship between input and output in a structured information processing network, consisting of hidden layers of nodes between input and output. The learning process is actually adapting the model to the training data by changing the structure of the network. To adopt the NN for muti-class classification, we build K feed-forward neural networks with a hidden layer of 25 neurons. The k-th neural network NN_k is trained by regressing the input features $\mathbf{x}_i$ to the indicator variable $I(y_i = k)$ on the training data. The decision on a test point is made by $C(\mathbf{x}) = \arg\max_k \mathrm{NN}_k(\mathbf{x})$, where $\mathrm{NN}_k(\mathbf{x})$ gives the prediction value on $\mathbf{x}$.

Decision Tree [5] is a widely used non-linear model for both regression or classification. A decision tree could be either a classification tree or a regression tree depending on the type of the target variable and it is built upon the training data by recursively splitting the feature space with one feature and a splitting criterion that minimizes error in the two resulting sub-spaces. To classify a data point into one of K classes, we construct K regression trees with each tree T_k built on the training data $\mathbf{x}_i, i = 1, \ldots, n$ with binary indicator variables $I(y_l = k)$, and predict the class of a test point by $C(\mathbf{x}) = \arg\max_k T_k(\mathbf{x})$. In this work, we choose the most well-known implementation of decision tree, CART [5].

Logistic Regression (LR) [15] is a discriminative model for classification. We consider linear logistic regression model for multi-class classification, which defines the class conditional probability by $\Pr(y = k|\mathbf{x}) = \frac{\exp(\mathbf{w}_k^\top \mathbf{x})}{\sum_{l=1}^{K} \exp(\mathbf{w}_l^\top \mathbf{x})}$ and learns the K weights $\mathbf{w}_1, \ldots, \mathbf{w}_k$ by maximizing the log-likelihood on the training data. To avoid over-fitting, a regularization term $(\lambda/2) \sum_{k=1}^{K} \|\mathbf{w}_k\|_2^2$ is added to the objective.

3.2   Nearest Neighbour Search

Given multiple models denoted by $f_1, \ldots, f_m$ learned from the training data, the remaining question is to combine them into a single model for achieving a better performance. In this and next section, we address the question by nearest neighbour search using a distance metric for bias correction and learning an optimal combination weights for model fusion.

The raw prediction scores of model f_j on a given test example X are generated by $f_j(\mathrm{X}) \in \mathbb{R}^K$. The motivation of bias correction is to reduce the bias of individual models in predicting test data points. If we can accurately estimate the bias $b_j(\mathrm{X}) = f_j(\mathrm{X}) - \mathrm{Y}$, where Y is the unknown class label of the given example X, we can subtract the estimated bias $\widehat{b}_j(\mathrm{X})$ from the raw prediction scores $f_j(\mathrm{X})$ and obtain a more accurate classification decision based on $f_j(\mathrm{X}) - \widehat{b}_j(\mathrm{X})$. The question reduced to accurately estimation of the bias $\widehat{b}_j(\mathbf{X})$ for a given test point X. We take a non-parametric method, i.e., nearest neighbour estimation. A non-parametric method fits into the framework of directed inference [19], which is useful for constructing individual rules for test examples.

Let $\mathcal{N}(\mathrm{X})$ denote a small neighbourhood of X that contains the nearest training data points, which we assume shares the similar bias as the test data point X, then the bias of X can be estimated by

$$\widehat{b}_j(\mathrm{X}) = \frac{1}{|\mathcal{N}(\mathrm{X})|} \sum_{\mathbf{x}_i \in \mathcal{N}(\mathrm{X})} (f_j(\mathbf{x}_i) - y_i) \tag{1}$$

It still remains a problem how to retrieve a nearest neighbourhood of the test point X. A simple method is to define a nearest neighbourhood by using the Euclidean distance metric $\|\mathbf{x}_i - \mathrm{X}\|_2$. However, in some cases it may not reflect the underlying manifold of the bias function $b_j(\mathbf{x}) = f_j(\mathbf{x}) - y$, which depends on the model prediction $f(\mathbf{x})$ and the ground-truth y. A simple example that provides a negative evidence of using the Euclidean distance is given in Figure 1, where for the green test point, the bias of the nearest training data points (in the

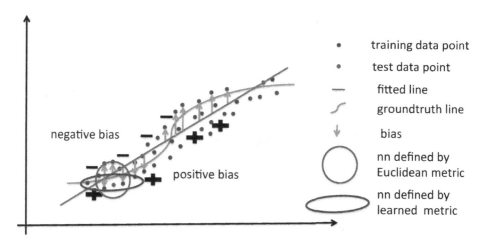

Fig. 1. An Illustration of nearest neighbourhood (nn) defined by the Euclidean metric and the learned metric for estimating the bias on a test point (green dot)

green circle) defined by Euclidean distance metric are mixed with positive values and negative values. As a consequence, by averaging the biases of the nearest training data points may yield a poor estimation of bias on the test data point. In contrast, if we define a nearest neighbourhood by a distance metric (e.g. the blue elliptical circle) that is consistent with the ground truth, i.e. data points with the same class labels have small distances and data points with different labels have large distances, then the estimation of bias can be improved. There exist many methods to formulate the distance metric learning [23,28,24]. In our empirical study, we choose a simple and effective method, relevant component analysis (RCA) [17,1], which is briefly described below.

RCA is originally proposed for learning a distance metric from partially labelled similar data points. Let $C_1, \ldots, C_K$ denote a set of K chunklets, where a chunklet is defined as a set of data points that share the same class labels. In our settings, each chunklet corresponds to one class. Then a positive semidefinite distance metric $A \in \mathbb{R}^{d \times d}$ is learned by minimizing the within class distances, i.e.,

$$\min_{A \in \mathbb{S}_+^{d \times d}} \sum_{k=1}^{K} \frac{1}{n_k} \sum_{y_i = k} (\mathbf{x}_i - \mathbf{c}_k)^\top A (\mathbf{x}_i - \mathbf{c}_k) - \log \det A \qquad (2)$$

where $\mathbb{S}_+^{m \times m} \subseteq \mathbb{R}^{m \times m}$ denotes a PSD cone, $\mathbf{c}_k$ is the center of the kth chunklet and n_k is the number of data points in C_k. The negative log-determinant term is added to avoid a trivial solution, which also has an information theoretic and Bayesian interpretation [1]. Finally, one can easily show that the optimal solution to (2) can be computed as

$$A = \left(\sum_{k=1}^{K} \frac{1}{n_k} \sum_{y_i=k} (\mathbf{x}_i - \mathbf{c}_k)(\mathbf{x}_i - \mathbf{c}_k)^\top \right)^{-1}$$

Equipped with a distance metric A, either the Euclidean metric or the learned metric, we can retrieve k nearest neighbors of the test sample X with k shortest distance $(\mathbf{x}_i - X)^\top A(\mathbf{x}_i - X)$ to form $\mathcal{N}(X)$.

3.3   Learning of an Optimal Combination Weights

In the previous section, we describe a nearest neighbour search for estimating the bias on a given test point X. Given the estimated bias, the prediction of each model is corrected by $f_j(X) - \hat{b}_j(X)$, and the corrected score will be combined by a weighted summation. In this section, we present a convex approach for learning a globally optimal combination weights. Let $\omega_1, \ldots, \omega_m$ denote the weights to be learned, the combined prediction is computed by

$$\hat{f}(X) = \sum_{j=1}^{m} \omega_j \left(f_j(X) - \hat{b}_j(X) \right) \tag{3}$$

The combination weights are global in the sense that all test points share the same weights. The optimal combination weights $\omega = (\omega_1, \ldots, \omega_m)^\top$ are learned following the spirit of cross-validation. To this end, we let $(\mathbf{x}_i^v, y_i^v), i = 1, \ldots, N$ denote a separate set of N validation data points sampled from the same distribution of the training data points, and then we optimize the following objective

$$\min_{\omega \in \Delta_+} \sum_{i=1}^{N} \ell \left(\sum_{j=1}^{m} \omega_j (f_j(\mathbf{x}_i^v) - \hat{b}_j(\mathbf{x}_i^v)), y_i^v \right) \tag{4}$$

where $\Delta_+ = \{\omega : \omega \geq 0, \sum_{j=1}^{m} \omega_j = 1\}$ is a simplex, $\hat{b}_j(\mathbf{x}_i^v)$ is the estimated bias from the nearest neighbors and $\ell(\mathbf{z}, y)$ is a hinge loss for multi-class defined as

$$\ell(\mathbf{z}, y) = \max_{k \neq y} ([\mathbf{z}]_k - [\mathbf{z}]_y + b)_+$$

where b is a specified margin parameter and $[s]_+ = \max(0, s)$

To optimize the objective in (4), we can employ the widely adopted gradient descent method that iteratively updates $\omega_t = \omega_{t-1} - \eta \nabla L(\omega_{t-1})$, where η is a step size. However, the standard gradient decent method suffers from a low convergence rate of $O(1/\sqrt{T})$ for the non-smooth hinge loss function, i.e., $L(\hat{\omega}_T) \leq \min_{\omega \in \Delta_+} L(\omega) + O(1/\sqrt{T}))$, where $\hat{\omega}_T = \sum_{t=1}^{T} \omega_t / T$. In this paper, we extend the primal dual prox method proposed in [29] to optimize $L(\omega)$ that enjoys a convergence rate of $O(1/T)$. To this end, we write the objective in (4) into a min-max formulation:

$$\min_{\omega \in \Delta_+} \max_{\alpha \in \Omega_+^N} \underbrace{\frac{1}{N} \sum_{i=1}^{N} \sum_{k \neq y} \alpha_k^i \left([\hat{f}(\mathbf{x}_i^v)]_k - [\hat{f}(\mathbf{x}_i^v)]_{y_i^v} + b \right)}_{F(\omega, \alpha)}$$

Algorithm 1. Pdprox algorithm for optimizing structured hinge loss over a simplex (Pdprox-shs)

1: **Input**: step size γ
2: **Initialization**: $\theta_0 = 1/m, \alpha_0 = \mathbf{0}$
3: **for** $t = 1, 2, \ldots$ **do**
4: $\quad \omega_t = P_{\theta_{t-1}}(\gamma \nabla_\omega(\theta_{t-1}, \alpha_{t-1})) = \dfrac{\theta_{t-1} \circ \exp(-\gamma \nabla_\omega(\theta_{t-1}, \alpha_{t-1}))}{\sum_{j=1}^m [\theta_{t-1} \circ \exp(-\gamma \nabla_\omega(\theta_{t-1}, \alpha_{t-1}))]_j}$
5: $\quad \alpha_t = \Pi_{\Omega_+^N}[\alpha_{t-1} + \gamma \nabla_\alpha(\omega_t, \alpha_{t-1})]$
6: $\quad \theta_t = P_{\theta_{t-1}}(\gamma \nabla_\omega(\omega_t, \alpha_t)) = \dfrac{\theta_{t-1} \circ \exp(-\gamma \nabla_\omega(\omega_t, \alpha_t))}{\sum_{j=1}^m [\theta_{t-1} \circ \exp(-\gamma \nabla_\omega(\omega_t, \alpha_t))]_j}$
7: **end for**
8: **Output** $\widehat{\omega}_T = \sum_{t=1}^T \omega_t / T$ and $\widehat{\alpha}_T = \sum_{t=1}^T \alpha_t / T$.

---

by observing that $\ell(\mathbf{z}, y) = \max_{\alpha \in \Omega_+} \sum_{k \neq y} \alpha_k([\mathbf{z}]_k - [\mathbf{z}]_y + b)$, where $\Omega_+ = \{\alpha \in \mathbb{R}^{K-1} : \alpha \geq 0, \sum_k \alpha_k \leq 1\}$. To present the algorithm, we let $\nabla_\alpha(\omega, \alpha)$ denote the partial gradient of $F(\omega, \alpha)$ in terms of ω, $\nabla_\alpha(\omega, \alpha)$ denote the partial gradient of $F(\omega, \alpha)$ in terms of α, and let $[\mathbf{u}]_j$ denote the jth element in $\mathbf{u}$. The detailed steps for updating the primal variable ω and the dual variables α are presented in Algorithm 1, which is a variant of Algorithm 2 proposed in [29]. The updating rule for the primal variable ω and the auxiliary primal variable θ is due to a proximal mapping $P_\theta(g) = \arg\min_{\omega \in \Delta_+} g^\top(\omega - \theta) + V(\omega, \theta)$, where $V(\omega, \theta) = \sum_j \omega_j \log(\omega_j / \theta_j)$ is the entropy distance function. The updating rule for the dual variables α is due to a projection $\Pi_{\alpha \in \Omega_+^N}[\widehat{\alpha}] = \arg\min_{\alpha \in \Omega_+^N} \|\alpha - \widehat{\alpha}\|_F^2$, which can be efficiently computed using the algorithm in [9]. Finally, we present the following theorem that states the convergence rate of Algorithm 1 for optimizing the structured hinge loss over a simplex. The proof can be easily duplicated following the analysis in [29].

Theorem 1. *Assuming* $\|[\widehat{f}(\mathbf{x})]\|_\infty \leq R$ *and setting* $\gamma = \sqrt{\frac{N}{8mR^2}}$, *by running Algorithm 1 with* T *steps, we have*

$$L(\widehat{\omega}_T) \leq \min_{\omega \in \Delta_+} L(\omega) + \frac{\log m + N}{2\gamma T}$$

4   Experiments

In this section, we present some preliminary experimental results. The data sets we choose for study include open benchmarks, *DNA, letter, pendigits, protein, satimage*, in **UCI** data repositories. We also adopted a jet engine fault classification data, which contains a total of $19,635$ instances. Each instance corresponding to a case of engine has 11 attributes from sensors and also is labelled to one of seven classes which indicates one of the seven fault types including normal. We refer to the data as aircraft engine fault diagnosis (AEFD) data.

More details can be found in [27]. The data is split into a training set of 15, 708 instances and a testing set of 3, 927 instances. Table 1 summarizes the statistics of the chosen datasets.

Table 1. Statistics of datasets

Name	instances	features	source	class	type
dna	3,186	180	statlog	3	multi-class
letter	20,000	16	Statlog	26	multi-class
segment	2,310	19	Statlog	7	mult-class
protein	24,387	357	JYW02a	3	multi-class
satimage	6,435	36	Statlog	6	multi-class
AEFD	19,635	11	GE	7	multi-class

Table 2. Prediction performance of individual models with/without bias correction, where nbs and bs indicate performances without and with bias correction, respectively. The reported results of bias correction is using the Euclidean distance metric.

	DNA				letter				segment			
	SVM	NN	CART	LR	SVM	NN	CART	LR	SVM	NN	CART	LR
nbs	0.9625	0.9475	0.9740	0.9765	0.8158	0.9350	0.8250	0.7532	0.9450	0.9850	0.9750	0.9350
bs	0.9740	0.9645	0.9800	0.9765	0.9436	0.9596	0.9148	0.8234	0.9750	0.9850	0.9800	0.9500

	protein				satimage				AEFD			
	SVM	NN	CART	LR	SVM	NN	CART	LR	SVM	NN	CART	LR
nbs	0.6709	0.6849	0.4948	0.6892	0.8575	0.8885	0.8320	0.8170	0.7675	0.8296	0.7056	0.7833
bs	0.6324	0.6130	0.5369	0.6801	0.9040	0.8995	0.8845	0.8170	0.8273	0.8442	0.7904	0.8182

Table 3. Prediction performance of multiple model fusion with bias correction using equal weights and optimal combination weights

DNA		letter		segment		protein		satimage		AEFD	
average	opt	average	opt	average	opt	average	opt	average	opt	average	opt
0.9795	0.9850	0.9126	0.9680	0.9950	0.9850	0.6730	0.6683	0.8528	0.8622	0.8180	0.8745

We use the default splitting of training, validation and testing if there exists a validation data, otherwise we manually generate a validation data set by sampling from the training data with the same size of the testing data set. For the purpose of demonstration, we train 4 classification models (SVM, NN, CART, LR) on all training data points, and report the metric of overall accuracy computed based on the confusion table [27]. The parameters in models are tuned on the validation data set. The number of nearest neighbours for estimating the bias is set to 5. The margin parameter in the structured hinge loss is set to 0.5. Both the bias correction and the model fusion are done on the previously listed dataset.

We first demonstrate the effectiveness of bias correction on individual models. The results are summarized in Table 2. From the results, we can observe

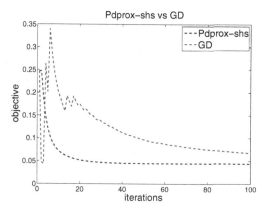

Fig. 2. Comparison of Pdprox-shs vs Gradient Descent (GD) method for optimizing the structured hinge loss on dna data set.

that bias correction can improve the prediction substantially. Furthermore, we compare the performance by using the Euclidean distance and a learned distance metric by RCA. We observed on several data sets that the distance metric learned from the ground truth can improve the performance of Euclidean distance metric, e.g. on AEFD data sets the performances of the four models are improved to $(0.8533, 0.8630, 0.8024, 0.8215)$, on letter data set the performances are improved to $(0.9546, 0.9656, 0.9216, 0.8302)$. On other data sets, the learned distance metric by RCA is comparable to the Euclidean distance metric.

Next, we demonstrate the effectiveness of model fusion. We compare the proposed convex approach for learning an optimal combination weights to the equal weighting fusion. The results are reported in Table 3. Among the six selected benchmark datasets for multi-class classification, the proposed optimal fusion approach significantly outperforms the equal weighting method on four datasets, and performs almost the same on the segment and protein data. By checking these two data sets, we find all individual classifiers perform almost equally. Thus, we draw a conclusion that the proposed fusion approach significantly outperforms simple fusion method when outputs of individual classifiers are diverse.

Finally, we show the efficiency of Pdprox-shs algorithm compared to gradient descent (GD) method for optimizing the structured hinge loss over a simplex. Both the initial step size of GD and the step size of Pdprox-shs are set to the same value 100. We plot the objective value versus the number of iterations on DNA data in Figure 2. It clearly shows that Pdprox-shs performs better than GD, which verifies our theoretical analysis on the convergence rate.

5 Conclusion

In this paper, we propose a directed inference approach for multi-class multi-model fusion. Different from traditional approaches, directed inference approach constructs a principle concept in training stage and individual (ad-hoc) rules for

classifying test samples. The presented approach consists of three key components: (i) learning of individual models, (ii) nearest neighbour search for estimating the bias of a given test sample, and (iii) learning of an optimal combination
weights for fusing the bias corrected scores of multiple models. We demonstrate
the effectiveness of the proposed approach on extensive data sets. In the future
work, we plan to extend the work to other tasks (e.g. regression and binary classification) and conduct the experiments in real distributed learning framework.

Acknowledgement. We sincerely thank Dr. Shengzhuo Zhu for pointing out
the connection to DAHI.

References

1. Bar-Hillel, A., Hertz, T., Shental, N., Weinshall, D.: Learning distance functions
using equivalence relations. In: Proc. of ICML, pp. 11–18 (2003)
2. Bonissone, P.P.: Lazy meta-learning: creating customized model ensembles on de-
mand. In: Liu, J., Alippi, C., Bouchon-Meunier, B., Greenwood, G.W., Abbass,
H.A. (eds.) WCCI 2012. LNCS, vol. 7311, pp. 1–23. Springer, Heidelberg (2012)
3. Bonissone, P.P., Varma, A., Aggour, K.S., Xue, F.: Design of local fuzzy models
using evolutionary algorithms. Comput. Stat. Data Anal. 51(1), 398–416 (2006)
4. Bonissone, P.P., Xue, F., Subbu, R.: Fast meta-models for local fusion of multiple
predictive models. Appl. Soft Comput. 11(2), 1529–1539 (2011)
5. Breiman, L., Friedman, J., Olshen, R., Stone, C.: Classification and Regression
Trees. Wadsworth and Brooks, Monterey (1984)
6. Breiman, L.: Bagging predictors. Machine Learning 24(2), 123–140 (1996)
7. Chang, C.-C., Lin, C.-J.: LIBSVM: A library for support vector machines. ACM
Transactions on Intelligent Systems and Technology 2, 27:1–27:27 (2011)
8. Cortes, C., Vapnik, V.: Support-vector networks. Machine Learning, 273–297
(1995)
9. Duchi, J., Shalev-Shwartz, S., Singer, Y., Chandra, T.: Efficient projections onto
the l1-ball for learning in high dimensions. In: Proc. of ICML, pp. 272–279 (2008)
10. Grove, A.J., Schuurmans, D.: Boosting in the limit: Maximizing the margin of
learned ensembles (1998)
11. Ho, T.K.: Random decision forest. In: Proc. of the 3rd International Conference
on Document Analysis and Recognition, pp. 278–282. IEEE (1995)
12. Ho, T.K., Hull, J.J., Srihari, S.N.: Decision combination in multiple classifier systems. IEEE TPAMI 16(1), 66–75 (1994)
13. Hopfield, J.J.: Neural networks and physical systems with emergent collective computational abilities. Pro. of the National Academy of Sciences 79(8), 2554–2558
(1982)
14. Lam, L., Suen, S.Y.: Application of majority voting to pattern recognition: an
analysis of its behavior and performance. Trans. Sys. Man Cyber. Part A 27(5),
553–568 (1997)
15. Mayers, J.H., Forgy, E.W.: The development of numerical credit evaluation systems. Journal of the American Statistical Association, 799–806 (1963)
16. Schapire, R.: The boosting approach to machine learning: An overview (2003)
17. Shental, N., Hertz, T., Weinshall, D., Pavel, M.: Adjustment learning and relevant
component analysis. In: Heyden, A., Sparr, G., Nielsen, M., Johansen, P. (eds.)
ECCV 2002, Part IV. LNCS, vol. 2353, pp. 776–790. Springer, Heidelberg (2002)

18. Vapnik, V.: Problems of empirical inference in machine learning and philosophy of science. Invited Talk at Tenth International Conference on Rough Sets, Fuzzy Sets, Data Mining and Granular Computing, Regina, Saskatchewan (2005)
19. Vapnik, V.: Estimation of Dependences Based on Empirical Data: Springer Series in Statistics. Springer-Verlag New York, Inc., Secaucus (1982)
20. Verikas, A., Lipnickas, A., Malmqvist, K., Bacauskiene, M., Gelzinis, A.: Soft combination of neural classifiers: A comparative study. Pattern Recognition Letters 20(4), 429–444 (1999)
21. Wolpert, D.H.: Stacked generalization. Neural Networks 5, 241–259 (1992)
22. Woods, K., Philip Kegelmeyer Jr., W., Bowyer, K.: Combination of multiple classifiers using local accuracy estimates. IEEE TPAMI 19(4), 405–410 (1997)
23. Wu, L., Hoi, S.C.H., Jin, R., Zhu, J., Yu, N.: Learning bregman distance functions for semi-supervised clustering. IEEE TKDE 24(3), 478–491 (2012)
24. Xing, E.P., Ng, A.Y., Jordan, M.I., Russell, S.: Distance metric learning, with application to clustering with side-information. In: Proc. of NIPS, pp. 505–512. MIT Press (2002)
25. Xu, L., Krzyzak, A., Suen, C.Y.: Methods of combining multiple classifiers and their applications to handwriting recognition. IEEE Transactions on Systems, Man, and Cybernetics 22(3), 418–435 (1992)
26. Xue, F., Subbu, R., Bonissone, P.P.: Locally weighted fusion of multiple predictive models. In: Proc. of IJCNN, pp. 2137–2143. IEEE (2006)
27. Yan, W., Xue, F.: Jet engine gas path fault diagnosis using dynamic fusion of multiple classifiers. In: Proc. of IJCNN, pp. 1585–1591. IEEE (2008)
28. Yang, T., Jin, R., Jain, A.K.: Learning from noisy side information by generalized maximum entropy model. In: Proc. of ICML (2010)
29. Yang, T., Mahdavi, M., Jin, R., Zhu, S.: An efficient primal-dual prox method for non-smooth optimization, arxiv (2012)

A New Feature Fusion Approach Based on LBP and Sparse Representation and Its Application to Face Recognition

He-Feng Yin and Xiao-Jun Wu

School of IoT Engineering, Jiangnan University, Wuxi 214122, China
yinhefeng@126.com, wu_xiaojun@yahoo.com.cn

Abstract. In this paper, we propose a new feature fusion approach based on lo-cal binary pattern (LBP) and sparse representation (SR). Firstly, local features are extracted by LBP and global features are sparse coefficients which are ob-tained via decomposing samples based on the over-complete dictionary. Then the global and local features are fused in a serial fashion. Afterwards PCA is used to reduce the dimensionality of the fused vector. Finally, SVM is em-ployed as a classifier on the reduced feature space for classification. Experi-mental results obtained on publicly available databases show that the proposed feature fusion method is more effective than other methods like LBP+PCA, Gabor+PCA and Gabor+SR in terms of recognition accuracy.

Keywords: Feature fusion, local binary pattern, sparse representation, support vector machine, face recognition.

1 Introduction

Automatic face recognition [1] remains one of the most visible and challenging re-search topics in computer vision, machine learning and biometrics. It is widely ap-plied to different fields including biometric authentication, security applications and human computer interaction. Compared with other biometrics, such as fingerprint identification and palm identification, face recognition has the advantages of being convenient, immediate and well accepted.

The question of which low-dimensional features of an object image are the most relevant or informative for classification is a central issue in face recognition. Con-ventional facial features can be roughly divided into global features (PCA [2], LDA [3], LPP [4], etc.) and local features (LBP [5], SIFT [6], etc.). However, both the global and local features are not rich enough to capture all of the classification infor-mation available in the image, in addition, researches have shown that different features have different classification capabilities and a fusion scheme that harnesses various features is likely to improve the overall performance.

There are three levels of information fusion, i.e. pixel level, feature level and deci-sion level. The decision level fusion, represented by multi-classifier combination, has been one of the hot research topics on pattern recognition [7-10]. In recent years,

Z.-H. Zhou, F. Roli, and J. Kittler (Eds.): MCS 2013, LNCS 7872, pp. 364–373, 2013.
© Springer-Verlag Berlin Heidelberg 2013

some feature level fusion methods have been proposed, for instance, Sun et al. [11] proposed a novel feature fusion method. Firstly, two groups of feature vectors are extracted with the same pattern, then a correlation criterion function is established between the two groups of feature vectors, finally their canonical correlation features are extracted to form effective discriminant vectors for recognition. Huang [12] put forward an efficient face representation and recognition method, which combines the both information between rows and those between columns from two-directional 2DPCA on fusion face image and the optimal discriminative information from column-directional 2DLDA. Song [13] provided a method based on the feature fusion of the local and global features, local features are extracted from sub-images and global features are obtained via PCA. Chowdhury et al. [14] presented a fusion method, first of all, face images are divided into a number of non-overlapping sub-images, the G-2DFLD method is applied to each of these sub-images as well as to the whole image to extract local as well as global discriminant features respectively. These extracted local and global features are fused to form a large feature vector and FLD method is applied on it to reduce its dimensionality. Nevertheless, the above fusion methods are largely dependent on the dimensionality of features, and in low-dimensional feature space, recognition accuracy of these methods is not that high.

However, within the framework of sparse representation, the precise choice of feature space is no longer critical. What is crucial is that the dimensionality of the feature space is sufficiently large and that the sparse representation is correctly computed [15]. In addition, according to related researches about local binary pattern (LBP), features coded by LBP have highly discriminative power [16], this property makes it suitable for image classification tasks. Inspired by these findings, we intend to use the fused features of sparse coefficient and local features extracted by LBP to improve the recognition performance.

The remainder of this paper is organized as follows: LBP and sparse representation are reviewed in Section 2 and Section 3 respectively. Section 4 presents the proposed method. Experiments are conducted on publicly available databases to verify the effectiveness of the proposed method in Section 5. Finally, conclusions are drawn in Section 6.

2   Local Binary Pattern

The LBP operator was first introduced by Ojala [17] and used as texture descriptor. Then Ahonen [5] applied it to face recognition and obtained outstanding results, which demonstrates that LBP is able to well describe face images.

The original LBP operator was defined as a window of size 3×3. This operator uses the value of the center pixel as a threshold, and the 8 surrounding pixels whose value is higher than or equal to the value of the threshold is assigned a binary value 1, otherwise the value is 0. When this process is accomplished, 8 values can be read start from the top left corner in the clockwise direction. The 8-bit binary number or its equivalent decimal number can be assigned to the center pixel and it can describe the texture information of an image. The basic LBP operator is illustrated in Fig. 1.

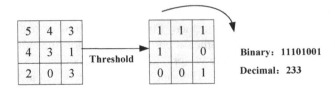

Fig. 1. The original LBP operator

In order to facilitate the analysis of textures with different scales, the basic LBP operator is extended by combining neighborhoods with different radius. In this case, P points on the edge of a circle, whose radius is R, are sampled and compared with the value of the center pixel. For ease of presentation, the notation (P,R) is employed to formulate P sampling points on a circle of radius of R. See Fig. 2 for an example of circular neighborhoods.

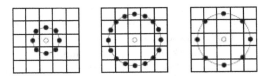

Fig. 2. The circular (8,1), (16,2) and (8,2) neighborhoods

Another extension of the original LBP operator is the definition of so called uniform patterns. A local binary pattern is called uniform if the binary pattern contains at most two bitwise transitions from 0 to 1 or vice versa when the bit pattern is considered circular [18]. Experimental results have demonstrated uniform patterns can describe most of the texture information, at the same time, they have strong ability to do classification tasks.

Generally, when we extract features from face images, we can divide the face image into small blocks. And features are extracted from each block independently. The descriptors are then concatenated to form a global description of the face image. In this way we can obtain a description of the face image on local and holistic levels. In this paper, uniform patterns of (8,1) are applied to extracted LBP features.

3   Sparse Representation (SR)

Theoretical results show that well-aligned images of a convex, Lambertian object lie near a low-dimensional feature space of the high-dimensional image space [19]. This is the only prior knowledge about the training samples in SR. The idea of SR is presented as follows [15].

Suppose we have C distinct classes, given sufficient training samples of the i-th object class, the size of face images is $w \times h$, and the total number of samples of i-th class is n_i. We stack the n_i training images from the i-th class as columns of a matrix $A_i = [v_{i,1}, \ldots, v_{i,n_i}] \in R^{m \times n_i}$ $(m = w \times h)$. For a test sample $y \in R^m$ belongs to this class, according to linear subspace theory, y can be approximated by the linear combination of the samples within A_i, i.e.

$$y \approx \alpha_{i,1} v_{i,1} + \alpha_{i,2} v_{i,2} + \cdots + \alpha_{i,n_i} v_{i,n_i} \tag{1}$$

$\alpha_{i,j} \in R, j = 1, 2, \cdots, n_i$.

Since the initial identity of the test sample y is unknown, let A be the concatenation of the n training samples from all the C classes, where $\sum_{i=1}^{C} n_i = n$, then we can define a new matrix A:

$$\begin{aligned} A &= [A_1, A_2, \cdots, A_C] \\ &= [v_{1,1}, \ldots, v_{1,n_1}, \ldots, v_{i,1}, v_{i,2}, \ldots, v_{i,n_i}, \ldots, v_{C,1,\ldots,} v_{C,n_C}] \end{aligned} \tag{2}$$

If we use the new matrix A to represent the test image y, that is

$$y = A x_0 \in R^m \tag{3}$$

where $x_0 = [0, \ldots, 0, \ldots, \alpha_{i,1}, \alpha_{i,2}, \ldots, \alpha_{i,n_i}, \ldots, 0, \ldots, 0]^T \in R^n$ is a coefficient vector whose entries are zero except those associated with the i-th class, and A is referred to as dictionary.

In robust face recognition, the system $y = Ax$ is always ill-determined, so its solution is not unique, but we just need to find a locally optimal solution. Conventionally, this problem is settled by choosing the minimum l^2-norm solution. However, the solution is non-sparse and it has no discriminative information. This motivates us to seek the sparsest solution to $y = Ax$, leading to the following optimization problem:

$$(l^0) \ x_0 = \arg \ \min \ \| x \|_0, \ subject \ \ to \ Ax = y \tag{4}$$

where $\|\cdot\|_0$ denotes the l^0-norm, which counts the number of nonzero elements in a vector.

However, the problem of finding the sparsest solution of an ill-determined system of linear equations is NP-hard. Recent progress in the theory of sparse representation and compressed sensing reveals that if the solution x_0 is sparse enough, the solution to the l^0-minimization problem (4) is equal to the following l^1-minimization problem [20]:

$$(l^1) x_1 = \arg \ \min \ \| x \|_1, \ subject \ \ to \ Ax = y \tag{5}$$

To solve the l^1-minimization problem, one can use gradient projection method [21], homotopy algorithm [22], iterative shrinkage-thresholding [23] etc.

In order to guarantee the coefficient vector x has the form $[0, \ldots, 0, \alpha, 0, \ldots, 0]$ where all the non-zero entries are together, we solve this optimization problem:

$$\min_{x} \| y - Ax \|_2 + \lambda_1 \| x \|_2^2 + \lambda_2 \| x \|_1 \tag{6}$$

The l_1 penalty in the above expression promotes sparsity of the coefficient vector x, while the quadratic l_2 penalty encourages grouping effect, i.e. selection of a group of correlated training samples.

4   Proposed Feature Fusion Method

Wavelet transform has been introduced in our method to perform the preprocessing of the face images, it can reduce noise of images, and the low frequency component is a coarser approximation to the original image. Thus the wavelet image should be more suitable for recognition.

Given all that, the procedure of the proposed method is presented as follows:

1. Perform wavelet transform to the original image and obtain its 1-level low-frequency component L.

2. Divide the 1-level low frequency component into small blocks, then extract LBP features for each small block.

3. Concatenate the LBP features of all the small blocks to form the local feature of the original image.

4. Based on the over-complete dictionary (which contains all the training samples), the same original image can be decomposed to obtain its sparse coefficient, i.e., the global feature.

5. Then the local and global features are fused in a serial fashion [24], after the dimensionality of the fused feature is reduced, it can be used for recognition.

Framework of the proposed method and other methods that will be compared with in this paper is depicted in Fig. 3.

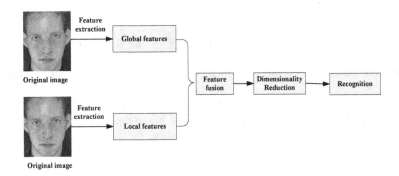

Fig. 3. Framework of the methods considered in this paper

5   Experiments and Analysis

In this section, we conduct experiments on publicly available databases for face recognition. The ORL and XM2VTS databases are used to verify the performance of the proposed method and its competing methods:

PCA: global features extract by PCA.
LBP: local features extract by LBP.
SR: sparse representation of the sample, i.e. sparse coefficient.
Gabor+PCA: fused features extracted by Gabor filter and PCA.
Gabor+SR: fused features extracted by Gabor filter and sparse coefficient.
LBP+PCA: fused features extracted by LBP and PCA.

When extracting local features based on LBP, the original face image is preprocessed by wavelet transform. In this experiment, the basis function of wavelet transform is *coif4*. In SR, the error tolerance ε is 0.05. We use Gabor filter at five different scales and eight orientations, thus we obtain 40 Gabor filters. The global and local features are fused in a serial fashion. Then PCA is utilized to do dimensionality reduction. Finally, linear SVM is employed for classification and the strategy for multi-class classification is one-against-one approach [25].

5.1   Experiments on the ORL Database

The ORL database contains images from 40 individuals, each providing 10 different images. For some subjects, the images were taken at different times. The facial expressions (open or closed eyes, smiling or non-smiling) and facial details (glasses or no glasses) also vary. The images were taken with a tolerance for some tilting and rotation of the face of up to 20 degrees. Moreover, there is also some variation in the scale of up to about 10 percent. All images are gray-scale and have a resolution of 92×112 pixels. Half of the images per subject are chosen as training samples, the reminder for testing, and the face image is divided into 4×4 blocks when extracting the LBP features. Fig. 4 shows the recognition performance for various methods, in conjunction with different feature dimensionality. Table 1 shows the detailed recognition accuracy of the methods considered and Table 2 records the computation time of Gabor+SR and the proposed method.

Table 1. Recognition rate(%) of different methods on the ORL database and the associated dimensionality of feature

Dimensionality	10	30	50	70	90
PCA	93.5%	92%	94.5%	93.5%	91.5%
LBP	83.5%	93.5%	95.5%	97%	97%
SR	82%	94%	93.5%	93.5%	93%
Gabor+PCA	92%	94%	95%	95.5%	95%
Gabor+SR	82%	96%	96.5%	97.5%	98%
LBP+PCA	92%	94%	95%	95.5%	95%
Proposed	97%	97%	97%	97%	97%

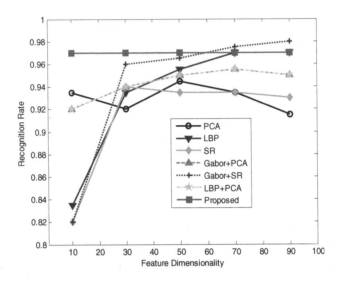

Fig. 4. Curves of recognition rate by different methods versus feature dimensionality on the ORL database

Table 2. Computation time(s) of Gabor+SR and the proposed method on the ORL database and the associated dimensionality of feature

Dimensionality	10	30	50	70	90
Gabor+SR	115.43s	116.03s	116.66s	117.42s	118.04s
Proposed	16.37s	16.42s	16.47s	16.55s	16.61s

5.2   Experiments on the XM2VTS Database

The XM2VTS database is a multi-modal database which consists of video sequences of talking faces recorded for 295 subjects at one month intervals. The data has been recorded in 4 sessions with 2 shots taken per session. From each session two facial images have been extracted to create an experimental face database of size 55×51. In our experiment, we chose a subset of the dataset consisting of 100 subjects. For each subject, four images are used as training samples, the rest for testing, and the face image is divided into 8×8 blocks when extracting the LBP features. The comparison of competing methods is given in Fig. 5 and Table 3. Computation time of Gabor+SR and the proposed method is recorded in Table 4.

Table 3. Recognition rate(%) of different methods on the XM2VTS database and the associated dimensionality of feature

Dimensionality	5	10	20	30	35
PCA	33.5%	58.75%	78.25%	86.5%	87.75%
LBP	35.75%	58.25%	81%	87.25%	88.5%
SR	27%	62.5%	80.5%	85.75%	88%
Gabor+PCA	45%	69%	80.5%	83.25%	83.75%
Gabor+SR	52.75%	79.75%	92%	95%	94.75%
LBP+PCA	45%	68.75%	80.25%	83.25%	83.75%
Proposed	96%	96%	96%	96%	96%

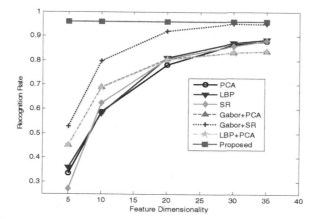

Fig. 5. Curves of recognition rate by different methods versus feature dimensionality on the XM2VTS database

Table 4. Computation time(s) of Gabor+SR and the proposed method on the XM2VTS database and the associated dimensionality of feature

Dimensionality	5	10	20	30	35
Gabor+SR	470.45s	470.61s	471.80s	473.14s	473.73s
Proposed	72.38s	72.66s	73.22s	73.77s	74.15s

Based on the above experimental results obtained on ORL and XM2VTS databases, we have the following observations:

1. As feature dimensionality increases, performance of LBP is better than that of PCA, this indicates that local features may contain more discriminative information.

2. When we fuse global features (e.g. features extracted by SR) with local features (e.g. Gabor features), performance of global features is boosted. This demonstrates that fused features can improve the overall performance.

3. By and large, the proposed method is more competitive than other methods, not only the performance of the proposed method remains stable, but the computation time is acceptable. Though performance of Gabor+SR is better than that of LBP+SR on ORL database, it is computationally expensive, and its computation time is about 7 times that of our method.

6   Conclusions

In this paper, we propose a new feature fusion approach based on LBP and sparse representation. Firstly, local features are extracted by LBP and global features are sparse coefficients which are obtained via decomposing samples based on the over-complete dictionary. Then the global and local features are fused in a serial fashion. Experiments conducted on the ORL and XM2VTS databases show the feasibility and effectiveness of the new method. However, in this paper, we do not explore other feature fusion methods, so in future, we will investigate other methods and come up with a better approach for robust face recognition.

References

1. Chellappa, R., Wilson, C.L., Sirohey, S.: Human and Machine Recognition of Faces: A Survey. Proceedings of the IEEE 83(5), 705–741 (1995)
2. Turk, M., Pentland, A.: Face Recognition Using Eigenfaces. In: CVPR, pp. 586–591 (1991)
3. Belhumeur, P., Hespanha, J., Kriegman, D.: Eigenfaces versus Fisherfaces: Recognition Using Class Specific Linear Projection. IEEE PAMI 9(7), 711–720 (1997)
4. He, X., Yan, S., Hu, Y., Niyogi, P., Zhang, H.: Face Recognition Using Laplacianfaces. IEEE Trans. Pattern Analysis and Machine Intelligence 27(3), 328–340 (2005)
5. Ahonen, T., Hadid, A., Pietikäinen, M.: Face Recognition with Local Binary Patterns. In: Pajdla, T., Matas, J. (eds.) ECCV 2004. LNCS, vol. 3021, pp. 469–481. Springer, Heidelberg (2004)
6. Bicego, M., Lagorio, A., Grosso, E., Tistarelli, M.: On the Use of SIFT Features for Face Authentication. In: Proc. of IEEE Conf. on Biometrics, in Association with CVPR Biometrics, p. 35 (2006)
7. Huang, Y.S., Suen, C.Y.: Method of Combining Multiple Experts for the Recognition of Unconstrained Handwritten Numerals. IEEE Trans. Pattern Anal. Mach. Intell. 7(1), 90–94 (1995)
8. Constantinidis, A.S., Fairhurst, M.C., Rahman, A.F.R.: A New Multi-Expert Decision Combination Algorithm and Its Application to the Detection of Circumscribed Masses in Digital Mammograms. Pattern Recognition 34(8), 1528–1537 (2001)
9. Jin, X.-Y., Zhang, D., Yang, J.-Y.: Face Recognition Based on a Group Decision-Making Combination Approach. Pattern Recognition 36(7), 1675–1678 (2003)
10. Kittler, J., Hatef, M., Duin, R.P.W., Matas, J.: On Combining Classifiers. Pattern Analysis and Machine Intelligence 20(3), 226–239 (1998)

11. Sun, Q.-S., Zeng, S.-G., Liu, Y., Heng, P.-A., Xia, D.-S.: A New Method of Feature Fusion and Its Application in Image Recognition. Pattern Recognition 38(12), 2437–2448 (2005)
12. Huang, G.H.: Fusion (2D)^{2}PCALDA: A New Method for Face Recognition. Applied Mathematics and Computation 216(11), 3195–3199 (2010)
13. Song, L.: Face Recognition Based on Feature Fusion. In: Cross Strait Quad-Regional Radio Science and Wireless Technology Conference (CSQRWC), pp. 1524–1527 (2011)
14. Chowdhury, S., Sing, J.K., Basu, D.K., Nasipuri, M.: Face Recognition by Fusing Local and Global Discriminant Features. In: 2nd International Conference on Emerging Applications of Information Technology (EAIT), pp. 102–105 (2011)
15. Wright, J., Yang, A.Y., Ganesh, A., Sastry, S.S., Ma, Y.: Robust Face Recognition via Sparse Representation. IEEE PAMI 31(2), 210–227 (2009)
16. Ojala, T., Pietikainen, M., Maenpaa, T.: Multiresolution Gray-Scale and Rotation Invariant Texture Classification with Local Binary Patterns. IEEE Trans. Pattern Analysis and Machine Intelligence 24(7), 971–987 (2002)
17. Ojala, T., Pietikinen, M.: A Comparative Study of Texture Measures with Classification Based on Feature Distribution. Pattern Recognition 29(1), 51–59 (1996)
18. Ahonen, T., Hadid, A., Pietikainen, M.: Face Description with Local Binary Patterns: Application to Face Recognition. PAMI 28(12), 2037–2041 (2006)
19. Basri, R., Jacobs, D.: Lambertian Reflectance and Linear Subspaces. IEEE Trans. Pattern Analysis and Machine Intelligence 25(2), 218–233 (2003)
20. Donoho, D.: For Most Large Underdetermined Systems of Linear Equations the Minimal l^1-Norm Solution Is Also the Sparsest Solution. Comm. Pure and Applied Math. 59(6), 797–829 (2006)
21. Figueiredo, M., Nowak, R., Wright, S.: Gradient Projection for Sparse Reconstruction: Application to Compressed Sensing and Other Inverse Problems. IEEE Journal of Selected Topics in Signal Processing 1(4), 586–597 (2007)
22. Drori, I., Donoho, D.: Solution of l^1-Minimization Problems by LARS/Homotopy Methods. In: ICASSP, pp. 636–639 (2006)
23. Hale, E., Yin, W., Zhang, Y.: A Fixed-Point Continuation Method for l^1-Regularized Minimization with Applications to Compressed Sensing. Technical Report, Rice University (2007)
24. Yang, J., Yang, J.-Y., Zhang, D., Lu, J.: Feature Fusion: Parallel Strategy vs. Serial Strategy. Pattern Recognition 36(6), 1369–1381 (2003)
25. Guo, G.-D., Li, S.-Z., Chan, K.-L.: Support Vector Machines for Face Recognition. Image and Vision Computing 19(9-10), 631–638 (2001)

Binary Decision Trees for Melanoma Diagnosis

Yu Zhou[1] and Zhuoyi Song[2]

[1] School of Computing, University of Leeds, Leeds, UK
Y.Zhou@leeds.ac.uk
[2] Department of Biomedical Science, University of Sheffield, S10 3TN, UK

Abstract. Although computer aided diagnosis of melanoma is an active research area for more than two decades, its clinical application is still just on horizon. To speed up its clinical application, two critical challenges need to be solved: the data gap and the decision-making gap. Ideally, these two issues shall be attacked simultaneously. However, in the literature, most current methods designing melanoma diagnosis classifiers adopt a biased approach by either focusing on the data gap or on the decision-making gap while neglecting the other. In this article, we present one prototype system covering both the data gap and the decision-gap. Performance of this new method is presented and comparisons with respect to alternative approaches, including the conventional one, are also included.

Keywords: computer aided diagnosis, decision tree, classification.

1 Introduction

Computer based early diagnosis of melanoma has been studied for more than two decades [1]. One of the key aims of this research field is to build a digital system for clinical diagnosis applications. This task is important because the manual inspection, while common in clinical practice, has undesirable features such as repetitiveness and subjectivity. Computer based methods, however, have huge potential in alleviating these shortcomings and providing an important clinical alternative when second opinion is needed.

For a computerized melanoma diagnosis system, there are two key components: data and decision-making. As shown in the flowchart in Fig. 1, the data are mainly images while the decision-making process means applying certain machine learning techniques to label the sample as either benign or malignant. In the literature, a variety of imaging protocols, such as digital dermoscopy, infrared imaging, multispectral imaging, confocal microscopy, have been applied in collecting digital data of lesions [2].

Though data is the inalienable part of a computer aided diagnosis (CAD) system, the decision-making is of utmost importance, especially with the abundance of numerous imaging devices on the market. Here for a CAD system, decision-making refers to carrying a diagnosis based on certain algorithms/classifiers. In other words, a CAD system mimics the clinical diagnosis process, which typically involves computing techniques such as feature extraction and classification.

Z.-H. Zhou, F. Roli, and J. Kittler (Eds.): MCS 2013, LNCS 7872, pp. 374–385, 2013.
© Springer-Verlag Berlin Heidelberg 2013

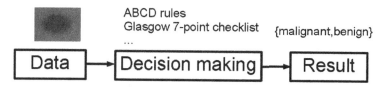

Fig. 1. General melanoma diagnosis process

Although CAD based melanoma diagnosis has been an active research area, its clinical application is still stagnating currently. Before bringing CAD based melanoma inspection into real clinical practices, there are two critical challenges:

- Data gap: The feature descriptors describing pigmented lesion properties may include both metric and non-metric values. Some attributes might involve physical meanings such as the diameter, some other features, such as Euler number of the lesion area, might not have any unit. However, typical learning paradigms like support vector machines [5] and neural networks [6] treat these data as purely numerical inputs without paying attention to the physical/clinical meanings of these attributes.
- Decision-making gap: This refers to the decision making style in clinical diagnosis and early CAD based diagnosis of melanoma. For instance, [10, 11] only used border attributes in melanoma diagnosis. Instead of applying a set of clinical heuristics such as the ABCD rules[7], this decision-making style focuses on just one property of pigmented lesions. Hence there is a gap between these CAD systems and the clinical diagnosis. Although the one-feature based CAD is an effective methodology in justifying the usefulness of certain descriptors, comprehensive features covering both geometric and colormetric properties should be incorporated in the decision-making process, particularly for CAD systems targeting clinical applications[13]. However, this combinatorial approach like [4] improves the diagnosis accuracy at the cost of increasing the complexity, i.e., jeopardizing the interpretability of the diagnosis system to a certain extent.

To build a successful diagnosis system, we believe both the data gap and the decision-making gap shall be solved. However, publications covering the decision gap [6, 13], albeit of varying performance, are neglecting the data gap by and large. Publications like [9, 10] are free of data gap while they have the decision-making gap when compared with clincial heuristics in diagnosis.

In this paper, a prototype system is designed, aiming at covering both the data gap and the decision-making gap. Section 2 discusses the geometrical and colormetric features extracted at first. Then the prototype is presented in line with the conventional design. Section 3 shows experimental results on a test dataset.

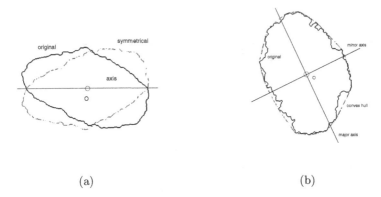

(a)   (b)

Fig. 2. (a) Extracting asymmetrical indices; (b) Extracting diameter indices

2   Method

Since a typical decision-making process involves feature extraction and classification, this section is divided into two subsections. In Section 2.1, the shape, size and colour features for the pigmented lesions are examined, which are inspired by the ABCD rules [2, 7]. In Section 2.2, unlike [4] that uses three heterogeneous classifiers, the decision tree is the only classifier employed here to conduct diagnosis.

2.1   Feature Descriptors

This is a classical topic in CAD based melanoma diagnosis. Here a set of features are extracted following the ABCD rules.

Asymmetry. In [7], asymmetry rule means the more asymmetrical the lesion is, the more likely it is malignant. [8, 13] use principal axis based methods to describe the asymmetry features. In this paper, a four-axis method is utilized to describe the asymmetry features. Fig. 2(a) shows the idea of how to construct the asymmetry features. In Fig. 2(a), the solid line is the first axis adopted and the original lesion area lies within the region enclosed by the solid boundary. The dashed line encloses an area which, given the axis, is symmetrical with respect to (w.r.t.) the original lesion area. Apart from the horizontal line in Fig. 2(a) serving as the symmetrical axis, the diagonal line, off-diagonal line and vertical line can also be used as symmetrical axes.

To construct one symmetrical area as shown in Fig. 2(a), there are two steps. Firstly, one detects the centroid of the lesion area. Then a reference axis, i.e., the symmetrical axis, is selected. These symmetrical axes are 0, 45, 90, 135 degrees w.r.t. the horizontal axis. Denote the original lesion area as A_0, and the symmetrical lesion area as A_i, the asymmetry indices are defined as:

$$a_i = \frac{XOR(A_0,\ A_i)}{2A_0},\ i = 1, 2, 3, 4, \tag{1}$$

where XOR means exclusive-or operation between two areas. Thus these four attributes are all non-metric variables.

Border Irregularity. To maintain the equivalence in terms of units for the border irregularity descriptors, indentation/protrusion index proposed in [10] is used here.

The method in [10] constructs an area based irregularity index in several steps: Firstly, a smoothed outline of the lesion is extracted via a series of multiscale Gaussian filters. Then the area enclosed within the smoothed outline is compared with the original lesion area. Denote the original lesion area as A_0 and the smoothed area as A_s, the indentation area lies within A_s while outside A_0. The protrusion area is the opposite: it lies within A_0 while outside A_s. So the indentation and protrusion maps can be obtained as follows:

$$I_{ind} = A_0 - A_s, \tag{2}$$

$$I_{pro} = A_s - A_0, \tag{3}$$

where I_{ind} and I_{pro} represents the indentation and protrusion images respectively. Specifically, for pixels in I_{ind} and I_{pro}, the definitions are as follows:

$$I_{ind}(x, y) = \begin{cases} 1, & A_0(x,y) == 1 \& A_s(x,y) == 0 \\ 0, & otherwise \end{cases}, \tag{4}$$

$$I_{pro}(x, y) = \begin{cases} 1, & A_0(x,y) == 0 \& A_s(x,y) == 1 \\ 0, & otherwise \end{cases}. \tag{5}$$

With the above definitions, one can extract the border irregularity features accordingly. In [10], these features are non-metric and they lie within [0,1].

Colour Variation. The colour variations are extracted via four different channels: red, green, blue and intensity channels. The intensity channel is obtained by fusing the RGB channels as follows:

$$I(x, y) = \sqrt{R^2(x, y) + G^2(x, y) + B^2(x, y)}, \tag{6}$$

where x and y denotes the coordinates of the pixels. By calculating the statistics of the images, the four features of colour variations are defined as follows:

$$\left[log \frac{\sigma(R)}{\mu(R)}, \ log \frac{\sigma(G)}{\mu(G)}, \ log \frac{\sigma(B)}{\mu(B)}, \ log \frac{\sigma(I)}{\mu(I)} \right], \tag{7}$$

where μ and σ denotes the mean value and standard deviation of the lesion area pixels of the given colour channel. Since the mean value and the standard deviation are of the same unit, i.e., the image intensity in different colour channels, the ratios between these two are free of metric unit. For computational conveniences, one can limit the range for the above colour variation descriptors within a reasonable region, e.g., [-10, 10].

Diameter. Unlike the original diameter feature expressed in [7] which states that 6mm diameter is the critical threshold in judging a suspicious lesion as malignant, here a 4-element diameter feature vector is extracted, which includes the following attributes:

$$d_1 := 2\sqrt{\sharp(A_0)/\pi}, \tag{8}$$
$$d_2 := 2\sqrt{\sharp(A_c)/\pi}, \tag{9}$$
$$d_3 := svd(S,1), \tag{10}$$
$$d_4 := svd(S,2), \tag{11}$$

where A_c is the convex hull extracted with the given lesion area and '$\sharp$' is the operator to calculate the number of non-zero elements lying within. Fig. 2(b) gives an example of a lesion area and its convex hull. Therefore in (8), d_1 is the equivalent diameter of the original lesion area. Likewise, d_2 in (9) is the equivalent diameter of the convex hull. S in (10)(11) is the covariance matrix constructed by the coordinates of the lesion area pixels. Since the image plane is 2D, S is a 2 by 2 positive definite matrix. In (10)(11), $svd(S,1)$ and $svd(S,2)$ represent the first and the second singular values of S respectively.

Clearly the four diameter feature descriptors above are of length units expressed in image pixels. In addition, the numerical values of these four attributes are non-negative.

Therefore, in the above subsections, the ABCD rules have been implemented in a way aiming at reducing the data gap between attributes within each feature group. As there are 4 elements for every feature group, altogether there are 16 attributes extracted for one 2D image sample of pigmented skin lesions.

2.2   Decision Trees

In the literature, there are a few protocols proposed for decision-making in melanoma diagnosis.

In [6], a neural network was proposed to carry diagnosis of melanoma. The feature descriptors cover asymmetry/border/colour properties and overall there are 14 entries in the feature vector. In experiments, one of the tested neural networks includes 14 input neurons, 7 hidden neurons and 1 output neuron. This neural network was trained with the well-known back-propagation method [14].

[4] formulated a multiple classifier system, including linear disciminant analysis (LDA), decision trees and k-nearest neighbor. The feature descriptors cover geometrical and colormetric domains, resembling the ABCD rules proposed in [7].

In [13], another ABCD rules based system for melanoma diagnosis was tested. In this system, there are 8 feature descriptors: skin line direction, skin line intensity, asymmetry, border irregularity, red/green/blue component variegation and diameter of lesion. After dimension reduction for the feature vectors, the first two principal components of these 8 components are selected for designing

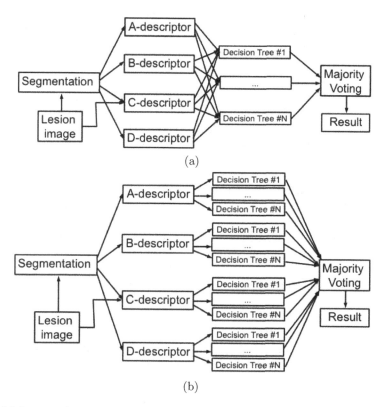

Fig. 3. Melanoma diagnosis using multiple decision trees. (a) with data gap; (b) data gap free.

a linear classifier. The area under the receiver operating characteristic (ROC) curve obtained is 0.94.

[13] represents a typical approach in designing computer aided diagnosis of melanoma, which involves dimension reduction after feature extraction. In [4, 6, 13], there is one hidden assumption behind applying dimension reduction methods: different features representing different properties are numerically computable. Here by computable, it means they are allowed to be numerically mixed together, including addition/deduction etc. Due to the non-homogeneity of the attributes, this operation, though numerically accepted commonly, neglects the *data gap* between different attributes. The associated undesirable risks include: Firstly, the result depends on the chosen metrics. Secondly, the result only has numerical meaning but the physical/clinical meaning might be elusive. For instance, suppose the skin line direction, blue component variegation and diameter of lesion are f_1, f_7 and f_8 respectively, adding them together as $f_1 + f_7 + f_8$ gives a numerical number, which is hardly of any clinical significance.

Unlike the above mentioned references, here decision trees [14] and only decision trees are used for decision making. They are selected not only because of the learning capability, but also because of its interpretability and its intuitiveness [3].

Recently, [3] proposes a decision tree based classification system which allows the end-users to tune the trees manually. This technique, called Visually Tuned Decision Tree (VTDT), illustrates that decision tree not only can generate comprehensible rules to domain experts, it also has the potential to allow human experts to embed their domain knowledge in designing the decision making models.

In utilizing decision trees to design the CAD system for melanoma diagnosis, there exist two approaches. As shown in Fig. 3(a), this method stacks the feature vectors together and feeds the overall feature vector into decision trees. The final output is generated by majority voting of the member decision trees. This approach has the data gap problem since the diameter attributes are in pixels while the other values are non-metric.

Fig. 3(b) offers an alternative which eliminates both the data gap and the decision-making gap. This is achieved by forwarding only homogeneous data into one given decision tree, i.e., no mixture or stacking of variables bearing different metric units. In addition, for the decision making part, since Fig. 3(b) combines both the geometric and colormetric properties of skin lesion in diagnosis, the decision-making gap is avoided as well, making it more similar to the clinical diagnosis than using a single diagnosis rule. Also compared with metric learning[15], Fig. 3(b) is a specific approach not only driven by the data, domain knowledge such as physical meanings of features also play important roles.

3   Experiments and Results

This section presents the CAD experimental results, which were obtained by testing the diagnosis systems in Fig. 3(a)(b) via cross-validations.

3.1   Experimental Setup

In the collected 2D image dataset, there are 110 malignant samples and 125 benign lesions. The lesion areas in these data were segmented by using the online graph-cut based algorithm [16–19]. Feature descriptors were extracted with the above methods. To run the experiments, Matlab 7.12 is used as the experimental platform.

In standard k-fold cross validation, the data are divided into k-subsets first. Then k-1 subsets are used to train the classifier and the remaining subset is left for testing. As multiple decision trees are used in both Fig. 3(a) and (b), employing standard k-fold cross-validation will introduce significant training data overlaps among different decision trees. Thus in training the multiple decision trees, only 1 of the k subsets was selected to train the classifier. Hence for

Table 1. Different setups for evaluating multiple decision trees

	Nodes (0/1 output)		Nodes (real number output)	
	Data Gap	ROC curve	Data Gap	ROC curve
Fig. 3(a)	No	-	Yes	Available
Fig. 3(b)	No	-	No	Available
RF	No	-	Yes	Available

different decision trees, the probability for two of them holding the same training data is generally small ($1/k^2$). In the following experiments, k is chosen as 5 and for multiple decision trees, N is set as 3. In this case, the Fig. 3(b) design has 12 trees. When making a majority voting for Fig. 3(b), if there is a draw, the outputs of the decision trees using colour attributes will be chosen to make a majority voting, leading to a final diagnosis.

For the design of nodes in the binary decision trees in Fig. 3, one can use either univariate nodes, or multivariate nodes with linear discriminant analysis to split a mother node into two daughter nodes. Specifically, for a univariate node, there is no data gap since no direct numerical operation between attributes arises. However, it also limits the flexibility in choosing the decision boundaries [14]. For multivariate nodes with linear classifiers, as shown in Table 1, receiver operating characteristic (ROC) curves can be generated, which offer a qualitative as well as quantitative perspective in evaluating the performance.

To give a third opinion, the random forest (RF) [20] is also incorporated in the experiments. In this paradigm, a series of feature-set are randomly selected via a uniform distribution over different descriptors. For every set, one decision tree is trained as described above. Here two random forests are tested: one with 3 trees and one with 12 trees so that the number of trees is equal to the Fig. 3(a) and Fig. 3(b) respectively with $N = 3$.

3.2 ROC Analysis

Fig. 4 shows the ROC curves for the ABCD features. Each ROC curve corresponds to one feature group processed via linear discriminant analysis based multivariate nodes. The areas under curves (AUCs) in Fig 4 are 0.688, 0.723, 0.758 and 0.644 respectively.

Fig. 5(a) presents the ROC curve by stacking the descriptors directly as one vector and then feeding the vector into a decision tree with multivariate nodes. As can be seen from Fig. 5(a), the AUC is 0.885. Fig. 5(b) shows the ROC curve for decision system in Fig. 3(b). This AUC for Fig 5(b) is 0.923, slightly higher than Fig. 5(a).

In addition, from Fig. 4 to Fig. 5, it can be seen that by applying multiple decision trees to mimic the ABCD rules, the diagnostic performance can be enhanced effectively. Although there exists data gap in Fig. 3(a), Fig. 5(a) still suggests that the direct stacking of features deserves certain efforts. The multiple

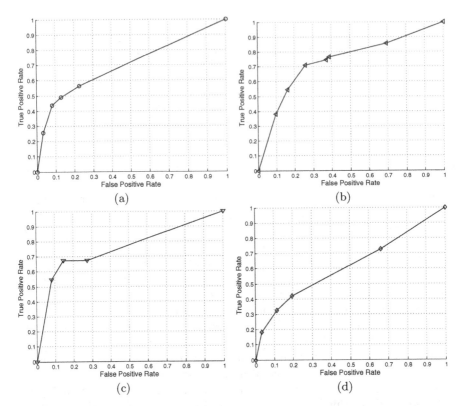

Fig. 4. ROC curves for ABCD descriptors respectively. (a) asymmetry (AUC=0.688); (b) border irregularity (AUC=0.723); (c) colour variations (AUC=0.758); (d) diameter (AUC=0.644).

decision trees in Fig. 3(b) eliminate the data gap while performing even better, at least qualitatively as shown in Fig. 5's ROC analysis.

For random forests, Fig. 5(c) and (d) show that 12-tree gives better results than 3-tree. This indicates an approach to improving the performance of random forests though overfitting could be a potential problem. Also the 12-tree's performance is comparable to the data-gap free design in Fig. 3(b). However, there are several disadvantages inherent in 12-tree random forests: Firstly, the probability of overlaps between different trees' training data is non-zero and sometime significant, albeit depending on the details of the sampling scheme. Also the computational cost of 12-tree random forests is higher than Fig. 3(b) due to the reuse of certain features via re-sampling. In addition, because of the randomness introduced in sampling features, interpretability of the classification process is sacrificed, i.e., it is not data-gap-free.

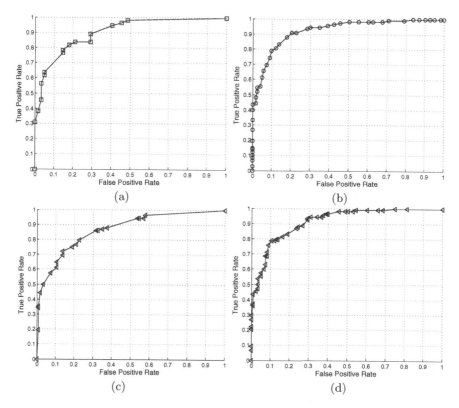

Fig. 5. ROC curves for multiple decision trees. (a) with data gap - Fig. 3(a): AUC = 0.885; (b) without data gap - Fig. 3(b): AUC = 0.923; (c)random forest - 3-tree: AUC = 0.867; (d) random forest - 12-tree: AUC=0.916.

Table 2. Simulation results with different multiple decision tree designs (%)

	Accuracy	Sensitivity	Specificity	PPV	NPV
Fig.3(a)	79.57(±0.39)	73.81(±0.50)	84.69(±0.39)	81.09(±0.44)	78.43(±0.42)
Fig.3(b)	82.86(±0.24)	77.22(±0.51)	87.80(±0.01)	84.76(±0.08)	81.45(±0.34)
RF*	76.28(±0.47)	71.11(±0.77)	80.81(±0.45)	76.49(±0.50)	76.11(±0.52)
RF**	82.94(±0.39)	76.11(±0.41)	88.94(±0.73)	85.81(±0.80)	80.92(±0.27)

(PPV: positive predictive value; NPV: negative predictive value; *3-tree; **12-tree)

3.3   Classification Result

Apart from the ROC analysis, the two multiple decision tree systems in Fig. 3 are also evaluated in classification. Table 2 shows the classification results, in which the performance indicators, including accuracy/sensitivity/specificity, positive predictive value (NPV) and negative predictive value (NPV), are measured.

From Table 2, it can be seen that Fig. 3(b) is better than Fig. 3(a) in terms of mean values of the above indicators, though with varying levels. The standard deviations of these indicators for Fig. 3(b)'s design are generally less than results obtained from using Fig. 3(a), indicating a slightly better consistency in performance.

Table 2 also shows the results via random forests. Here the performance of RF with 12 trees is better than the 3-tree version while similar to the results of Fig. 3(b). These observations from Table 2 confirms the observations made in Section 3.2.

4   Conclusion

By working on covering both the data gap and the decision-making gap, this paper examined the application of multiple binary decision trees in computer aided diagnosis of melanoma. Two structurally different designs are highlighted: the conventional design (Fig. 3(a)) and the new design (Fig. 3(b)). In experimental studies on a dataset containing 235 samples (125 benign and 110 malignant), while the first design gives good results in ROC analysis and classification tests, the second design can perform even better. The only downside of the new design is that its structure appears less straightforward than its conventional alternative. However, if one needs a high performance CAD system free of both the data gap and the decision-making gap, the system in Fig. 3(b) merits certain consideration.

Though Fig. 3(a) and Fig. 3(b) are different, in fact Fig. 3(b) can be derived from Fig. 3(a) in two steps. Firstly, one can set the number of decision trees in Fig. 3(a) as 12, i.e., N=12. Then by putting feature selections prior to feeding the data into the decision trees, one can obtain the design in Fig. 3(b).

References

1. Korotkov, K., Garcia, R.: Computerized analysis of pigmented skin lesions: A review. Artificial Intelligence in Medicine 56(2), 69–90 (2012)
2. Rigel, D., Russak, J., Friedman, R.: The evolution of melanoma diagnosis: 25 years beyond the ABCDs. CA: A Cancer Journal for Clinicians 60(5), 301–316 (2010)
3. Stiglic, G., Kocbek, S., Pernek, I., Kokol, P.: Comprehensive decision tree models in bioinformatics. PLoS ONE 7(3) (2012)
4. Sboner, A., Eccher, C., Blanzieri, E., Bauer, P., Cristofolini, M., Zumiani, G., et al.: A multiple classifier system for early melanoma diagnosis. Artificial Intelligence in Medicine 27(1), 29–44 (2003)
5. Gilmore, S., Hofmann-Wellenhof, R., Soyer, H.: A support vector machine for decision support in melanoma recognition. Experimental Dermatology 19(9), 830–835 (2010)
6. Ercal, F., Chawla, A., Stoecker, W., Lee, H., Moss, R.: Neural network diagnosis of malignant melanoma from color images. IEEE Transactions on Biomedical Engineering 41(9), 837–845 (1994)

7. Friedman, R., Rigel, D., Kopf, A.: Early detection of malignant melanoma: the role of physician examination and self-examination of the skin. CA: A Cancer Journal for Clinicians 35(3), 130–151 (1985)
8. Celebi, M.E., Kingravi, H.A., Uddin, B., Iyatomi, H., Aslandogan, Y.A., Stoecker, W.V., Moss, R.H.: A methodological approach to the classification of dermoscopy images. Computerized Medical Imaging and Graphics 31(6), 362–373 (2007)
9. Lee, T.K., McLean, D.I., Atkins, M.S.: Irregularity index: A new border irregularity measure for cutaneous melanocytic lesions. Medical Image Analysis 7(1), 47–64 (2003)
10. Lee, T.K., Claridge, E.: Predictive power of irregular border shapes for malignant melanomas. Skin Research and Technology 11(1), 1–8 (2005)
11. Zhou, Y., Smith, M., Smith, L., Warr, R.: A new method describing border irregularity of pigmented lesions. Skin Research and Technology 16(1), 66–76 (2010)
12. Abbas, Q., Celebi, M.E., Garcia, I.F., Rashid, M.: Lesion border detection in dermoscopy images using dynamic programming. Skin Research and Technology 17(1), 91–100 (2011)
13. She, Z., Liu, Y., Damatoa, A.: Combination of features from skin pattern and ABCD analysis for lesion classification. Skin Research and Technology 13(1), 25–33 (2007)
14. Duda, R.O., Hart, P.E., Stork, D.G.: Pattern Classification, 2nd edn. John Wiley, New York (2001)
15. Ying, Y., Li, P.: Distance Metric Learning with Eigenvalue Optimization. Journal of Machine Learning Research 13, 1–26 (2012)
16. Boykov, Y., Veksler, O., Zabih, R.: Efficient Approximate Energy Minimization via Graph Cuts. IEEE Transactions on Pattern Analysis and Machine Intelligence 20(12), 1222–1239 (2001)
17. Boykov, Y., Kolmogorov, V.: An Experimental Comparison of Min-Cut/Max-Flow Algorithms for Energy Minimization in Vision. IEEE Transactions on Pattern Analysis and Machine Intelligence 26(9), 1124–1137 (2004)
18. Kolmogorov, V., Zabih, R.: What Energy Functions can be Minimized via Graph Cuts? IEEE Transactions on Pattern Analysis and Machine Intelligence 26(2), 147–159 (2004)
19. Bagon, S.: Matlab Wrapper for Graph Cut (December 2006)
20. Breiman, L.: Random forests. Machine Learning 45(1), 5–32 (2001)

ECOC Matrix Pruning Using Accuracy Information

Cemre Zor, Terry Windeatt, and Josef Kittler

Centre for Vision, Speech and Signal Processing (CVSSP), University of Surrey,
Guildford, GU2 7XH, United Kingdom
{c.zor,t.windeatt,j.kittler}@surrey.ac.uk

Abstract. The target of ensemble pruning is to increase efficiency by reducing the ensemble size of a multi classifier system and thus computational and storage costs, without sacrificing and preferably enhancing the generalization performance. However, most state-of-the-art ensemble pruning methods are based on unweighted or weighted voting ensembles; and their extensions to the Error Correcting Output Coding (ECOC) framework is not strongly evident or successful. In this study, a novel strategy for pruning ECOC ensembles which is based on a novel accuracy measure is presented. The measure is defined by establishing the link between the accuracies of the two-class base classifiers in the context of the main multiclass problem. The results show that the method outperforms the ECOC extensions of the state-of-the-art pruning methods in the majority of cases and that it is even possible to improve the generalization performance by only using 30% of the initial ensemble size in certain scenarios.

1 Introduction

Ensemble pruning aims to decrease the number of base classifiers of an existing ensemble system without sacrificing performance and brings about the benefits of reduced complexity and storage requirements, as well as increase in performance in some cases. Due to the computational complexity of finding the optimum classifier combination through exhaustive search, various sub-optimal techniques have been proposed in the literature for ensemble pruning. Among these approaches, search and ordering based methods can be listed as the most straightforward methods. While carrying out the search or the ordering (ranking), the idea of having diverse and/or accurate base classifiers is usually required as a key for success.

Error Correcting Output Coding (ECOC) [6] is a powerful ensemble technique, in which multiple base classifiers are trained according to the information obtained from a pre-set binary code matrix. The idea is to solve the original multiclass problem by decomposing it into simpler two-class decompositions. The pruning of the ECOC matrix using state-of-the-art pruning methods, which have mainly been defined for majority / weighted voting ensembles, is not appropriate due to the specific requirements of the ECOC framework: The link

Z.-H. Zhou, F. Roli, and J. Kittler (Eds.): MCS 2013, LNCS 7872, pp. 386–397, 2013.

between the accuracies of the base classifiers solving different two-class problems and the ensemble accuracy obtained for the main multiclass problem is not taken into account. In order to overcome these deficiencies and further improve the existing approaches, we propose a method for pruning ECOC which is based on a novel accuracy measure.

Section 2 summarizes the state-of-the-art pruning methods found in the literature and Sec. 3 gives insight about the ECOC framework. In Sec. 4, the proposed novel pruning method is explained and finally in Sec. 5-6 details of experimentation and the conclusions drawn are presented.

2   Pruning Background

Existing ensemble pruning algorithms in the literature have mainly been developed in the context of majority or weighted voting ensembles. Taxonomies and detailed analysis of the existing strategies can be found in [14,10,18]. Below, the descriptions of some of the most popular pruning algorithms are given. Note that these algorithms do not have straightforward extensions to the ECOC framework.

Ordered aggregation pruning methods aim to rank all the classifiers according to a desired measure and then select the first n many desired components. The uth classifier to be added to the set C_{u-1}, which contains the first $u-1$ classifiers of the ordered sequence, is selected based on a measure gauging maximum improvement on the ensemble.

Reduce Error Pruning (REP) [9] is an ordered aggregation pruning method in which the first classifier selected is the one having the lowest classification error. In each subsequent iteration, the classifier which provides maximum improvement in the current subensemble accuracy is added to the ensemble. Backfitting, which aims to interchange an already selected classifier with a new one from the pool of unselected, can be applied if the new classifier reduces the subensemble error.

Margin Distance Minimization (MDM), as initially defined in [11], is an ordered aggregation pruning method based on the base classifiers' average success in correctly classifying patterns belonging to a selection set, S. For a given S of size μ, the mth component of the signature vector c for the classifier i is defined to be equal to 1 if the classification decision is correct for the pattern S_m and -1 otherwise. That is,

$$c_m^{(i)} = 2I\left(d_i\left(S_m\right) = \phi\left(S_m\right)\right)\text{-}1 \tag{1}$$

where $I\left(\text{true}\right) = 1$, $I(\text{false}) = 0$; $d_i\left(S_m\right)$ is the decision of the base classifier i for the pattern S_m, and $\phi\left(S_m\right)$ is the actual label.

The aim is then to select a subensemble whose average signature vector, $< c >$, over all classifiers is as close as possible to a reference vector, o, in the first quadrant. The reference vector is designed arbitrarily to consist of equal components, $o_{m,\,m\epsilon(1,\alpha)} = p$, in which the choice of p is advised to be sufficiently small for the algorithm to progressively focus on examples that are more difficult to classify. When the Euclidean distance, d_{Eucl}, is used as the distance metric, the classifier selected in the uth iteration is

$$s_u = \operatorname{argmin}_t \operatorname{dist}_{\text{Eucl}} \left(o, \frac{1}{T} \left(c^{(t)} + \sum_{i=1}^{u-1} c^{(i)} \right) \right) \tag{2}$$

in which t is the index of a yet unselected base classifier and T is the total number of base classifiers.

Boosting-based Pruning (OB) [12] is also an ordered aggregation pruning method in which base classifiers are ordered according to their performance in boosting. In each iteration of boosting, the classifier with the lowest weighted training error is selected from the initial pool of classifiers. OB does not halt even when the selected classifier has zero training error and continues running even if all the classifiers have training error more than 50%, by resetting the weights. OB has later been combined with Instance-based Pruning [7] in [13]; however the results have shown improvement over OB in speed rather than accuracy.

Different from the above mentioned commonly used ordering and search based approaches, Zhang et al. [17] have proposed an optimization framework for ensemble pruning. Using the finding that the success of the ensemble depends on the individual classification powers and complementarities of the base classifiers [9,3,4], an optimization problem which maximizes accuracy and diversity at the same time is formulated.

In general, the more accurate the base classifiers are, the less different and therefore diverse they become. Hence, in order to end up with the optimal accuracy-diversity trade-off, a matrix P is formed by using the data points in a selection set S such that $P_{mi} = 0$ if i^{th} classifier is correct on data point S_m; and $P_{mi} = 1$ otherwise. Thus, $G = P^T P$ is the matrix in which the diagonal entry G_{ii} is the total number of errors made by classifier i, and the off-diagonal G_{ij} is the number of common errors of the classifiers i and j. Then, $\sum_i \tilde{G}_{ii}$ is supposed to be a measure for the overall ensemble strength in the sense of accuracy and $\sum_{ij,i\neq j} \tilde{G}_{ij}$ in the sense of diversity, where $\tilde{G}$ is obtained after normalizing each element of G into $[0,1]$. Therefore, the overall $\sum_{ij} \tilde{G}_{ij}$, incorporating both accuracy and diversity, is considered to be a good approximation for the ensemble error and the optimization problem is formulated as

$$\begin{aligned} &\min_x x^T \tilde{G} x \\ &\text{s.t} \sum_t x_t = p \\ &\quad x_t \epsilon \{0,1\} \end{aligned} \tag{3}$$

where x is a vector with elements $x_t = 1$ if t^{th} classifier is chosen as a result of pruning and 0 otherwise; and p is the desired input size of the pruned ensemble.

This problem is NP-hard, and the suboptimal solution is found by transforming it into the form of the max-cut problem with size p and using semidefinite programming (SDP).

3   Error Correcting Output Coding (ECOC)

Error Correcting Output Coding (ECOC) [6] is a powerful multiclass classification technique, in which multiple two-class base classifiers are trained using re-labeled subsets of the training data, determined by a preset code matrix.. The main idea behind this procedure is to solve the original multiclass problem by combining the decision boundaries obtained from simpler two-class decompositions.

In an ECOC matrix C, a particular element $C_{ji} \in \{-1, +1\}$ indicates the desired label for class j, to be used in training the base classifier i. An example ECOC matrix is provided in Figure 1 for a 4 class problem to be solved using 5 base classifiers. During decision making, namely decoding, the decisions of each base classifier for a given test sample are located in a vector consecutively, and the similarity between this vector and the codeword for each class (the row array C_j is the codeword for class j) is measured using a distance metric such as the Hamming distance or $L1$ norm.

As the name implies, ECOC can handle incorrect base classification results up to a certain degree. Specifically, if the minimum Hamming Distance (HD) between any pair of codewords is h, then up to $\lfloor (h-1)/2 \rfloor$ single bit errors can be corrected. Thus, in order to help with the error correction, the code matrix is suggested to be designed to have large HD between the codewords of different classes.

There are various data dependent and independent methods proposed for the design, namely encoding of the ECOC matrix. Most importantly, it has been theoretically and experimentally proven that the randomly generated long or deterministic equi-distant code matrices give close to optimum performance when used with strong base classifiers [8,16]. This is why random codes have also been used for the experiments in this study. Finally for encoding, note that the use of ternary ECOC matrices [1], where $C_{ji} = 0$ is introduced to leave a class out of the consideration of a base classifier, has also been investigated.

	Base Cl. 1	Base Cl. 2	Base Cl. 3	Base Cl. 4	Base Cl. 5
Class 1	1	-1	1	-1	1
Class 2	1	1	1	-1	-1
Class 3	-1	-1	1	1	-1
Class 4	-1	1	-1	-1	-1

Fig. 1. An example ECOC matrix for a 4 class problem with 5 base classifiers

4 ECOC Pruning

The algorithms presented in Section 2 have initially been proposed for majority/weighted voting ensembles like Bagging. In our experiments, we extend their scope to the ECOC framework and make use of them to prune the base classifiers that have been trained on the two-class decompositions determined by a given ECOC matrix. However, the following problems arise during the extensions of these methods to the ECOC framework: The connection between the base classifier accuracies measured on two-class decompositions of the ECOC matrix and the target multiclass accuracy is not evident; and neither a complex diversity measure, nor the HD information which is crucial in the design of ECOC matrices are taken into account during pruning.

To overcome these shortcomings and further improve the pruning efficiency, we introduce a novel pruning method based on a novel accuracy measure for ECOC, called *AcEc*. Using *AcEc*, the problem that the dichotomies within ECOC contain indirect information about the multiclass problem is overcome by taking into account the ECOC matrix structure while calculating the base accuracies.

4.1 Pruning Using the ECOC Accuracy Measure ($AcEc - P$)

When we focus on the base classifier accuracy for different two-class decompositions in the ECOC framework, the fact that the ultimate objective is to solve a multiclass problem is usually overlooked. We propose a novel approach to investigate a given base classifier's effectiveness by measuring its accuracy k times with respect to each individual class of a k class problem, and averaging the results. We shall refer to this measure as *AcEc*, and the pruning method achieved by using ordered aggregation based on it as $AcEc - P$.

Consider an ECOC ensemble with N base classifiers and k classes, and a selection set S consisting of μ training patterns. The desired label for the pattern S_m belonging to class j, to be used in training the base classifier i within the ECOC framework can be denoted by

$$\theta_{m,i} = \psi_i(j) = \psi_i(\phi(S_m)) \tag{4}$$

where $1 \leq m \leq \mu$, $1 \leq i \leq N$, ϕ is the target label function for the multiclass problem and ψ_i is the binary decomposition function defined by the ith column ECOC matrix.

To calculate $AcEc_i$ for a given classifier i, each pattern S_m of the selection set S is relabeled and target-mapped k times, with respect to each of the k classes. In each run l of k, the relabeling function r is defined as

$$\begin{aligned} r_l(S_m) &= +1, \text{ if } \phi(S_m) = l \\ r_l(S_m) &= -1, \text{ otherwise} \end{aligned} \tag{5}$$

whereas the target-mapping function f is formulated as

$$f_{i,l}(S_m) = \psi_i(l)\, d_i(S_m) \tag{6}$$

where $d_i(S_m)$ is the decision of the base classifier i for the pattern S_m.

For each run l, the function r creates a $1 - vs - the\ rest$ relabeling by assigning value 1 to the patterns from the class of interest, l, and -1 to the rest. On the other hand, the mapping function f maps the base classifier's decision based on the information obtained from the ECOC matrix. Any pattern S_m, which is at the same bi-partition with class l and is therefore sharing the same ECOC labeling $\theta_{m,i} = \psi_i(l)$ is mapped to the label $f_{i,l}(S_m) = +1$ if they are correctly classified by i; and to label $f_{i,l}(S_m) = -1$ if they are misclassified by i. Conversely, the rest of the patterns lying in the opposite bi-partition (as a result of having opposite ECOC labeling) with respect to class l are mapped to $f_{i,l}(S_m) = -1$ if they are correctly classified by i; and are mapped to label $f_{i,l}(S_m) = 1$ if they are misclassified. Note that f is a function of i, whereas r is not.

The final accuracy measure for classifier i on set S at the lth run, which might be referred to as the $class\ l - vs - the\ rest$ analysis for i on S, is given by

$$AcEc_{i,l} = \sum_{m=1}^{\mu} f_{i,l}(S_m)\, r_l(S_m) \tag{7}$$

and $AcEc$ over all runs $l = 1...k$ is defined as the average

$$AcEc_i = \frac{1}{k}\sum_{l=1}^{k}\left(\sum_{m=1}^{\mu} f_{i,l}(S_m)\, r_l(S_m)\right). \tag{8}$$

Figure 2 presents an example $AcEc$ calculation for the base classifier 1 of the ECOC matrix C given in Figure 1. The pattern column indicates the patterns from all classes which have been correctly or incorrectly classified by classifier 1; for example cl2 denotes the patterns from class 2 correctly classified by classifier 1, and $\widetilde{cl2}$ denotes those incorrectly classified. The relabeling and mapping columns show the results of r and f functions respectively, for each one of the k cases. Finally, the $AcEc$ columns denote the final accuracy calculation for each pattern group. Thus, in this example,

$$AcEc_{1,1} = no(cl1)\text{-}no(\widetilde{cl1})\text{-}no(cl2)+no(\widetilde{cl2})+no(cl3)\text{-}no(\widetilde{cl3})+no(cl4)\text{-}no(\widetilde{cl4}),$$

$$AcEc_{1,2} = \text{-}no(cl1)+no(\widetilde{cl1})+no(cl2)\text{-}no(\widetilde{cl2})+no(cl3)\text{-}no(\widetilde{cl3})+no(cl4)\text{-}no(\widetilde{cl4}),\ ...$$

where $no(cll)$ denotes the number of patterns belonging to class l. The resulting $AcEc_i$ is the average over all column $AcEc_{i,l}$'s.

The pseudo-code for the $AcEc$ can be found in Algorithm 1.

k	1		2		3		4	
Pattern	Relabeling	Mapping	Relabeling	Mapping	Relabeling	Mapping	Relabeling	Mapping
cl1	1	1	-1	1	-1	-1	-1	-1
cl1	1	-1	-1	-1	-1	1	-1	1
cl2	-1	1	1	1	-1	-1	-1	-1
cl2	-1	-1	1	-1	-1	1	-1	1
cl3	-1	-1	-1	-1	1	1	-1	1
cl3	-1	1	-1	1	1	-1	-1	-1
cl4	-1	-1	-1	-1	-1	1	1	1
cl4	-1	1	-1	1	-1	-1	1	-1

k	1	2	3	4
Pattern	$AcEc_{1,1}$	$AcEc_{1,2}$	$AcEc_{1,3}$	$AcEc_{1,4}$
cl1	1	-1	1	1
cl1	-1	1	-1	-1
cl2	-1	1	1	1
cl2	1	-1	-1	-1
cl3	1	-1	1	-1
cl3	-1	-1	-1	1
cl4	1	1	-1	1
cl4	-1	-1	1	-1

Fig. 2. *AcEc* calculation given the base classifier 1 of the ECOC matrix C

Interpretation of *AcEc*. *AcEc* evaluates the average strength of a given base classifier i in coping with $1 - vs - the\ rest$ problems for each class in the original multiclass problem. This can only be elucidated by making use of the base classifier's decisions together with information derived from the ECOC matrix. Below is a summary interpretation of an *AcEc* column for any base classifier i and for the run l (class $l - vs - the\ rest$) given without loss of generality.

1. Patterns from class l are rewarded with +1 if they have been correctly classified; or penalized with −1 otherwise.
2. Patterns from the opposite ECOC bi-partition of class l are rewarded with +1 if correctly classified, and penalized with −1 otherwise.
3. Patterns from the same bi-partition with class l (but not belonging to class l) are rewarded with +1 if they have been misclassified, and −1 otherwise.

With reference to 3, note that as the strength of the base classifier in run l is measured in terms of its ability to separate class l from the rest, any pattern from the same bi-partition but not from the same class as l is being penalized for the correct classification.

Therefore, for a given class $l - vs - the - rest$ analysis for i,

$$AcEc_{i,l} = (no(TP_l)+no(TN_l))-(no(FP_l)-no(FN_l))$$
$$AcEc_{i,l} = 2\,(no(TP_l)+no(TN_l)) - \mu$$

where μ is the number of training patterns; $no(TP_l)$, $no(TN_l)$, $no(FP_l)$ and $no(FN_l)$ are the numbers of true positives, true negatives, false positives and false negatives for class l, respectively. Thus, it can be observed that $AcEc_{i,l}$ is an indicative of accuracy for class l given classifier i.

Algorithm 1: The $AcEc$ Measure Calculation

Input: a k class problem
a base classifier i
a selection set S consisting of α patterns
Output: $AcEc_i$, the $AcEc$ accuracy for i

1 **for** $l \leftarrow 1$ **to** k **do**
2   **for** $j \leftarrow 1$ **to** α **do**
3   calculate $\mathbf{mapVector}_{i,l}[Y_j] \leftarrow f_{i,l}(Y_j)$;
4   calculate $\mathbf{relabelVector}_l[Y_j] \leftarrow r_l(Y_j)$;
5     $AcEc_{i,l} \leftarrow \left(\mathbf{mapVector}_{i,l}\right)^T \cdot \mathbf{relabelVector}_l$;
6 $AcEc_i \leftarrow \mathrm{mean}_{l=1}^{k} AcEc_{i,l}$;

---

5 Experiments

Experiments have been carried out on 2 artificial and 7 UCI Machine Learning Repository (MLR) [2] datasets summarized in Table 3.

Pruning algorithms have been analyzed on ECOC ensembles of pruned CART (CART-P) trees , unpruned CART trees and Neural Networks (NNs). As it has been theoretically and experimentally proven that the randomly generated long ECOC matrices give close to optimum performance when used with strong base classifiers [8], randomly generated ECOC matrices of 50 and 150 base classifiers are used to embrace both short (lower ensemble accuracy) and long codes (higher ensemble accuracy). ECOC decoding is carried out using HD.

All experiments have been repeated 15 times using different ECOC matrices, and additional random perturbation for the base classifiers is obtained by use of bootstrapping. In each run, patterns have random 50/50 train/test split unless they have already been provided as separate sets by UCI MLR. Finally note that the training set is used as the selection set for the pruning techniques in this study.

Tables 1 and 2 show pruned ensemble accuracies calculated on the datasets for pruning rates of 50% and 70% respectively. Pruning methods of MDM, REP, PSDP, OB, $AcEc$-P are analyzed together with two more pruning algorithms: ordered aggregation pruning based on base classifier accuracy (BaseAc) and random selection which is repeated 20 times within each of the 15 runs (RAND). The unpruned ensemble accuracy is also provided under the name "FULL EN". In each table, the results obtained using CART-P trees, NNs with Levenberg-Marquardt backpropagation for 16 nodes and 15 epochs, and CART trees as base classifiers are presented. For each base classifier type, initial ECOC matrix sizes of 50 (block (a) in each table) and 150 (block (b)) columns are evaluated.

The pruning method which obtains the highest accuracy in each case is marked in bold, whereas the one having the second rank is underlined.

Table 1. Test error rates (%) obtained using different pruning methods, base classifier types and ensemble sizes; for prune rate = 0.50

		MDM	RAND	BaseAc	AcEc-P	PSDP	REP	OB	FULL EN	MDM	RAND	BaseAc	AcEc-P	PSDP	REP	OB	FULL EN
CART	art1	73.52	72.85	73.31	73.91	72.96	73.83	73.49	74.66	74.51	74.4	74.46	74.7	74.47	74.39	74.6	74.82
	art2	78.37	76.92	79.04	79.61	78.33	79.29	78.83	80.71	80.38	80.13	80.3	80.8	80.38	80.56	80.68	81.02
	derma	93.03	94.5	95.62	95.39	92.36	95.99	95.73	96.67	94.53	96.46	96.55	96.29	94.46	96.6	96.18	97.3
	glass	65.27	64.81	67.11	68.44	64.57	68.19	69	68.83	69.78	70.92	72.38	72.25	69.59	71.94	72.38	72.89
	vehicle	70.51	72.72	72.07	73.93	68.72	73.73	73.55	74.83	69.86	74.1	73.4	74.69	69.04	74.09	74.96	74.85
	yeast	49.99	52.53	51.44	56.14	49.63	55.65	55.5	57.81	53.55	58.31	53.76	58.95	53.16	58.82	58.99	59.77
	optdig	92.64	90.72	93.14	93.89	92.64	93.08	93.25	95.39	95.56	95.14	95.69	95.8	95.56	95.26	95.93	96.29
	segm	89.75	90.15	93.22	92.94	89.19	92.47	92.94	93.82	92.54	93.8	94.52	94.49	92.44	94.35	94.56	94.63
	sat	88.37	87.84	88.2	89.11	88.03	88.83	88.95	89.72	90.02	89.98	89.78	90.4	89.96	89.65	90.05	90.41
CARTP	art1	72.66	72.31	73.05	73.61	72.42	72.8	72.84	73.99	73.9	73.79	74.11	74.3	73.89	73.86	74.02	74.34
	art2	77.51	76.37	79.03	78.86	77.2	78.5	78.69	80.05	79.9	79.79	80.06	80.1	79.89	80.11	80.03	80.5
	derma	94.02	94.93	95.36	95.92	93.68	96.26	95.56	97.02	95.02	96.77	96.25	96.25	94.49	97	95.84	97.27
	glass	67.14	66.44	68.06	69.21	67.43	69.24	70	71.29	70.16	70.59	71.05	72.06	69.33	72.8	72.13	72.44
	vehicle	69.68	71.67	71.33	73.73	68.33	73.03	72.65	74.06	69.92	74.08	72.81	74.9	69.53	74.09	74.44	74.6
	yeast	49.78	52.81	53.06	55.68	49.07	56.06	55.72	58.65	54.52	58.02	54.73	58.8	53.89	58.09	58.75	59.57
	optdig	92.33	90.36	92.89	93.45	92.52	92.64	92.89	95.11	95.83	95.2	95.59	95.98	95.83	95.82	96	96.38
	segm	89.31	90.11	92.88	92.7	88.98	92.85	92.87	93.59	92.18	93.72	94.54	94.54	92.17	94.31	94.7	94.7
	sat	87.97	87.8	88.28	88.87	87.98	88.59	88.77	89.89	89.83	89.94	89.75	90.3	89.81	89.62	90.12	90.38
NN	art1	76.16	75.61	75.9	76.18	75.94	75.77	75.97	76.45	76.66	76.61	76.72	76.7	76.61	76.52	76.72	76.79
	art2	81.27	80.36	81.72	81.9	81.17	81.59	81.68	82.47	82.76	82.68	82.89	83	82.75	82.72	82.8	83.1
	derma	95.54	95.95	93.86	96.18	95.28	96.63	96.07	96.74	95.81	96.43	95.54	96.8	95.73	96.7	96.7	96.7
	glass	62.67	63.27	64	65.14	61.71	65.78	65.71	66.22	65.33	64.94	64.44	65.4	64.63	66.3	65.59	65.71
	vehicle	76.29	79.99	79.08	80.44	74.69	81.31	81.4	81.61	76.4	81.59	81.03	81.97	75.45	82.2	81.96	82.15
	yeast	57.98	56.74	53.92	58.49	56.27	58.19	58.8	59.31	59	58.94	55.16	59.4	57.65	58.47	58.99	59.6
	optdig	91.7	92.4	95.58	95.24	91.69	95.61	95.24	95.66	94.37	95.74	96.9	96.66	94.47	96.64	96.61	96.45
	segm	90.35	89.78	91.14	91.08	89.92	91.75	91.64	92.03	91.93	92.14	93	92.39	91.82	92.97	93.08	92.64
	sat	88.51	88.39	88.68	89.14	88.32	89.42	89.01	89.3	88.94	89.18	88.91	89.3	88.96	89.3	89.19	89.27

(a)   (b)

5.1   Discussion of the Results

When all datasets, base classifier types and pruning rates are taken into account; $AcEc - P$, REP and OB are found to be the most successful pruning algorithms in the sense of pruned ensemble accuracy, in the order given. As a result of pairwise comparisons between $AcEc - P$ and the rest of the algorithms, the ratio of cases where $AcEc - P$ is found to reveal ensemble accuracy better than or equal to REP and OB over all cases is 63.8% and 75% respectively.

As the number of base classifiers decrease and the pruning rate increases, the results obtained using REP start getting better than the rest. Though, even when REP is at the peak of its performance (case of 50 base classifiers & 70% pruning rate), the performance of $AcEc - P$ follows closely, being at most 2% worse than the results obtained by REP. Due to the above mentioned overall superior performance of $AcEc - P$ and the high time complexity of REP, $AcEc - P$ still comes out to be the most favorable pruning algorithm in general.

Contrary to the findings for pruned Bagging ensembles [10], pruned ECOC ensembles are not observed to commonly improve the generalization performance of the unpruned ones. This is explained by the characteristics of the ECOC framework such as its resistance to overfitting and better capacity for error correction, especially when used with long code matrices and strong base classifiers, making it hard to improve its performance via pruning. Table 4 shows both the significant differences between the results obtained by $AcEc-P$ pruned and unpruned

Table 2. Test error rates (%) obtained using different pruning methods, base classifier types and ensemble sizes; for prune rate= 0.70

(a)

Clf	Data	MDM	RAND	BaseAc	AcEc-P	PSDP	REP	OB	FULL EN
CART	art1	71.12	71.13	71.74	72.8	69.67	72.22	71.84	74.66
	art2	74.01	73.91	75.09	77.54	73.97	77.68	75.81	80.71
	derma	89.36	92.89	89.89	94.46	88.13	95.73	92.96	96.67
	glass	59.75	62.19	60	65.21	58.54	66.41	64.83	68.83
	vehicle	69.05	70.99	68.39	72.07	67.2	72.54	72.23	74.83
	yeast	44.38	49.89	45.83	53.21	43.5	53.28	52.35	57.81
	optdig	88.67	86.96	88.15	90.75	88.58	90.47	89.39	95.39
	segm	84.43	87.62	89.46	91.49	83.95	91.29	90.24	93.82
	sat	85.44	86.12	75.59	87.66	84.63	87.86	87.39	89.72
CARTP	art1	70.66	70.68	71.81	72.08	69.54	71.88	71.69	73.99
	art2	72.78	73.38	76.65	76.74	72.8	76.8	75.95	80.05
	derma	89.47	93.28	89.1	95.17	88.71	94.75	91.77	97.02
	glass	63.05	64.58	59.3	66.35	62.19	66.43	65.05	71.29
	vehicle	67.43	70.42	67.77	71.63	65.81	71.97	71.46	74.06
	yeast	43.79	49.6	46.45	53.23	42.83	53.67	52.79	58.65
	optdig	87.86	86.81	88.02	90.33	88.1	90.51	89.29	95.11
	segm	84.03	87.06	90.42	91.65	83.43	91.83	90.63	93.59
	sat	85.02	86.26	75.51	87.58	84.85	87.74	87.27	89.89
NN	art1	75.27	74.98	72.47	75.6	74.67	75.43	75.49	76.45
	art2	78.69	78.31	79.89	80.76	78.12	80.55	80.14	82.47
	derma	92.96	94.55	86.4	95.88	91.87	95.13	88.84	96.74
	glass	58.16	61.17	49.84	63.49	56.51	65.33	63.37	66.22
	vehicle	75.01	78.79	73.1	80.47	72.5	80.35	78.89	81.61
	yeast	54.24	54.94	48.28	57.51	51.42	57.35	56.46	59.31
	optdig	86.91	89.6	94.44	93.31	86.94	94.61	93.5	95.66
	segm	87.57	88.13	86.76	90.56	86.93	90.82	89.9	92.03
	sat	86.49	87.56	75.8	88.73	86.44	88.97	88.4	89.3

(b)

Clf	Data	MDM	RAND	BaseAc	AcEc-P	PSDP	REP	OB	FULL EN
CART	art1	73.85	74.15	73.64	74.45	73.67	73.63	73.98	74.82
	art2	79.23	79.41	79.41	80.08	79.1	79.72	79.84	81.02
	derma	92.36	96.26	92.66	95.81	92.25	96.8	95.43	97.3
	glass	67.75	70.17	64.76	70.67	66.6	71.3	71.17	72.89
	vehicle	69.12	73.72	68.37	73.73	68.4	73.8	73.76	74.85
	yeast	49.84	56.99	50.08	58.22	49.64	57.04	57.76	59.77
	optdig	94.48	94.19	94.68	95.07	94.49	94.5	94.96	96.29
	segm	89.91	93.1	93.59	94.15	89.76	93.82	93.79	94.63
	sat	88.63	89.5	79.68	89.99	88.53	89.14	89.48	90.41
CARTP	art1	73.23	73.49	73.17	73.86	73.01	73.09	73.44	74.34
	art2	78.47	79.16	79.34	79.49	78.52	79.31	79.38	80.5
	derma	92.81	96.28	91.87	96.29	92.36	96.29	94.94	97.27
	glass	66.86	69.33	65.33	71.49	65.78	71.49	71.11	72.44
	vehicle	68.8	73.61	68.8	74.17	68.01	72.92	73.81	74.6
	yeast	50.28	56.86	49.75	58.03	49.17	56.98	57.73	59.57
	optdig	89.95	92.97	93.83	94.29	89.83	93.79	94.06	94.7
	segm	88.27	89.54	84.48	89.78	88.18	89.18	89.53	90.38
NN	art1	76.42	76.48	76.47	76.62	76.32	76.44	76.7	76.79
	art2	82.31	82.3	82.46	82.79	82.29	82.12	82.43	83.1
	derma	94.64	96.44	87.75	96.89	94.76	96.52	94.08	96.7
	glass	62.79	64.6	53.9	65.65	61.9	65.4	64.7	65.71
	vehicle	76.03	81.26	72.45	81.93	75.02	81.64	81.64	82.15
	yeast	57.65	58.5	49.89	58.37	55.24	57.87	58.42	59.6
	optdig	92.66	95.08	96.9	96.44	92.68	96.47	96.33	96.45
	segm	90.97	91.64	92.25	92.53	90.76	92.57	92.7	92.64
	sat	88.31	89.08	83.23	89.38	88.3	89.35	89.12	89.27

Table 3. Summary of the datasets used in the experiments. Sizes of the datasets not having separate test sets are given under #Training.

	Type	#Training Samples	#Test Samples	#Attributes	#Classes
ArtMulti1	Artificial	300	18000	2	5
ArtMulti2	Artificial	300	18000	3	9
Dermatology	UCI MLR	358	-	34	6
Glass Identification	UCI MLR	214	-	9	6
Vehicle	UCI MLR	846	-	18	4
Yeast	UCI MLR	1484	-	8	10
Optdigits	UCI MLR	3823	1797	64	10
Segmentation	UCI MLR	210	2100	19	7
Satellite Image	UCI MLR	4435	2000	36	6

ECOC ensembles, and the number of datasets for which the pruned ensembles with the given rates perform better than the unpruned, over all datasets. For the significance tests, the procedure suggested by Demsar [5] is utilized by using the Friedman significance test with the Nemenyi post-hoc procedure, with $p < 0.05$. The toolbox used in the implementation is provided in [15]. Here, the ensembles pruned by $AcEc - P$ with the given pruning rates are assigned 0 if there is no statistical difference between them and the unpruned ensemble, and 1 otherwise. It can be deduced that it is possible to reduce the size of the ensemble down to 30% (using a pruning rate of 70%) of the initial size without a significant difference between the $AcEc - P$ pruned and the unpruned ensemble, and even improve the generalization performances in some cases.

Table 4. Significance test results between the pruned and unpruned ensembles / No. of datasets (over 9) for which pruned ensemble performs better than the unpruned

	30%	40%	50%	60%	70%
CART with 50 Col.	1 / 1	1 / 0	1 / 0	1 / 1	1 / 0
CART with 150 Col.	1 / 1	1 / 1	1 / 0	1 / 0	1 / 0
CARTP with 50 Col.	1 / 1	1 / 1	1 / 0	1 / 0	1 / 0
CARTP with 150 Col.	0 / 3	0 / 2	1 / 0	1 / 0	1 / 0
16N-15E NN with 50 Col.	1 / 1	1 / 0	1 / 0	1 / 0	1 / 0
16N-15E NN with 150 Col.	1 / 1	0 / 2	0 / 3	0 / 4	0 / 2

It should also be noted that according to findings in Table 1, the deterioration in the ensemble accuracy when $AcEc - P$ is used with a pruning rate of 50% is only within ~ 0.01 for all datasets and base classifier types, when an initial pool of 150 classifiers is used. That is to say, instead of using a 150 column ECOC matrix, using one with 75 will never degrade the performance by more than $\sim 1\%$, which is highly useful when applications requiring speed and accuracy at the same time are considered.

6   Conclusions

We have presented a novel ECOC pruning method, which works by establishing the link between the individual two class decompositions of an ECOC matrix and the main multiclass problem using accuracy information. It is found as a result of experimentation that this method, namely $AcEc - P$, yields superior results to those of the state-of-the art ensemble pruning methods applied to ECOC. It is also shown that in certain cases it is possible to use a pruning rate of 70% without a significant difference in the performance and even help increase the classification performance at times. Especially when used with longer codes and lower pruning rates, the difference in the performance of the pruned ensemble and the unpruned one can be kept within an upper limit as small as 0.01.

As for future work, the strength of the method is to be further investigated using theoretical bounds, and possible improvements are to be sought via utilizing further information such as HD and diversity in addition to accuracy.

References

1. Allwein, E.L., Schapire, R.E., Singer, Y.: Reducing multiclass to binary: A unifying approach for margin classifiers. JMLR 1, 113–141 (2000)
2. Asuncion, A., Newman, D.: UCI machine learning repository (2007), http://www.ics.uci.edu/~mlearn/MLRepository.html
3. Breiman, L.: Random forests. Machine Learning 45(1), 5–32 (2001)
4. Chandra, A., Yao, X.: Ensemble learning using multi-objective evolutionary algorithms. J. Math. Model. Algorithms 5(4), 417–445 (2006)
5. Demsar, J.: Statistical comparisons of classifiers over multiple data sets. JMLR 7, 1–30 (2006)

6. Dietterich, T.G., Bakiri, G.: Solving multiclass learning problems via error-correcting output codes. J. Artif. Intell. Res. (JAIR) 2, 263–286 (1995)
7. Hernandez-Lobato, D., Martinez-Munoz, G., Suarez, A.: Statistical instance-based pruning in ensembles of independent classifiers. IEEE Trans. Pattern Anal. Mach. Intell. 31(2), 364–369 (2009)
8. James, G., Hastie, T.: The error coding method and picts (1998)
9. Margineantu, D.D., Dietterich, T.G.: Pruning adaptive boosting. In: International Conference on Machine Learning, pp. 211–218 (1997)
10. Martinez-Munoz, G., Hernandez-Lobato, D., Suarez, A.: An analysis of ensemble pruning techniques based on ordered aggregation. IEEE Trans. Pattern Anal. Mach. Intell. 31(2), 245–259 (2009)
11. Martinez-Munoz, G., Suarez, A.: Aggregation ordering in bagging. In: Proc. of the IASTED ICAIA, pp. 258–263. Acta Press (2004)
12. Martinez-Munoz, G., Suarez, A.: Using boosting to prune bagging ensembles. Pattern Recognition Letters 28, 156–165 (2007)
13. Soto, V., Martínez-Muñoz, G., Hernández-Lobato, D., Suárez, A.: A double pruning algorithm for classification ensembles. In: El Gayar, N., Kittler, J., Roli, F. (eds.) MCS 2010. LNCS, vol. 5997, pp. 104–113. Springer, Heidelberg (2010)
14. Tsoumakas, G., Partalas, I., Vlahavas, I.: A taxonomy and short review of ensemble selection. In: ECAI 2008, Workshop SUEMA (2008)
15. Ulas, A., Yildiz, O.T., Alpaydin, E.: Cost-conscious comparison of supervised learning algorithms over multiple data sets. Pattern Recognition (2011)
16. Windeatt, T., Ghaderi, R.: Coding and decoding strategies for multi-class learning problems. Information Fusion 4(1), 11–21 (2003)
17. Zhang, Y., Burer, S., Street, W.N.: Ensemble pruning via semi-definite programming. JMLR 7, 1315–1338 (2006)
18. Zhou, Z.H.: Ensemble Methods: Foundations and Algorithms, 1st edn. CRC Press, Boca Raton (2012)

Author Index